STUDENT SOLUTIONS MANUAL
AND STUDY GUIDE FOR
DROOYAN AND FRANKLIN'S

Algebra for College Students

BERNARD FELDMAN
Los Angeles Pierce College

Wadsworth Publishing Company
Belmont, California
A Division of Wadsworth, Inc.

Preface

This student supplement has been prepared to accompany *Algebra for College Students*. Detailed solutions, including graphs, are included for the even-numbered exercises for each section.

Many of the exercises can be solved in different ways. To keep this supplement to a convenient size, alternate solutions are in general not shown. It is anticipated that sometimes you will devise a way to solve a problem that is simpler than the way shown. However, if you have difficulty with a particular exercise, you should refer to the solution given to see one way it can be solved. Of course, it is best for you to do this only after you have made a diligent attempt to work the problem yourself.

The Test Problems section at the end of this manual is intended as a study guide to help you determine areas in which you might need more practice. Work the problems in each Test Problems chapter on your own first, then check the answers in the back of the manual. For more help in particular chapters, refer to the Chapter Reviews in the main textbook.

Bernard Feldman

Printed in the United States of America

1 2 3 4 5 6 7 8 9 10---92 91 90 89 88

ISBN 0-534-08155-X

Mathematics Publisher: Kevin Howat
Assistant Editor: Barbara Holland
Typist: Joyce A. Gaiser
Artist: Robert Gaiser
Proofreader: Rene Lynch

Contents

1 Review of the Real Number System

EXERCISE 1.1

2. $\{-2,-1,0,1,2,3\}$

4. $\{1,2,3\}$

6. $\{\ldots,-1,0,1\}$

8. $\{-6,-2,-4\}$

10. Infinite

12. Infinite

14. Infinite

16. 0 is not a natural number; hence, $0 \notin$ {natural numbers}.

18. -3 is not a whole number; hence, $-3 \notin$ {whole numbers}.

20. -6 is an integer; hence, $-6 \in$ {integers}.

22. {8}; 8 is the only member of A that is a natural number.

24. $\{-5, -3.44, -\frac{2}{3}, 0, \frac{1}{5}, \frac{7}{3}, 6.1, 8\}$; $\sqrt{15}$ is the only member in A that is not a rational number.

26. $\{x | x \in A\}$; all the members in A are real numbers.

28. No. Since $A \neq B$, B will have members which A does not have.

30. No. Consider: $A = \{1,2\}$, $B = \{1\}$, $C = \{1,2\}$; $A \neq B$ and $B \neq C$, yet $A = C$.

EXERCISE 1.2

2. $t + 3$

4. $4 = y$

6. a

8. 4

10. $2 + x$

12. $-5 > -8$

14. $-3 < 4$

16. $x - 3 > 0$

18. $x + 2 \geq 0$

20. $-4 < y < 0$

22. $0 \leq 3t \leq 4$

24. $<$

26. $>$

28. $<$

30. $=$

32. $>$; $>$

34. $<$; $<$

36.
-5
0
5

38.
-5
0

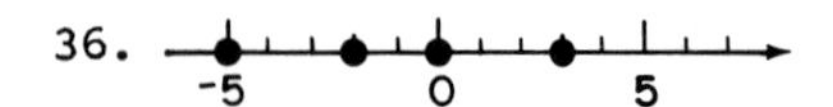

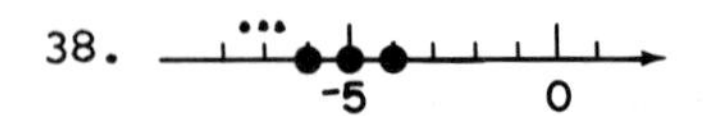

The 3 dots above the number line mean that the graph continues indefinitely to the left.

40.
0
5

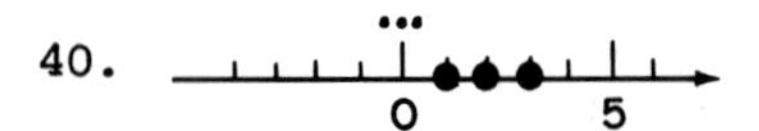

See the comment in the solution to Problem 38.

42.
-5
0

44.
0
5

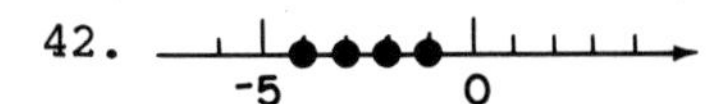

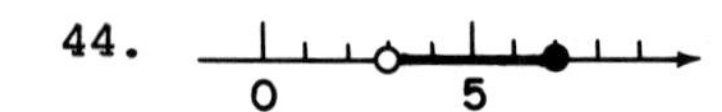

46.
0
5
10

48.
-5
0
5
10

50.
-30
-20
-10
0

52.
0
10
20
30

54.
0
20
40
60

EXERCISES 1.3

2. $(2m)n = 2(mn)$

4. $(2 + z) + 3 = 2 + (z + 3)$

6. (-7)

8. 0

10. r

12. $\frac{1}{6} \cdot 6$

14. $(y + z)$

16. 0

18. t; 3

20. -7

22. $x + 2$

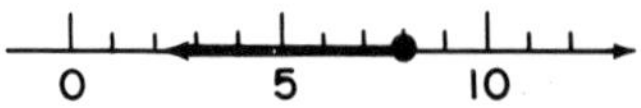

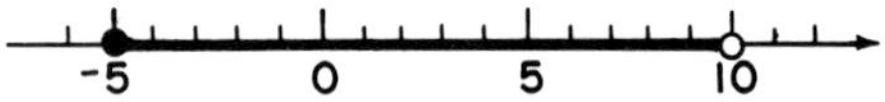

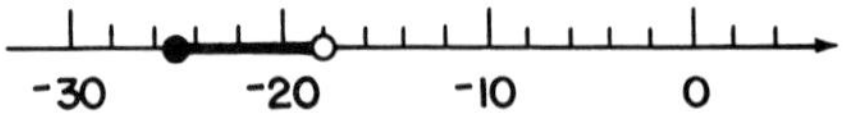

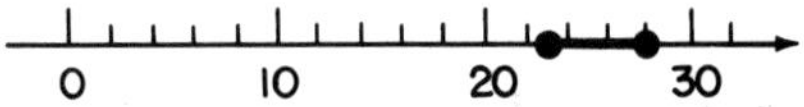

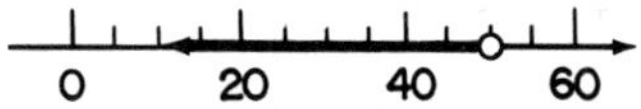

24. Negative number; because $-(-x) = x$.

26. 7

28. 6

30. -4

32. -7

34. $-x$, if $-x \geq 0$
$-(-x)$ or x, if $-x < 0$

36. $x + 3$, if $x + 3 \geq 0$
$-(x + 3)$, if $x + 3 < 0$

38. Not closed with respect to addition, since $1 + 1 = 2$ and $2 \notin \{-1,0,1\}$. Closed with respect to multiplication, since $(-1) \cdot (-1)$, $-1 \cdot 0$, $0 \cdot 0$, $0 \cdot 1$, and $1 \cdot 1$ are all elements of $\{-1,0,1\}$.

40. All of the sets are closed with respect to addition and with respect to multiplication.

42. No. If $x = 0$, they are not equal. If $x \neq 0$, the left member denotes a positive number while the right member denotes a negative number.

43. Yes. If $x \neq 0$ and $y \neq 0$, $|y|$ will always be less than $|x| + |y|$.
Consider for example, $x = -2$ and $y = 1$: $|1| < |-2| + |1|$ is true.

EXERCISE 1.4

2. 15

4. 2

6. -5

8. -12

10. $6 + (-1) = 5$

12. $8 + (-14) = -6$

14. $-5 + (-11) = -16$

16. $8 + [-(-4)] = 8 + 4 = 12$

18. $-8 + [-(-5)] = -8 + 5 = -3$

20. $8 + 3 + (-5) = 6$

22. $6 + (-4) + (-2) = 0$

24. $4 + [-(-3)] + 2 = 4 + 3 + 2 = 9$

26. $8 - (8) = 0$

28. $[3 + (-7)] + (-1)$
$= [-4] + (-1) = -5$

30. $4 - ([6 + 2] + [-11])$
$= 4 - ([8] + [-11])$
$= 4 - (-3) = 4 + 3 = 7$

32. $[5 + (-9)] + (2) = [-4] + (2)$
$= -2$

34. $(13) + (13 + [(-17) + (-2)])$
$= 13 + (13 + [-19])$
$= 13 + (-6) = 7$

36. $(-15 + [3 + (-1)]) - [(9 - 5) + 12] = (-15 + [2]) - [(4) + 12]$
$= (-13) - (16)$
$= (-13) + (-16) = -29$

38. $[(-27 + 3) + (-6)] - [19 + (-12)] = [-24 + (-6)] - [7]$
$= -30 + (-7) = -37$

40. Since $(4 - 3) - 2 = 1 - 2 = -1$ and $4 - (3 - 2) = 4 - 1 = 3$, it follows that $(4 - 3) - 2 \neq 4 - (3 - 2)$. Hence, subtraction is not associative.

42. a. If $a > -b$ or $b > -a$, $a + b$ is positive.
Consider $a = 2$ and $b = 3$ or $a = 2$ and $b = -1$.

b. If $a > b$, $a - b$ is positive.
Consider $a = 2$ and $b = 1$ or $a = 2$ and $b = -1$.

EXERCISE 1.5

2. -15

4. 40

6. $[(-5)(2)](-3) = [-10](-3) = 30$

8. $[(2)(-2)](6) = [-4](6)$
$= -24$

10. $[(5)(-2)][(-1)(6)] = [-10][-6]$
$= 60$

12. $[(-6)(0)][(2)(3)] = [0][6] = 0$

14. $-2(2 \cdot 6) = -2(12) = -24$

16. $(-5 \cdot 2)(-3) = (-10)(-3) = 30$

18. $(-2 \cdot 3)(3 \cdot 4) = (-6)(12) = -72$

20. $-3[(-2 \cdot 2) \cdot 4] = -3[-4 \cdot 4]$
$= -3[-16] = 48$

22. $2 \cdot 13$

24. $2 \cdot 3 \cdot 3$

26. $-1 \cdot 2 \cdot 2 \cdot 2 \cdot 2$

28. Prime

30. $5 \cdot 13$

32. $-1 \cdot 2 \cdot 2 \cdot 13$

34. -4

36. -3

38. 9

40. 0

42. Undefined

44. -4

46. $27\left(\frac{1}{7}\right)$

48. $9\left(\frac{1}{17}\right)$

50. $3\left(\frac{1}{13}\right)$

52. $3\left(\frac{1}{1000}\right)$

54. $\frac{8}{3}$

56. $2 \cdot \frac{1}{z} \cdot x = (2 \cdot x) \cdot \frac{1}{z}$
$= \frac{2x}{z}$

58. $(\frac{1}{5} \cdot 5) \cdot y = y$

60. $(\frac{1}{3} \cdot 3)(x - y) = x - y$

62. Let $x = 1$ and $y = 2$. Then $-3(xy) = -3[(1)(2)] = -6$ and $(-3x)(-3y) = [(-3)(1)][(-3)(2)] = 18$. Therefore, $-3(xy)$ and $(-3x)(-3y)$ are not equal for all x and y.

64. True if $x \neq 0$ or if $y = 0$. A fraction equals zero only if its numerator equals zero and the denominator doesn't. Also, a fraction is undefined if its denominator equals zero.

66. True if $x > 0$ and $y < 0$ or if $x < 0$ and $y > 0$. A fraction is negative if its numerator and denominator have opposite signs.

68. Since $(8 \div 4) \div 2 = 2 \div 2 = 1$ and $8 \div (4 \div 2) = 8 \div 2 = 4$,
$$(8 \div 4) \div 2 \neq 8 \div (4 \div 2).$$
Hence, division is not associative.

EXERCISE 1.6

2. $7 - 12 = -5$

4. $18 + (-10) = 8$

6. $5(-6) + 7 = -30 + 7$
$= -23$

8. $4 - 7(10) = 4 - 70$
$= -66$

10. $\frac{12}{6} - 5 = 2 - 5$
$= -3$

12. $\frac{5(-2)}{2} - \frac{18}{-3} = \frac{-10}{2} - (-6) = -5 + 6 = 1$

14. $6[5 - 3(-3)] + 3$
$= 6[5 + 9] + 3$
$= 6[14] + 3$
$= 84 + 3 = 87$

16. $(2)[5 + 7(-1)]$
$= (2)[5 - 7]$
$= (2)[-2]$
$= -4$

18. $27 \div (3[9 - 3(2)])$
$= 27 \div (3[9 - 6])$
$= 27 \div (3[3])$
$= 27 \div (9) = 3$

20. $-3[-2 + 5] \div [-9]$
$= -3[3] \div [-9]$
$= -9 \div [-9]$
$= 1$

22. $[-2 + 3(-3)] \cdot [-15 \div 3]$
$= [-2 + (-9)] \cdot [-5]$
$= [-11] \cdot [-5] = 55$

24. $\left[\frac{10}{-5}\right]\left[\frac{-9}{3}\right] = [-2][-3]$
$= 6$

26. $\left[7 + 3\left(\frac{-12}{6}\right) - 5\right] + 3$
$= [7 + 3(-2) - 5] + 3$
$= [(7 - 6) - 5] + 3$
$= [1 - 5] + 3$
$= [-4] + 3 = -1$

28. $\dfrac{12 + 3\left(\frac{-8}{2}\right) - 1}{-8 + 6\left(\frac{-18}{-3}\right) + 1}$
$= \dfrac{12 + 3(-4) - 1}{-8 + 6(6) + 1}$
$= \dfrac{[12 + (-12)] - 1}{-8 + (36 + 1)}$
$= \dfrac{0 - 1}{-8 + 37} = \dfrac{-1}{29}$

30. $$\frac{6 - 2\left(\frac{10}{5}\right) + 8}{3 - 6 + 8} = \frac{6 - 2(2) + 8}{-3 + 8} = \frac{(6 - 4) + 8}{5} = \frac{2 + 8}{5} = \frac{10}{5} = 2$$

32. $$\frac{2|-3|}{|-3|} - \frac{|-4| + 2}{|-2|} = \frac{2(3)}{3} - \frac{4 + 2}{2} = 2 - 3 = -1$$

34. $$\frac{12 + 2}{2} = \frac{14}{2} = 7$$

36. $$\frac{4 - 2(12)}{1 - 2} = \frac{4 - 24}{-1} = 20$$

38. $$2.5[1 + (0.05)(20)] = 2.5[1 + 1] = 2.5[2] = 5.0$$

40. $$A = \frac{1}{2}(14)(10 + 12) = 7(22) = 154$$

42. $$\text{Net } R = \frac{10(20)}{10 + 20} = \frac{200}{30} = \frac{20}{3}$$

2 Polynomials

EXERCISE 2.1

2. Trinomial; degree 2

4. Binomial; degree 1

6. Monomial; degree 3

8. Binomial; degree 2

10. $(-5)^2 = 25$

12. $-3^2 = -(3)^2$
$= -9$

14. $4^2 - (6)^2$
$= 16 - 36 = -20$

16. $\dfrac{4 \cdot 9}{6} + (12)^2 = \dfrac{36}{6} + 144$
$= 6 + 144 = 150$

18. $\dfrac{9 \cdot 4}{3} + \dfrac{-3 \cdot 8}{6} = \dfrac{36}{3} + \dfrac{-24}{6}$
$= 12 + (-4) = 8$

20. $\dfrac{49 - 36}{13} - \dfrac{64(-2)}{16} = \dfrac{13}{13} - \dfrac{-128}{16}$
$= 1 + 8 = 9$

22. $3 - (-2)^2 = 3 - (4)$
$= -1$

24. $[3 + 2(-2)]^2 = [3 + (-4)]^2$
$= [-1]^2 = 1$

26. $[3(-2)]^2 - 3(3) = (-6)^2 - 9$
$= 36 - 9 = 27$

28. $\dfrac{-3(-2)^2}{6} + 2(3)(-2)$
$= \dfrac{-3(4)}{6} + (-12)$
$= \dfrac{-12}{6} + (-12)$
$= -2 + (-12) = -14$

30. $[3 + (-2)]^2 + [3 - (-2)]^2$
$= (1)^2 + (5)^2$
$= 1 + 25$
$= 26$

32. $\frac{1}{2}(32)(3)^2 - 12(3)$
$= 16(9) - 36$
$= 144 - 36 = 108$

34. $\dfrac{(32)(12 - 4)^2}{32}$
$= \dfrac{(32)(8)^2}{32} = 64$

36. $\dfrac{(4) - (4)(2)^3}{1 - (2)}$

$= \dfrac{(4) - (4)(8)}{-1}$

$= \dfrac{4 - 32}{-1} = \dfrac{-28}{-1} = 28$

38. $P(3) = 2(3)^3 + (3)^2 - 3(3) + 4$
$= 2(27) + 9 - 9 + 4 = 58;$

$P(-3) = 2(-3)^3 + (-3)^2 - 3(-3) + 4$
$= 2(-27) + 9 + 9 + 4 = -32$

40. $D(4) = 11(4)^2 - 6(4) + 1$

$= 176 - 24 + 1 = 153$

$D(0) = 11(0)^2 - 6(0) + 1 = 1$

42. $P(0) = -3(0)^2 + 1$
$= 0 + 1 = 1;$

$Q(-1) = 2(-1)^2 - (-1) + 1$
$= 2(1) + 1 + 1 = 4$

44. $P(-1) = (-1)^6 - (-1)^5$
$= 1 - (-1) = 2;$

$Q(-1) = (-1)^7 - (-1)^6$
$= -1 - (1) = -2$

46. $P(1) = 1 + 1 = 2;$

$R(-2) = (-2)^2 + (-2) - 1$

$= 1;$

$Q(-1) = (-1)^2 - 1 = 0;$

$P(1) + R(-2) \cdot Q(-1)$
$= 2 + (1)(0)$
$= 2 + 0 = 2$

48. $P(3) = 3 + 1 = 4;$

$R(-1) = (-1)^2 + (-1) - 1$

$= -1;$

$Q(1) = 1^2 - 1 = 0;$

$P(3)[R(-1) + Q(1)]$
$= 4[-1 + 0]$
$= 4[-1] = -4$

50. $P(-1) = -1 + 1 = 0;$

$R[P(-1)] = R(0)$

$= 0^2 + 0 - 1$

$= -1$

52. $R(-2) = (-2)^2 + (-2) - 1$
$= 4 + (-2) - 1 = 1;$

$Q[R(-2)] = Q(1)$

$= 1^2 - 1 = 0$

54. $P(0) = 0 + 1 = 1;$

$Q[P(0)] = Q(1)$

$= 1^2 - 1 = 0$

56. $Q(-2) = (-2)^2 - 1 = 3;$

$P[Q(-2)] = P(3)$
$= 3 + 1 = 4$

58. $(-x)^n = (-x)(-x) \cdots (-x)$, n factors
$= (-1)(x)(-1)(x) \cdots (-1)(x)$, n factors of (-1) and of (x),
$= [(-1)(-1) \cdots (-1)][(x)(x) \cdots (x)]$ by repeated application of the commutative property and by the associative property of multiplication.

Hence, $(-x)^n = (-1)^n(x)^n$.

If n is odd, $(-1)^n = -1$, and $(-x)^n = (-1)(x)^n = -x^n$.

If n is even, $(-1)^n = 1$, and $(-x)^n = 1(x)^n = x^n$.

EXERCISE 2.2

2. $4x^3$

4. $-11y^2$

6. $-6z^3 + 5z^2$

8. $2xy^2 + 3y$

10. $r^2 + (-2r^2 - r) = -r^2 - r$

12. $(s^2 + s) + (-3s^2)$
$= -2s^2 + s$

14. $t^2 - 4t + 1 - 2t^2 + 2$
$= -t^2 - 4t + 3$

16. $2u^2 + 4u + 2 - u^2 + 4u + 1$
$= (2u^2 - u^2) + (4u + 4u) + (2 + 1)$
$= u^2 + 8u + 3$

18. $(4y^2 - 3y - 7) - (6y^2 - y + 2) = 4y^2 - 3y - 7 - 6y^2 + y - 2$
$= (4y^2 - 6y^2) + (-3y + y) + (-7 - 2)$
$= -2y^2 - 2y - 9$

20. $(4s^3 - 3s^2 + 2s - 1) - (s^3 - s^2 + 2s - 1)$
$= (4s^3 - 3s^2 + 2s - 1) + (-s^3 + s^2 - 2s + 1)$
$= (4s^3 - s^3) + (-3s^2 + s^2) + (2s - 2s) + (-1 + 1)$
$= 3s^3 - 2s^2$

22. $(7c^2 - 10c + 8) - (8c + 11) + (-6c^2 - 3c - 2)$
$= 7c^2 - 10c + 8 - 8c - 11 - 6c^2 - 3c - 2$
$= c^2 - 21c - 5$

24. $(m^2n^2 - 2mn + 7) - (-2m^2n^2 + mn - 3) - (3m^2n^2 - 4mn + 2)$
$= m^2n^2 - 2mn + 7 + 2m^2n^2 - mn + 3 - 3m^2n^2 + 4mn - 2$
$= mn + 8$

26. $[(2t^2 - 3t + 5) + (t^2 + t + 2)] - (2t^2 + 3t - 1)$
$= [2t^2 - 3t + 5 + t^2 + t + 2] - (2t^2 + 3t - 1)$
$= 3t^2 + (-2t) + 7 - 2t^2 - 3t + 1$
$= t^2 + (-5t) + 8 = t^2 - 5t + 8$

28. $(2c^2 + 3c + 1) - [(7c^2 + 3c - 2) + (3 - c - 5c^2)]$
$= (2c^2 + 3c + 1) - [2c^2 + 2c + 1]$
$= 2c^2 + 3c + 1 - 2c^2 - 2c - 1$
$= c$

30. $3a + [2a - (a + 4)] = 3a + [2a - a - 4]$
$= 3a + [a - 4]$
$= 3a + a - 4 = 4a - 4$

32. $5 - [3y + (y - 4) - 1] = 5 - [3y + y - 4 - 1]$
$= 5 - [4y - 5]$
$= 5 - 4y + 5 = 10 - 4y$

34. $-(x - 3) + [2x - (3 + x) - 2] = -x + 3 + [2x - 3 - x - 2]$
$= -x + 3 + [x - 5]$
$= -x + 3 + x - 5 = -2$

36. $[2y^2 - (4 - y)] + [y^2 - (2 + y)] = [2y^2 - 4 + y] + [y^2 - 2 - y]$
$= 2y^2 - 4 + y + y^2 - 2 - y$
$= 3y^2 - 6$

38. Let $x = 1$ and $y = 2$. Then

$$-(x - y) = -(1 - 2) = 1$$

and
$$-x - y = -1 - 2 = -3.$$

Therefore, $-(x - y)$ is not equivalent to $-x - y$.

40. $[x - (y + x)] - (2x - [3x - (x - y)] + y)$
$= [x - y - x] - (2x - [3x - x + y] + y)$
$= -y - (2x - [2x + y] + y)$
$= -y - (2x - 2x - y + y)$
$= -y - 0 = -y$

42. $-(2y - [2y - 4y + (y - 2)] + 1) + [2y - (4 - y) + 1]$
$= -(2y - [-2y + y - 2] + 1) + [2y - 4 + y + 1]$
$= -(2y - [-y -2] + 1) + [3y - 3]$
$= -(2y + y + 2 + 1) + [3y - 3]$
$= -(3y + 3) + [3y - 3]$
$= -3y - 3 + 3y - 3$
$= -6$

44. $P(x) - Q(x) + R(x) = x - 1 - (x^2 + 1) + (x^2 - x + 1)$
$= x - 1 - x^2 - 1 + x^2 - x + 1$
$= -1$

46. $Q(x) - [R(x) + P(x)] = x^2 + 1 - [(x^2 - x + 1) + (x - 1)]$
$= x^2 + 1 - [x^2]$
$= x^2 + 1 - x^2$
$= 1$

48. $P(x) - [R(x) - Q(x)] = x - 1 - [(x^2 - x + 1) - (x^2 + 1)]$
$= x - 1 - [x^2 - x + 1 - x^2 - 1]$
$= x - 1 - [-x]$
$= 2x - 1$

50. $Q(x) + [R(x) - P(x)] = x^2 + 1 + [(x^2 - x + 1) - (x - 1)]$
$= x^2 + 1 + [x^2 - x + 1 - x + 1]$
$= x^2 + 1 + [x^2 - 2x + 2]$
$= x^2 + 1 + x^2 - 2x + 2$
$= 2x^2 - 2x + 3$

EXERCISE 2.3

2. $(4c^3)(3c) = 4 \cdot 3c^{3+1}$
$= 12c^4$

4. $(-6r^2s^2)(5rs^3) = -6 \cdot 5r^{2+1}s^{2+3}$
$= -30r^3s^5$

6. $(-8abc)(-b^2c^3)$
$= (-8abc)(-1b^2c^3)$
$= -8 \cdot (-1)ab^{1+2}c^{1+3} = 8ab^3c^4$

8. $-5(ab^3)(-3a^2bc)$
$= -5(-3)a^{1+2}b^{3+1}c$
$= 15a^3b^4c$

10. $(-5mn)(2m^2n)(-n^3)$
$= (-5mn)(2m^2n)(-1 \cdot n^3)$
$= -5 \cdot 2 \cdot (-1)m^{1+2}n^{1+1+3}$
$= 10m^3n^5$

12. $(-3xy)(2xz^4)(3x^3y^2z)$
$= -3 \cdot 2 \cdot 3x^{1+1+3}y^{1+2}z^{4+1}$
$= -18x^5y^3z^5$

14. $-a^2(ab^2)(2a)(-3b^2)$
$= -1 \cdot 2 \cdot (-3)a^{2+1+1}b^{2+2}$
$= 6a^4b^4$

16. $(-2y^2)(y^2)(y)$
$= -2 \cdot y^{2+2+1}$
$= -2y^5$

18. If any one factor in a product equals zero, then the product equals zero. Therefore, $(-t)(2t^2)(-t)(0) = 0$.

20. x^{12}

22. x^2y^2

24. y^2z^8

26. $3^2(x^2)^2z^2 + 9x^4z^2$

28. $3^3(x^2)^3y^3(z^2)^3 = 27x^6y^3z^6$

30. $(-1)^3 3^3(x^2)^3y^3(z^3)^3$
$= -27x^6y^3z^9$

32. $2^4x^4(y^3)^4 + 3^2x^2y^2$
$= 16x^4y^{12} + 9x^2y^2$

34. $3^3(x^2)^3y^3z^3 - 4^2(x^2)^2(y^2)^2(z^2)^2$
$= 27x^6y^3z^3 - 16x^4y^4z^4$

36. $(x^2y)(x^2y^2) + x^2y^4$
$= x^4y^3 + x^2y^4$

38. $x^2y^2 + (x^4y^2)(-xy^2)$
$= x^2y^2 - x^5y^4$

40. $3x^4y^2 + x^3y - x^4y^2$
$= 2x^4y^2 + x^3y$

42. $2x^2y^3 + 4y(x^2y^2) - x^2y^3 = 2x^2y^3 + 4x^2y^3 - x^2y^3$
$= 5x^2y^3$

44. $b^{-n} \cdot b^{2n+1} = b^{-n+2n+1} = b^{n+1}$

46. $y^{2n+6} \cdot y^{4-n} = y^{2n+6+4-n} = y^{n+10}$

48. $b^{n+2} \cdot b^{2n-1} = b^{n+2+2n-1}$
$= b^{3n+1}$

50. $(xy^{3n})^2 = x^2y^{(3n)2}$
$= x^2y^{6n}$

52. $(xy^{n+2})^2 = x^2y^{2(n+2)}$
$= x^2y^{2n+4}$

54. $(x^{n-2}y^{2n+1})^2 = x^{2(n-2)}y^{2(2n+1)}$
$= x^{2n-4}y^{4n+2}$

EXERCISE 2.4

2. $3x(2x + y) = 3x(2x) + 3x(y)$
$= 6x^2 + 3xy$

4. $-2y(y^2 - 3y + 2)$
$= -2y(y^2) + (-2y)(-3y) + (-2y)(2)$
$= -2y^3 + 6y^2 - 4y$

6. $-(2y^2 - y + 3) = -1(2y^2 - y + 3)$
$= -2y^2 + y - 3$

8. $(y - 4)^2 = (y - 4)(y - 4)$
$= (y)(y) + (-4)(y) + (y)(-4) + (-4)(-4)$
$= y^2 + [(-4y) + (-4y)] + 16$
$= y^2 - 8y + 16$

10. $(3x + 2)^2 = (3x + 2)(3x + 2)$
$= (3x)(3x) + (2)(3x) + (3x)(2) + (2)(2)$
$= 9x^2 + [(6x) + (6x)] + 4$
$= 9x^2 + 12x + 4$

12. $(x - 7)(x + 7) = (x)(x) + (-7)(x) + (x)(7) + (-7)(7)$
$= x^2 + [(-7x) + (7x)] + (-49)$
$= x^2 - 49$

14. $(r - 1)(r - 6) = r \cdot r + [-1 \cdot r + (-6 \cdot r)] + (-1)(-6)$
$= r^2 - 7r + 6$

16. $(y - 2)(y + 3) = (y)(y) + [(-2)(y) + (y)(3)] + (-2)(3)$
$= y^2 + [-2y + 3y] + (-6) = y^2 + y - 6$

18. $(z - 3)(z - 5) = z^2 - 8z + 15$

20. $(3t - 1)(2t + 1) = 6t^2 + t - 1$

22. $(2z - 1)(3z + 5)$
$= 6z^2 + 7z - 5$

24. $(3t - 4s)(3t + 4s)$
$= 9t^2 - 16s^2$

26. $(t + 4)(t^2 - t - 1) = t(t^2 - t - 1) + 4(t^2 - t - 1)$
$= t^3 - t^2 - t + 4t^2 - 4t - 4$
$= t^3 + 3t^2 - 5t - 4$

28. $(x - 7)(x^2 - 3x + 1) = x(x^2 - 3x + 1) - 7(x^2 - 3x + 1)$
$= x^3 - 3x^2 + x - 7x^2 + 21x - 7$
$= x^3 - 10x^2 + 22x - 7$

30. $(y + 2)(y - 2)(y + 4) = [(y + 2)(y - 2)](y + 4)$
$= [y^2 - 4](y + 4)$
$= y^2(y + 4) - 4(y + 4)$
$= y^3 + 4y^2 - 4y - 16$

32. $(z - 5)(z + 6)(z - 1) = [(z - 5)(z + 6)](z - 1)$
$= [z^2 + z - 30](z - 1)$
$= z^2(z - 1) + z(z - 1) - 30(z - 1)$
$= z^3 - z^2 + z^2 - z - 30z + 30$
$= z^3 - 31z + 30$

34. $(3x - 2)(4x^2 + x - 2) = 3x(4x^2 + x - 2) - 2(4x^2 + x - 2)$
$= 12x^3 + 3x^2 - 6x - 8x^2 - 2x + 4$
$= 12x^3 - 5x^2 - 8x + 4$

36. $(b^2 - 3b + 5)(2b^2 - b + 1)$
$= b^2(2b^2 - b + 1) - 3b(2b^2 - b + 1) + 5(2b^2 - b + 1)$
$= 2b^4 - b^3 + b^2 - 6b^3 + 3b^2 - 3b + 10b^2 - 5b + 5$
$= 2b^4 - 7b^3 + 14b^2 - 8b + 5$

38. $3[2a - (a + 1) + 3] = 3[2a - a - 1 + 3]$
$= 3[a + 2]$
$= 3a + 6$

40. $-2a[3a + (a - 3) - (2a + 1)] = -2a[3a + a - 3 - 2a - 1]$
$= -2a[2a - 4] = -4a^2 + 8a$

42. $-[(a + 1) - 2(3a - 1) + 4] = -[a + 1 - 6a + 2 + 4]$
$= -[-5a + 7]$
$= 5a - 7$

44. $-4(4 - [3 - 2(a - 1) + a] + a) = -4(4 - [3 - 2a + 2 + a] + a)$
$= -4(4 - [-a + 5] + a)$
$= -4(4 + a - 5 + a)$
$= -4(2a - 1) = -8a + 4$

46. $x(4 - 2[3 - 4(x + 1)] - x) = x(4 - 2[3 - 4x - 4] - x)$
$= x(4 - 2[-4x - 1] - x)$
$= x(4 + 8x + 2 - x)$
$= x(7x + 6)$
$= 7x^2 + 6x$

48. $3[2x + (x + 2)^2] = 3[2x + x^2 + 4x + 4]$
$= 3[x^2 + 6x + 4]$
$= 3x^2 + 18x + 12$

50. $-2x + x[3 - (x + 4)^2] = -2x + x[3 - (x^2 + 8x + 16)]$
$= -2x + x[3 - x^2 - 8x - 16]$
$= -2x + x[-x^2 - 8x - 13]$
$= -2x - x^3 - 8x^2 - 13x$
$= -x^3 - 8x^2 - 15x$

52. $-x[2x - (2x + 1)^2 + 3] = -x[2x - (4x^2 + 4x + 1) + 3]$
$= -x[-4x^2 - 2x + 2]$
$= 4x^3 + 2x^2 - 2x$

54. $-3[2x^2 - 3(x - 2)(x + 3) + 3x]$
$= -3[2x^2 - 3(x^2 + x - 6) + 3x]$
$= -3[2x^2 - 3x^2 - 3x + 18 + 3x]$
$= -3[-x^2 + 18]$
$= 3x^2 - 54$

56. $(x + a)^2 = (x + a)(x + a) = x^2 + 2ax + a^2$

58. $(ax + by)(cx + dy) = ax(cx + dy) + by(cx + dy)$
$= acx^2 + adxy + bcxy + bdy^2;$
$= acx^2 + (ad + bc)xy + bdy^2$

Therefore, $(ax + by)(cx + dy) = acx^2 + (ad + bc)xy + bdy^2$.

60. $(x - a)(x^2 + ax + a^2) = x(x^2 + ax + a^2) - a(x^2 + ax + a^2)$
$= x^3 + ax^2 + a^2x - ax^2 - a^2x - a^3$
$= x^3 - a^3$

62. Let $x = 2$ and $y = 1$. Then
$(x - y)^2 = (2 - 1)^2 = 1$
and $x^2 - y^2 = 2^2 - 1^2 = 3$.
Therefore, $(x - y)^2$ is not equivalent to $x^2 - y^2$.

64. $3t^n(2t^n + 3)$
$= 3t^n(2t^n) + 3t^n(3)$
$= 6t^{n+n} + 9t^n$
$= 6t^{2n} + 9t^n$

66. $b^{n-1}(b + b^n)$
$= b^{n-1}(b) + b^{n-1}(b^n)$
$= b^{n-1+1} + b^{n-1+n}$
$= b^n + b^{2n-1}$

68. $b^{2n+2}(b^{n-1} + b^n) = b^{2n+2}(b^{n-1}) + b^{2n+2}(b^n)$
$= b^{2n+2+n-1} + b^{2n+2+n}$
$= b^{3n+1} + b^{3n+2}$

70. $(a^n - 3)(a^n + 2)$
$= a^{2n} - 3a^n + 2a^n - 6$
$= a^{2n} - a^n - 6$

72. $(a^{3n} - 3)(a^{3n} + 3)$
$= a^{6n} - 9$

74. $(a^{2n} - 2b^n)(a^{3n} + b^{2n}) = a^{2n}(a^{3n} + b^{2n}) - 2b^n(a^{3n} + b^{2n})$
$= a^{5n} + a^{2n}b^{2n} - 2a^{3n}b^n - 2b^{3n}$

3 Equations and Inequalities

EXERCISE 3.1

2.
$$\begin{aligned} 2 + 5x &= 37 \\ 2 + 5x + (-2) &= 37 + (-2) \\ 5x &= 35 \\ \frac{1}{5}(5x) &= \frac{1}{5}(35) \\ x &= 7; \quad \{7\} \end{aligned}$$

4.
$$\begin{aligned} 3x - 5 &= 7 \\ 3x - 5 + (5) &= 7 + (5) \\ 3x &= 12 \\ \frac{1}{3}(3x) &= \frac{1}{3}(12) \\ x &= 4; \quad \{4\} \end{aligned}$$

6.
$$\begin{aligned} y + (y - 160) &= 830 \\ y + y - 160 &= 830 \\ 2y - 160 &= 830 \\ 2y - 160 + (160) &= 830 + 160 \\ 2y &= 990 \\ \frac{1}{2}(2y) &= \frac{1}{2}(990) \\ y &= 495; \quad \{495\} \end{aligned}$$

8.
$$\begin{aligned} y + (y - 620) + (y - 810) &= 8630 \\ y + y - 620 + y - 810 &= 8630 \\ 3y - 1430 &= 8630 \\ 3y &= 10060 \\ y &= \frac{10060}{3}; \quad \{\frac{10,060}{3}\} \end{aligned}$$

10.
$$\begin{aligned} 2(z - 3) &= 15 \\ 2z - 6 &= 15 \\ 2z &= 21 \\ z &= \frac{21}{2}; \quad \{\frac{21}{2}\} \end{aligned}$$

12.
$$\begin{aligned} 5z - (z + 1) &= 14 \\ 5z - z - 1 &= 14 \\ 4z &= 15 \\ z &= \frac{15}{4}; \quad \{\frac{15}{4}\} \end{aligned}$$

14.
$$\begin{aligned} 3[3x - 2(x - 3) + 1] &= 8 \\ 3[3x - 2x + 6 + 1] &= 8 \\ 3[x + 7] &= 8 \\ 3x + 21 &= 8 \\ 3x &= -13 \\ x &= \frac{-13}{3}; \quad \{\frac{-13}{3}\} \end{aligned}$$

16.
$$\begin{aligned} 2[1 - 3(x + 1) - 2x] &= 12 \\ 2[1 - 3x - 3 - 2x] &= 12 \\ 2[-5x - 2] &= 12 \\ -10x - 4 &= 12 \\ -10x &= 16 \\ x &= \frac{-16}{10} = \frac{-8}{5}; \quad \{-\frac{8}{5}\} \end{aligned}$$

18.
$$\begin{aligned}
-[7 - (y - 3) - 4y] &= 0 \\
-[7 - y + 3 - 4y] &= 0 \\
-[-5y + 10] &= 0 \\
5y - 10 &= 0 \\
5y &= 10 \\
y &= 2; \quad \{2\}
\end{aligned}$$

20.
$$\begin{aligned}
-3[2y - (y - 2)] &= 2(y + 3) \\
-3[2y - y + 2] &= 2y + 6 \\
-3[y + 2] &= 2y + 6 \\
-3y - 6 &= 2y + 6 \\
-5y &= 12 \\
y &= \frac{-12}{5}; \quad \{\frac{-12}{5}\}
\end{aligned}$$

22.
$$\begin{aligned}
(z + 2)^2 - z^2 &= 8 \\
z^2 + 4z + 4 - z^2 &= 8 \\
4z + 4 &= 8 \\
4z &= 4 \\
z &= 1; \quad \{1\}
\end{aligned}$$

24.
$$\begin{aligned}
2 + (x - 2)(x + 3) &= x^2 + 12 \\
2 + x^2 + x - 6 &= x^2 + 12 \\
x - 4 &= 12 \\
x &= 16; \quad \{16\}
\end{aligned}$$

26.
$$\begin{aligned}
-3[y - (y + 2)^2] &= 3y^2 - 10 \\
-3[y - (y^2 + 4y + 4)] &= 3y^2 - 10 \\
-3[y - y^2 - 4y - 4] &= 3y^2 - 10 \\
-3[-y^2 - 3y - 4] &= 3y^2 - 10 \\
3y^2 + 9y + 12 &= 3y^2 - 10 \\
9y + 12 &= -10 \\
9y &= -22 \\
y &= \frac{-22}{9}; \quad \{\frac{-22}{9}\}
\end{aligned}$$

28.
$$\begin{aligned}
2[2x + 3(x + 1)^2 - x^2] &= 4x^2 + 1 \\
2[2x + 3(x^2 + 2x + 1) - x^2] &= 4x^2 + 1 \\
2[2x + 3x^2 + 6x + 3 - x^2] &= 4x^2 + 1 \\
2[2x^2 + 8x + 3] &= 4x^2 + 1 \\
4x^2 + 16x + 6 &= 4x^2 + 1 \\
16x + 6 &= 1 \\
16x &= -5 \\
x &= \frac{-5}{16}; \quad \{\frac{-5}{16}\}
\end{aligned}$$

30.
$$\begin{aligned}
-[(2x + 1)^2 + 2x] &= -(2x - 1)^2 - 2x \\
-[4x^2 + 4x + 1 + 2x] &= -(4x^2 - 4x + 1) - 2x \\
-4x^2 - 6x - 1 &= -4x^2 + 4x - 1 - 2x \\
-6x - 1 &= 2x - 1 \\
-8x &= 0 \\
x &= 0; \quad \{0\}
\end{aligned}$$

32. $0.60(y + 2) = 3.60$
$60(y + 2) = 360$
$60y + 120 = 360$
$60y = 240$
$y = 4;$ $\{4\}$

34. $0.12y + 0.08(y + 10{,}000) = 12{,}000$
$12y + 8(y + 10{,}000) = 1{,}200{,}000$
$12y + 8y + 80{,}000 = 1{,}200{,}000$
$20y = 1{,}120{,}000$
$y = 56{,}000;$ $\{56{,}000\}$

36. $0.10x + 0.12(x + 4000) = 920$
$10x + 12(x + 4000) = 92000$
$10x + 12x + 48000 = 92000$
$22x + 48000 = 92000$
$22x = 44000$
$x = 2000;$ $\{2000\}$

38. $18 = \frac{1}{2}b(9)$
$18 = \frac{9}{2}b$
$36 = 9b$
$4 = b;$ $\{4\}$

40. $48 = (4)w(2)$
$48 = 8w$
$6 = w;$ $\{6\}$

42. $2320 = 1000 + 1000(0.11)t$
$2320 = 1000 + 110t$
$1320 = 110t$
$12 = t$

44. $pv = k$
$v = \frac{k}{v}$

46. $v = lwh$
$\frac{v}{lw} = h$

48. $S = 3\pi d + \pi d$
$S = 4\pi d$
$\frac{S}{4\pi} = d$

50. $v = k + gt$
$v - k = gt$
$\frac{v - k}{g} = t$

52. $S(1 - r) = a$
$S - Sr = a$
$-Sr = a - S$
$r = \frac{a - S}{-S}$ or $\frac{S - a}{S}$

54. $2A = h(b + c)$
$2A = hb + hc$
$2A - hb = hc$
$\frac{2A - hb}{h} = c$

56. $P(1 + rt) = A$
$P + Prt = A$
$Prt = A - P$
$t = \frac{A - P}{Pr}$

EXERCISE 3.2

2. $2 + 5x = 37$

4. $3x - 5 = 7$

6. $3x + 2x = 40$

8. $\frac{x - 2}{5} = 3$

10. Total salary: x

$\frac{4}{5}x = 120$ (Note: $80\% = \frac{4}{5}$)

$4x = 5(120) = 600$

$x = 150$

Total salary is \$150.

12. Total number of gears manufactured: x
Number of gears defective: $0.03x$

$$x - 0.03x = 1200$$
$$0.97x = 1200$$
$$x \approx 1237.11$$

1238 gears must be manufactured to meet specifications.

14. Number of shots attempted: x

$0.40x = 8$

$x = \frac{80}{0.40} = 20$

He attempted 20 shots.

16. Number of transistors ordered: x

$x + 0.02x = 22400$

$1.02x = 22400$

$x = \frac{22400}{1.02} = 21960.784$

21961 transistors were ordered.

18. Original cost: x

$x - 0.40x = 384$

$0.60x = 384$

$x = \frac{384}{0.60} = 640$

The original cost was \$640.

20. The shorter piece: x
The longer piece: $2x$

$x + 2x = 36$

$3x = 36$

$x = 12$

$2x = 24$

The shorter piece is 12 feet.
The longer piece is 24 feet.

22. The number of votes for the winner: x
The number of votes for the loser: $x - 160$

$$x + x - 160 = 1830$$
$$2x - 160 = 1830$$
$$2x = 1990$$
$$x = 995$$
$$x - 160 = 835$$

The winner received 995 votes and the loser received 835 votes.

24. The number of votes for the first candidate: x
The number of votes for the second candidate: $x - 620$
The number of votes for the third candidate: $x - 810$

$$\begin{aligned} x + x - 620 + x - 810 &= 8650 \\ 3x - 1430 &= 8650 \\ 3x &= 10080 \\ x &= 3360 \\ x - 620 &= 2740 \\ x - 810 &= 2550 \end{aligned}$$

The first, second, and third candidates received 3360, 2740, and 2550 votes, respectively.

26. Measure of third angle: x
Measure of smallest angle: $x - 50$
Measure of remaining angle: $x - 25$

$$(x) + (x - 50) + (x - 25) = 180$$

$$\begin{aligned} 3x - 75 &= 180 \\ 3x &= 255 \\ x &= 85,\ x - 25 = 60,\ x - 50 = 35 \end{aligned}$$

The angles of the triangle measure 35°, 60°, and 85°.

28. Measure of "another" angle: x
Measure of "one" angle: $x + 20$
Measure of third angle: $6x$

$$\begin{aligned} x + x + 20 + 6x &= 180 \\ 8x + 20 &= 180 \\ 8x &= 160 \\ x &= 20,\ x + 20 = 40,\ 6x = 120 \end{aligned}$$

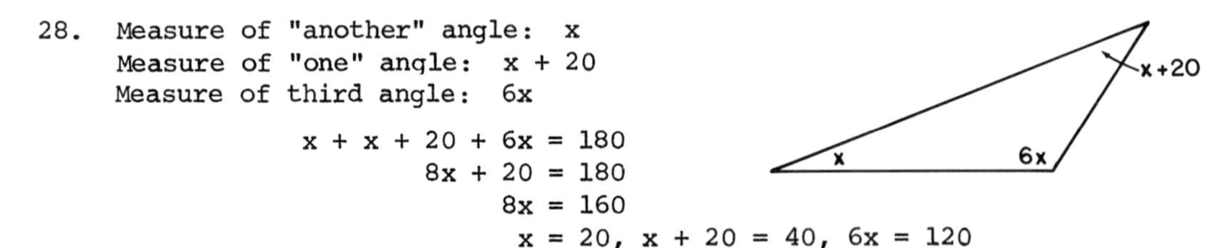

The angles of the triangle measure 20°, 40°, and 120°.

30. Length of shortest side: x
Length of longest side: $3x$
Length of third side: $x + 72$

$$\begin{aligned} x + 3x + x + 72 &= 272 \\ 5x + 72 &= 272 \\ 5x &= 200 \\ x &= 40,\ 3x = 120,\ x + 72 = 112 \end{aligned}$$

The lengths of the three sides are 40 cm, 120 cm, and 112 cm.

32. Length of side of original square: x
Length of side of final square: $x - 6$

$$\begin{pmatrix}\text{Area of}\\\text{Original Square}\end{pmatrix} - 84 = \begin{pmatrix}\text{Area of}\\\text{Final Square}\end{pmatrix}$$

$$x^2 - 84 = (x - 6)^2$$
$$x^2 - 84 = x^2 - 12x + 36$$
$$12x = 120$$
$$x = 10,\ x - 6 = 4$$

The length of side of the original square is 10 inches.

34. Width: x
Length: $2x + 24$

$$2x + 2(2x + 24) = 210$$
$$2x + 4x + 48 = 210$$
$$6x + 48 = 210$$
$$6x = 162$$
$$x = 27,\ 2x + 24 = 78$$

The width is 27 feet and the length is 78 feet.

36. Number of nickels: n
Number of dimes: $n + 22$

Denomination	Value of 1 Coin	Number of Coins	Value of Coins
Nickels	0.05	n	$0.05n$
Dimes	0.10	$n + 22$	$0.10(n + 22)$

$$\begin{pmatrix}\text{Value of}\\\text{nickels}\end{pmatrix} + \begin{pmatrix}\text{Value of}\\\text{dimes}\end{pmatrix} = \begin{pmatrix}\text{Value of}\\\text{collection}\end{pmatrix}$$

$$0.05n + 0.10(n + 22) = 4.90$$
$$5n + 10(n + 22) = 490$$
$$5n + 10n + 220 = 490$$
$$15n = 270$$
$$n = 18,\ n + 22 = 40$$

There are 18 nickels and 22 dimes.

38. Number of dimes: x
Number of quarters: $x - 3$
Number of nickels: $3x$

$$\begin{bmatrix}\text{value of}\\\text{dimes}\end{bmatrix} + \begin{bmatrix}\text{value of}\\\text{quarters}\end{bmatrix} + \begin{bmatrix}\text{value of}\\\text{nickels}\end{bmatrix} = \begin{bmatrix}\text{value of}\\\text{collection}\end{bmatrix}$$

$$0.10x + 0.25(x-3) + 0.05(3x) = 1.75$$
$$10x + 25(x-3) + 5(3x) = 175$$
$$10x + 25x - 75 + 15x = 175$$
$$50x - 75 = 175$$
$$50x = 250$$
$$x = 5,\ x - 3 = 2,\ 3x = 15$$

There are 5 dimes, 2 quarters, and 15 nickels.

40. Number of chocolate-covered bars: x
Number of sandwich bars: $50 - x$

$$\begin{bmatrix}\text{value of chocolate-}\\\text{covered bars}\end{bmatrix} + \begin{bmatrix}\text{value of}\\\text{sandwich bars}\end{bmatrix} = \begin{bmatrix}\text{value of all}\\\text{the bars}\end{bmatrix}$$

$$0.18x + 0.15(50-x) = 8.10$$
$$18x + 15(50-x) = 810$$
$$18x + 750 - 15x = 810$$
$$3x + 750 = 810$$
$$3x = 60$$
$$x = 20$$
$$50 - x = 30$$

He purchased 20 chocolate-covered bars and 30 sandwich bars.

42. Number of 16-cent bolts: x
Number of 12-cent bolts: $3x$

$$\begin{bmatrix}\text{value of}\\\text{16-cent bolts}\end{bmatrix} + \begin{bmatrix}\text{value of}\\\text{12-cent bolts}\end{bmatrix} = \begin{bmatrix}\text{total value}\\\text{of purchase}\end{bmatrix}$$

$$0.16x + 0.12(3x) = 7.80$$
$$16x + 12(3x) = 780$$
$$16x + 36x = 780$$
$$52x = 780$$
$$x = 15,\ 3x = 45$$

There were 15 16-cent bolts and 45 12-cent bolts purchased.

44. Number of first-class tickets: x
Number of tourist-class tickets: $48 - x$

$$\begin{bmatrix}\text{value of first-}\\\text{class tickets}\end{bmatrix} + \begin{bmatrix}\text{value of tourist-}\\\text{class tickets}\end{bmatrix} = \begin{bmatrix}\text{total value of}\\\text{all tickets}\end{bmatrix}$$

$$240x + 150(48 - x) = 8100$$

$$\begin{aligned} 240x + 7200 - 150x &= 8100 \\ 90x + 7200 &= 8100 \\ 90x &= 900 \\ x &= 10,\ 48 - x = 38 \end{aligned}$$

There were 10 first-class and 38 tourist-fare tickets sold.

46. Amount invested at 8%: A
Amount invested at 10%: $A + 3000$

Investment	Rate	Amount	Interest
8%	0.08	A	0.08A
10%	0.10	A + 3000	0.10 (A + 3000)

$$\begin{bmatrix}\text{Interest from}\\\text{8\% investment}\end{bmatrix} + \begin{bmatrix}\text{Interest from}\\\text{10\% investment}\end{bmatrix} = \begin{bmatrix}\text{Total interest}\end{bmatrix}$$

$$0.08A + 0.10(A + 3000) = 1110$$
$$8A + 10(A + 3000) = 111,000$$

$$\begin{aligned} 8A + 10A + 30000 &= 111,000 \\ 18A + 30000 &= 111,000 \\ 18A &= 81000 \\ A &= 4500 \\ A + 3000 &= 7500 \end{aligned}$$

\$4500 is invested at 8% and \$7500 is invested at 10%.

48. Amount invested at 8%: A
Amount invested at 12%: 2A

Investment Rate		Amount	Interest
8%	0.08	A	0.08A
12%	0.12	2A	0.12(2A)

$$\begin{bmatrix}\text{Interest from}\\ \text{8\% investment}\end{bmatrix} + \begin{bmatrix}\text{Interest from}\\ \text{12\% investment}\end{bmatrix} = \begin{bmatrix}\text{Total interest}\end{bmatrix}$$

$$0.08A + 0.12(2A) = 1280$$

$$\begin{aligned} 0.08A + 0.24A &= 1280 \\ 0.32A &= 1280 \\ A &= 4000 \\ 2A &= 8000 \end{aligned}$$

\$4000 is invested at 8% and \$8000 is invested at 12%.

50. Amount invested at 14%: x
Amount invested at 11%: 2400 - x
Return on 14% investment: 0.14x
Return on 11% investment: 0.11(2400 - x)

$$0.14x - 0.11(2400 - x) = 111$$

$$\begin{aligned} 14x - 11(2400 - x) &= 11{,}100 \\ 14x - 26{,}400 + 11x &= 11{,}100 \\ 25x &= 37{,}500 \\ x &= 1500 \\ 2400 - x &= 900 \end{aligned}$$

Hence, \$1500 is invested at 14% and \$900 is invested at 11%.

52. Amount invested at 12%: x
Interest at 9%: 0.09(6000)
Interest at 12%: 0.12x

$$\begin{bmatrix}\text{Interest from}\\ \text{9\% investment}\end{bmatrix} + \begin{bmatrix}\text{Interest from}\\ \text{12\% investment}\end{bmatrix} = \begin{bmatrix}\text{Interest from 11\%}\\ \text{of total investment}\end{bmatrix}$$

$$0.09(6000) + 0.12x = 0.11(6000 + x)$$

$$\begin{aligned} 9(6000) + 12x &= 11(6000 + x) \\ 54{,}000 + 12x &= 66{,}000 + 11x \\ x &= 12{,}000 \end{aligned}$$

Additional money necessary at 12% to give return of 11% is \$12,000.

54. Number of quarts of 30% solution: n

Percent salt in mixture		Number of Quarts	Amount of Salt
30%	0.30	n	0.30n
10%	0.10	50	0.10(50)
25%	0.25	n + 50	0.25(n + 50)

$$\begin{bmatrix}\text{Pure salt in}\\ \text{30\% solution}\end{bmatrix} + \begin{bmatrix}\text{Pure salt in}\\ \text{10\% solution}\end{bmatrix} = \begin{bmatrix}\text{Pure salt in}\\ \text{25\% solution}\end{bmatrix}$$

$$0.30n + 0.10(50) = 0.25(n + 50)$$

$$\begin{aligned} 30n + 10(50) &= 25(n + 50) \\ 30n + 500 &= 25n + 1250 \\ 5n &= 750 \\ n &= 150 \end{aligned}$$

150 quarts of a 30% solution must be added.

56. Number of pounds of 30% copper alloy: n

Percent copper in alloy		Number of pounds	Amount of copper
30%	0.30	n	0.30n
25%	0.25	12	0.25(12)
10%	0.10	12 - n	0.10(12 - n)

$$\begin{bmatrix}\text{Copper in}\\ \text{30\% alloy}\end{bmatrix} + \begin{bmatrix}\text{Copper in}\\ \text{10\% alloy}\end{bmatrix} = \begin{bmatrix}\text{Copper in}\\ \text{25\% alloy}\end{bmatrix}$$

$$0.30n + 0.10(12 - n) = 0.25(12)$$

$$\begin{aligned} 30n + 10(12 - n) &= 25(12) \\ 30n + 120 - 10n &= 300 \\ 20n &= 180 \\ n &= 9 \end{aligned}$$

9 pounds of 30% copper alloy are needed.

58. Number of grams of alloy containing 45% silver: n

Percent of silver in alloy		Amount of alloy	Amount of silver
45%	0.45	n	0.45n
60%	0.60	40 − n	0.60(40 − n)
48%	0.48	40	0.48(40)

$$\begin{bmatrix}\text{Silver in}\\ \text{45\% alloy}\end{bmatrix} + \begin{bmatrix}\text{Silver in}\\ \text{60\% alloy}\end{bmatrix} = \begin{bmatrix}\text{Silver in}\\ \text{48\% alloy}\end{bmatrix}$$

$$(0.45n) + 0.60(40-n) = 0.48(40)$$

$$\begin{aligned} 45n + 60(40 - n) &= 48(40) \\ 45n + 2400 - 60n &= 1920 \\ -15n &= -480 \\ n &= 32 \end{aligned}$$

32 grams of the 45% alloy are needed.

60. Number of liters of 20% sugar solution: n

Percent of sugar in solution		Amount of solution	Amount of sugar
20%	0.20	n	0.20n
32%	0.32	40	0.32(40)
28%	0.28	n + 40	0.28(n + 40)

$$\begin{aligned} 20n + 32(40) &= 28(n + 40) \\ 20n + 1280 &= 28n + 1120 \\ 8n &= 160 \\ n &= 20 \end{aligned}$$

20 liters of the 20% sugar solution should be added.

EXERCISE 3.3

2. $x + 7 + \boxed{(-7)} > 8 + \boxed{(-7)}$

$x > 1$

$\{x|x > 1\}$ or $(1,+\infty)$

4. $2x - 3 + \boxed{(3)} < 4 + \boxed{(3)}$

$2x < 7$

$\left(\frac{1}{2}\right)2x < \left(\frac{1}{2}\right)7$

$x < \frac{7}{2}$

$\left\{x|x < \frac{7}{2}\right\}$ or $\left(-\infty,\frac{7}{2}\right)$

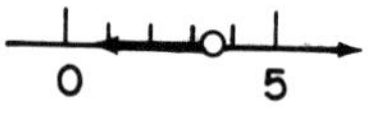

6. $2x + 3 + \boxed{(-x) + (-3)} \le x - 1 + \boxed{(-x) + (-3)}$

$x \le -4$; $\{x|x \le -4\}$ or $(-\infty,-4]$

8. $(2)\frac{2x - 3}{2} \le (2)5$

$2x - 3 \le 10$

$2x \le 13$

$x \le \frac{13}{2}$

$\left\{x|x \le \frac{13}{2}\right\}$ or $\left(-\infty,\frac{13}{2}\right]$

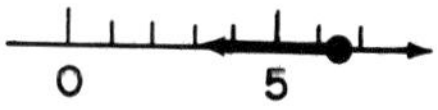

10. $\frac{-2x}{5} \le 6$

$-2x \le 30$

$\left(-\frac{1}{2}\right)(-2)x \ge \left(-\frac{1}{2}\right)30$

$x \ge -15$

$\{x|x \ge -15\}$ or $[-15,+\infty)$

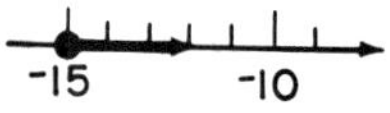

12. $\frac{-5x}{2} < -20$

$(2)\frac{-5x}{2} < (2)(-20)$

$-5x < -40$

$\left(-\frac{1}{5}\right)(-5)x > \left(-\frac{1}{5}\right)(-40)$

$x > 8$; $\{x|x > 8\}$ or $(8,+\infty)$

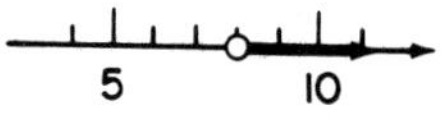

14. $6x \geq 5x - 9$

$x \geq -9$

$\{x|x \geq -9\}$ or $[-9,+\infty)$

-10 -5 0

16. $-2x < 5x - 6$

$-7x < -6$

$x > \frac{6}{7}$

$\left\{x|x > \frac{6}{7}\right\}$ or $\left(\frac{6}{7},+\infty\right)$

0 $\frac{6}{7}$ 5

18. $\left(\frac{1}{2}\right)0 \leq \left(\frac{1}{2}\right)2x \leq \left(\frac{1}{2}\right)12$

$0 \leq x \leq 6$; $\{x|0 \leq x \leq 6\}$ or $[0,6]$

0 5

20. $2 + (4) \leq 3x - 4 + (4) \leq 8 + (4)$

$6 \leq 3x \leq 12$

$\left(\frac{1}{3}\right)6 \leq \left(\frac{1}{3}\right)3x \leq \left(\frac{1}{3}\right)12$

$2 \leq x \leq 4$; $\{x|2 \leq x \leq 4\}$ or $[2,4]$

0 5

22. $-3 + (-3) < 3 - 2x + (-3) < 9 + (-3)$

$-6 < -2x < 6$

$\left(-\frac{1}{2}\right)(-6) > \left(-\frac{1}{2}\right)(-2x) > \left(-\frac{1}{2}\right)6$

$3 > x > -3$ or $-3 < x < 3$

$\{x|-3 < x < 3\}$ or $(-3,3)$

-3 0 3

24. $\{x|x \leq 5\}$

$\{x|x \geq 1\}$

0 5

$\{x|x \leq 5\} \cap \{x|x \geq 1\}$ or $\{x|1 \leq x \leq 5\}$

26. $2x - 3 < 5$, $-2x + 3 < 5$

$2x < 8$ $\quad -2x < 2$

$x < 4$ $\quad x > -1$

The graph of the given set is the same as the graph of

$\{x|x < 4\} \cap \{x|x > -1\}$.

$\{x|x < 4\}$

$\{x|x > -1\}$

-1 0 5

$\{x|x < 4\} \cap \{x|x > -1\}$ or $\{x|-1 < x < 4\}$

28. $9 \le 3x < 21,\quad -10 < 5x \le 15$
$3 \le x < 7 \qquad -2 < x \le 3$

The graph of the given set is the same as the graph of

$\{x \mid 3 \le x < 7\} \cap \{x \mid -2 < x \le 3\}$.

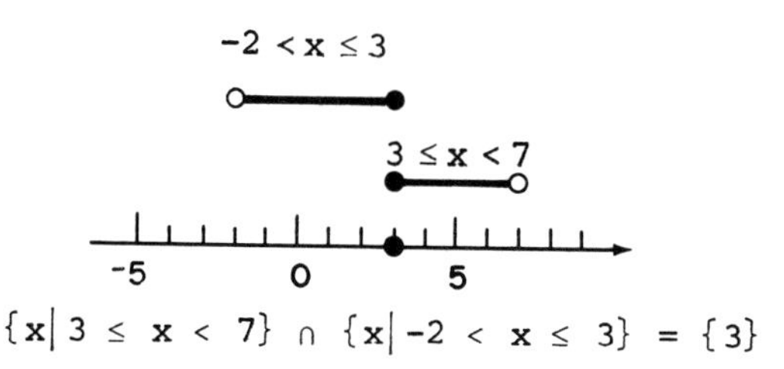

$\{x \mid 3 \le x < 7\} \cap \{x \mid -2 < x \le 3\} = \{3\}$

30. Grade on final exam: x

$$80 \le \frac{72 + 68 + 84 + 70 + 2x}{6} < 90$$
$$480 \le 72 + 68 + 84 + 70 + 2x < 540$$
$$480 \le 294 + 2x < 540$$
$$186 \le 2x < 246$$
$$93 \le x < 123$$

Hence, any grade greater than or equal to 93 and less than or equal to 100 would provide a grade of B for the course.

32. Temperature in degrees Fahrenheit: F

$$-10 < \frac{5}{9}(F - 32) < 20$$
$$-90 < 5(F - 32) < 180$$
$$-18 < F - 32 < 36$$
$$14 < F < 68$$

The temperature in Fahrenheit must be between 14° and 68°.

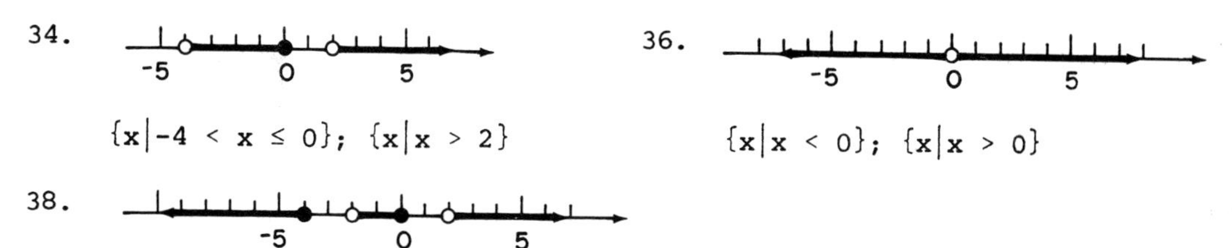

34. $\{x \mid -4 < x \le 0\};\ \{x \mid x > 2\}$

36. $\{x \mid x < 0\};\ \{x \mid x > 0\}$

38. $\{x \mid x \le -4\};\ \{x \mid -2 < x \le 0\};\ \{x \mid x > 2\}$

For each of the Exercises 40-46, if necessary graph the given sets on the same number line to determine the required intersection.

40. $[-1, 0]$

42. Disjoint

44. $(-4,0] \cap \emptyset = \emptyset$

Hence the given sets are disjoint.

46. $(-5,-3)$

EXERCISE 3.4

2. $x = 7$ or $-x = 7$

$\{7\} \cup \{-7\} = \{7,-7\}$

4. $x - 3 = 7$ or $-(x - 3) = 7$

$x = 10$ $\quad -x + 3 = 7$

$-x = 4$

$\{10\} \cup \{-4\} = \{10,-4\}$

6. $3x - 1 = 5$ or $-(3x - 1) = 5$

$3x = 6$ $\quad -3x + 1 = 5$

$x = 2$ $\quad -3x = 4$

$x = \frac{-4}{3}$

$\{2\} \cup \left\{\frac{-4}{3}\right\} = \left\{2,\frac{-4}{3}\right\}$

8. $6 - 5x = 4$ or $-(6 - 5x) = 4$

$-5x = -2$ $\quad -6 + 5x = 4$

$x = \frac{2}{5}$ $\quad 5x = 10$

$x = 2$

$\left\{\frac{2}{5}\right\} \cup \{2\} = \left\{\frac{2}{5},2\right\}$

10. $3x - 7 = 0$

$3x = 7$

$\left\{\frac{7}{3}\right\}$

12. $3x - 2 = 2$ or $3x - 2 = -2$

$3x = 4$ $\quad 3x = 0$

$x = \frac{4}{3}$ $\quad x = 0$

$\left\{\frac{4}{3}\right\} \cup \{0\} = \left\{0,\frac{4}{3}\right\}$

14. $-5 < x < 5$

$\{x|-5 < x < 5\}$ or $(-5,5)$

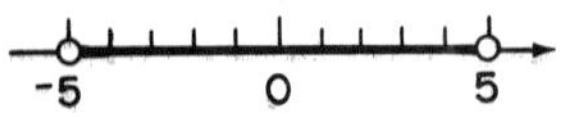

16. $-8 \le x + 1 \le 8$

$-8 + (-1) \le x + 1 + (-1) \le 8 + (-1)$

$-9 \le x \le 7$

$\{x|-9 \le x \le 7)$ or $[-9,7]$

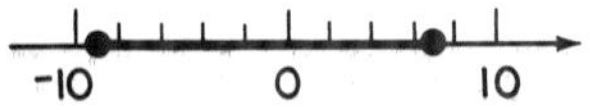

18. $-6 < 2x + 4 < 6$

$-6 + (-4) < 2x + 4 + (-4) < 6 + (-4)$

$-10 < 2x < 2$

$\left(\frac{1}{2}\right)(-10) < \left(\frac{1}{2}\right)2 < \left(\frac{1}{2}\right)2$

$-5 < x < 1$; $\{x \mid -5 < x < 1\}$ or $(-5,1)$

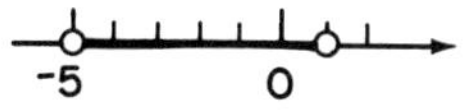

-5 0

20. $-15 \le 5 - 2x \le 15$

$-15 + (-5) \le (-5) + 5 - 2x \le 15 + (-5)$

$-20 \le -2x \le 10$

$\left(-\frac{1}{2}\right)(-20) \ge \left(-\frac{1}{2}\right)(-2x) \ge \left(-\frac{1}{2}\right)10$

$10 \ge x \ge -5$ or $-5 \le x \le 10$

$\{x \mid -5 \le x \le 10\}$ or $[-5,10]$

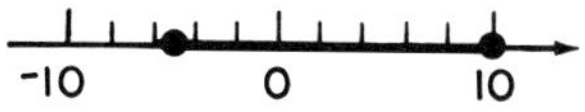

-10 0 10

22. $x \ge 5$ or $-x \ge 5$

$x \le -5$; $\{x \mid x \le -5\} \cup \{x \mid x \ge 5\}$

or $(-\infty,-5] \cup [5,+\infty)$

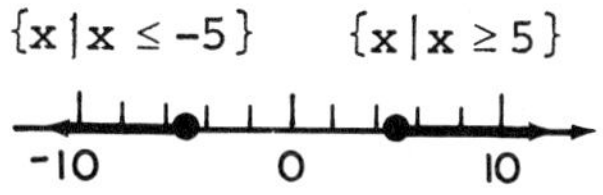

$\{x \mid x \le -5\}$ $\{x \mid x \ge 5\}$

-10 0 10

24. $x + 5 > 2$ or $-(x + 5) > 2$

$x > -3$ $\quad -x - 5 > 2$

$-x > 7$

$x < -7$; $\{x \mid x < -7\} \cup \{x \mid x > -3\}$

or $(-\infty,-7) \cup (-3,+\infty)$

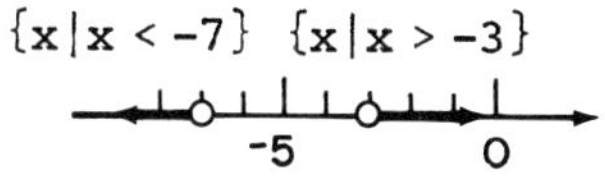

$\{x \mid x < -7\}$ $\{x \mid x > -3\}$

-5 0

26. $4 - 3x > 10$ or $-(4 - 3x) > 10$

$-3x > 6$ $\quad -4 + 3x > 10$

$x < -2$ $\quad 3x > 14$

$x > \frac{14}{3}$; $\{x \mid x < -2\} \cup \left\{x \mid x > \frac{14}{3}\right\}$

or $(-\infty,-2) \cup \left(\frac{14}{3},+\infty\right)$

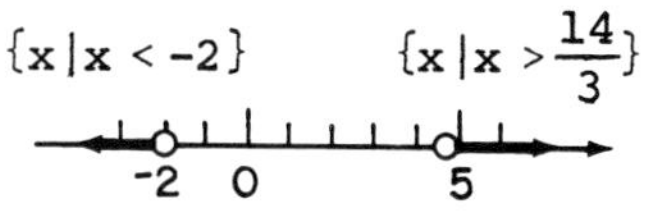

$\{x \mid x < -2\}$ $\left\{x \mid x > \frac{14}{3}\right\}$

-2 0 5

28. $-5 < x - 2 < 5$

$-3 < x < 7$

The solution set is $(-3,7)$.

30. $3x + 5 \geq 2$ or $-(3x + 5) \geq 2$

$3x \geq -3$ $\qquad -3x - 5 \geq 2$

$x \geq -1$ $\qquad -3x \geq 7$

$x \leq -\frac{7}{3}$

$\left(-\infty,-\frac{7}{3}\right] \cup [-1,+\infty)$

32. $2x - 6 = 0$

$2x = 6$

$x = 3$

$\{3\}$

34. $3x + 2 > 5$ or $-(3x + 2) > 5$

$3x > 3$ $\qquad -3x - 2 > 5$

$x > 1$ $\qquad -3x > 7$

$x < -\frac{7}{3}$

$\left(-\infty,\frac{-7}{3}\right) \cup (1,+\infty)$

For each of the Exercises 36-40, if necessary graph the given sets on the same number line to determine the required set.

36. $[-5,-2] \cup [-3,4] = [-5,4]$. Hence,

$$([-5,-2] \cup [-3,4]) \cap [0,6] = [-5,4] \cap [0,6] = [0,4]$$

38. $(-7,0) \cup (-2,1) = (-7,1)$. Hence,

$$((-7,0) \cup (-2,1)) \cap [-3,2] = (-7,1) \cap [-3,2] = [-3,2]$$

40. $(-1,5) \cup (-3,-1] = (-3,5)$. Hence,

$$((-1,5) \cup (-3,-1]) \cap [-4,4] = (-3,5) \cap [-4,4] = (-3,4]$$

4 Factoring Polynomials

EXERCISE 4.1

2. $3x - 9 = 3(? - ?)$
   $= 3(x - 3)$; because $3(x - 3) = 3x - 9$

4. $3xy(? + ?) = 3xy(x + 2)$; because
   $3xy(x + 2) = 3x^2y + 6xy$

6. $x(? - ? + ?) = x(x^2 - x + 1)$; because
   $x(x^2 - x + 1) = x^3 - x^2 + x$

8. $3rs(? + ? - ?) = 3rs(5r + 6s - 1)$; because
   $3rs(5r + 6s - 1) = 15r^2s + 18rs^2 - 3rs$

10. $3n^2(? - ? + ?) = 3n^2(n^2 - 2n + 4)$; because
    $3n^2(n^2 - 2n + 4) = 3n^4 - 6n^3 + 12n^2$

12. $x(? - ? + ?) = x(2xy^2 - 3y + 5x)$; because
    $x(2xy^2 - 3y + 5x) = 2x^2y^2 - 3xy + 5x^2$

14. $xz(? + ? - ?) = xz(xy^2z + 2y - 1)$; because
    $xz(xy^2z + 2y - 1) = x^2y^2z^2 + 2xyz - xz$

16. $3x(? - ? + ?) = 3x(2xy - 3y^2 + 4)$; because
    $3x(2xy - 3y^2 + 4) = 6x^2y - 9xy^2 + 12x$

18. $7xy(? + ? - ?) = 7xy(2 + 3xy - 4z)$; because
    $7xy(2 + 3xy - 4z) = 14xy + 21x^2y^2 - 28xyz$

20. $(a - 2)(? + ?) = (a - 2)(b + a)$; because
    $(a - 2)(b + a) = b(a - 2) + a(a - 2)$

22. $(y - 2)(? + ?) = (y - 2)(y - 3x)$; because
$(y - 3x)(y - 2) = y(y - 2) - 3x(y - 2)$

24. $(2a - b)(? + ?) = (2a - b)(3x + 4y)$; because
$(3x + 4y)(2a - b) = 3x(2a - b) + 4y(2a - b)$

26. $(x - 6)[? + ?] = (x - 6)[(x - 6) + 1]$
$= (x - 6)[x - 5]$

28. $3(x + 1)[? - ?] = 3(x + 1)[2 - (x + 1)]$
$= 3(x + 1)(-x + 1)$

30. $(2x + 3)^2[? + ?] = (2x + 3)^2[1 + x]$

32. $x(x + 3)[? - ?] = x(x + 3)[x - (x + 3)]$
$= x(x + 3)(-3)$ or $-3x(x + 3)$

34. $3(x + 2)(x - 4) + 6(x + 2)(x + 1)$
$= 3(x + 2)[? + ?]$
$= 3(x + 2)[(x - 4) + 2(x + 1)]$
$= 3(x + 2)(3x - 2)$

36. $(x + 2)^2 - (x + 2)(x - 1) = (x + 2)[? - ?]$
$= (x + 2)[(x + 2) - (x - 1)]$
$= (x + 2)(3)$ or $3(x + 2)$

38. $-(-3m + 2n) = -(2n - 3m)$

40. $-(-r^2 + s^2t^2) = -(s^2t^2 - r^2)$

42. $-3(? + ?) = -3(2x + 3)$

44. $?(a - b) = -a(a - b)$

46. $-(? + ?) = -(-x^2 + 3)$
$= -(3 - x^2)$

48. $-(? + ? + ?)$
$= -(-3x - 3y + 2z)$

50. $x^{2n} \cdot x^{2n} + x^{2n} \cdot 1 = x^{2n}(x^{2n} + 1)$

52. $y^{2n} \cdot y^{2n} + y^{2n} \cdot y^n + y^{2n} \cdot 1 = y^{2n}(y^{2n} + y^n + 1)$

54. $x^n \cdot x^2 + x^n \cdot x + x^n \cdot 1 = x^n(x^2 + x + 1)$

56. $(-x^{2n})x^{3n} - (-x^{2n}) \cdot 1 = -x^{2n}(x^{3n} - 1)$

58. $(-y^2)y^a - (-y^2) \cdot 1 = -y^2(y^a - 1)$

EXERCISE 4.2

2. $(x + 1)(x + 4)$

4. $(y - 2)(y - 5)$

6. $(x - 5)(x + 3)$

8. $(y - 3)(y + 7)$

10. $(3x - 1)(x - 2)$

12. $(3x - 1)(2x - 1)$

14. $(4x - 1)(x - 1)$

16. $(5x + 6)(2x - 3)$

18. $(8x - 3)(x + 1)$

20. $(2y + 1)(2y + 1)$

22. $(3x - 2a)(3x + 5a)$

24. $(2xy + 3)(2xy + 3)$

26. $(x - 6)(x + 6)$

28. $(xy - 2)(xy + 2)$

30. $(y^2 - 7)(y^2 + 7)$

32. $(3x - y)(3x + y)$

34. $(4x + 3y)(4x - 3y)$

36. $(8xy - 1)(8xy + 1)$

38. $2(x^2 + 3x - 10)$
$= 2(x + 5)(x - 2)$

40. $a(2a^2 + 15a + 7)$
$= a(a + 7)(2a + 1)$

42. $5(4a^2 + 12ab + 9b^2)$
$= 5(2a + 3b)(2a + 3b)$
$= 5(2a + 3b)^2$

44. $x^2(1 - 4y^2)$
$= x^2(1 + 2y)(1 - 2y)$

46. $x^2(1 - 2x + x^2) = x^2(1 - x)^2$, or
$x^2(1 - 2x + x^2) = x^2(x^2 - 2x + 1)$
$= x^2(x - 1)^2$

48. $xy(x^2 - y^2) = xy(x + y)(x - y)$

50. $(a^2 + 2)(a^2 + 3)$

52. $(4x^2 + 1)(x^2 - 3)$

54. $(x^2 - 9)(x^2 + 3) = (x + 3)(x - 3)(x^2 + 3)$

56. $(y^2 - 4)(y^2 - 9) = (y + 2)(y - 2)(y + 3)(y - 3)$

58. $(x^2 - 4)(3x^2 + 1) = (x + 2)(x - 2)(3x^2 + 1)$

60. $(x^2 - 9a^2)(4x^2 + 3a^2) = (x + 3a)(x - 3a)(4x^2 + 3a^2)$

62. $4^2 - (y^{2n})^2 = (4 - y^{2n})(4 + y^{2n})$
$= [2^2 - (y^n)^2](4 + y^{2n})$
$= (2 - y^n)(2 + y^n)(4 + y^{2n})$

64. $(x^{2n} - 1)(x^{2n} - 1) = (x^n - 1)(x^n + 1)(x^n - 1)(x^n + 1)$

or

$(x^n - 1)^2(x^n + 1)^2$

66. $6(y^{2n} + 5y^n - 150) = 6(y^n - 10)(y^n + 15)$

EXERCISE 4.3

2. $a(5 + b) + b(5 + b)$
$= (5 + b)(a + b)$

4. $a(1 + b) + b(1 + b)$
$= (1 + b)(a + b)$

6. $(x - y)x^2 + (x - y)y$
$= (x - y)(x^2 + y)$

8. $c(2a - b) + d(2a - b)$
$= (2a - b)(c + d)$

10. $5z(x - y) - 1 \cdot (x - y)$
$= (x - y)(5z - 1)$

12. $2x^2(3x - 2) + 1 \cdot (3x - 2)$
$= (3x - 2)(2x^2 + 1)$

14. $a(2a + 3) - b(2a + 3)$
$= (2a + 3)(a - b)$

16. $a(2b^2 + 5) - 4(2b^2 + 5)$
$= (2b^2 + 5)(a - 4)$

18. $4(3 - y^3) - x^2(3 - y^3)$
$= (3 - y^3)(4 - x^2)$
$= (3 - y^3)(2 + x)(2 - x)$

20. $x^2(x + 3) + 3(x + 3)$
$= (x + 3)(x^2 + 3)$

22. $x^2(2x + 7) + 2(2x + 7)$
$= (2x + 7)(x^2 + 2)$

24. $x^2(x - 2) - 3(x - 2)$
$= (x - 2)(x^2 - 3)$

26. $[x - (y - 3)][x + (y - 3)]$
$= (x - y + 3)(x + y - 3)$

28. $(x^2 - 6x + 9) - y^2$
$= (x - 3)^2 - y^2$
$= [(x - 3) - y][(x - 3) + y]$
$= (x - 3 - y)(x - 3 + y)$

or

$(x - y - 3)(x + y - 3)$

30. $y^2 - (x^2 - 4x + 4)$
$= y^2 - (x - 2)^2$
$= [y - (x - 2)][y + (x - 2)]$
$= (y - x + 2)(y + x - 2)$

32. $(9x^2 - 6x + 1) - 9y^2$
$= (3x - 1)^2 - (3y)^2$
$= [(3x - 1) - 3y][(3x - 1) + 3y]$
$= (3x - 3y - 1)(3x + 3y - 1)$

34. $y^3 + 3^3 = (y + 3)(y^2 - 3y + 9)$

36. $y^3 - (3x)^3 = (y - 3x)[y^2 + y(3x) + (3x)^2]$
$= (y - 3x)(y^2 + 3xy + 9x^2)$

38. $(3a)^3 + b^3 = (3a + b)[(3a)^2 - (3a)b + b^2]$
$= (3a + b)(9a^2 - 3ab + b^2)$

40. $2^3 + (xy)^3 = (2 + xy)[2^2 - 2(xy) + (xy)^2]$
$= (2 + xy)(4 - 2xy + x^2y^2)$

42. $a^3 - (5b)^3 = (a - 5b)[a^2 + a(5b) + (5b)^2]$
$= (a - 5b)(a^2 + 5ab + 25b^2)$

44. $(2ab)^3 + (1)^3 = (2ab + 1)[(2ab)^2 - (2ab)(1) + (1)^2]$
$= (2ab + 1)(4a^2b^2 - 2ab + 1)$

46. $[(x + y) - z][(x + y)^2 + (x + y)z + z^2]$
$= (x + y - z)(x^2 + 2xy + y^2 + xz + yz + z^2)$

48. $(x^2)^3 + (x - 2y)^3$
$= [(x^2) + (x - 2y)][(x^2)^2 - (x^2)(x - 2y) + (x - 2y)^2]$
$= (x^2 + x - 2y)(x^4 - x^3 + 2x^2y + x^2 - 4xy + 4y^2)$

50. $[(2y - 1) + (y - 1)][(2y - 1)^2 - (2y - 1)(y - 1) + (y - 1)^2]$
$= (2y - 1 + y - 1)[4y^2 - 4y + 1 - (2y^2 - 3y + 1) + y^2 - 2y + 1]$
$= (3y - 2)(4y^2 - 4y + 1 - 2y^2 + 3y - 1 + y^2 - 2y + 1)$
$= (3y - 2)(3y^2 - 3y + 1)$

52. $a^2 - b^2 - c^2 + 2bc = a^2 - b^2 + 2bc - c^2$
$= a^2 - (b^2 - 2bc + c^2)$
$= (a)^2 - (b - c)^2$
$= [(a) - (b - c)][(a) + (b - c)]$
$= (a - b + c)(a + b - c)$

54. $x^4 - 8x^2y^2 + 4y^4 = x^4 - 4x^2y^2 + 4y^4 - 4x^2y^2$
$= (x^4 - 4x^2y^2 + 4y^4) - 4x^2y^2$
$= (x^2 - 2y^2)^2 - (2xy)^2$
$= [(x^2 - 2y^2) - 2xy][(x^2 - 2y^2) + 2xy]$
$= (x^2 - 2xy - 2y^2)(x^2 + 2xy - 2y^2)$

56. $$\begin{aligned}4a^4 - 5a^2b^2 + b^4 &= 4a^4 - 4a^2b^2 + b^4 - a^2b^2\\ &= (4a^4 - 4a^2b^2 + b^4) - a^2b^2\\ &= (2a^2 - b^2)^2 - (ab)^2\\ &= [(2a^2 - b^2) - ab][(2a^2 - b^2) + ab]\\ &= (2a^2 - ab - b^2)(2a^2 + ab - b^2)\end{aligned}$$

EXERCISE 4.4

2. $x + 3 = 0;\ x - 4 = 0$
$\{-3,4\}$

4. $x + 1 = 0;\ 3x - 1 = 0$
$3x = 1$
$\left\{-1,\frac{1}{3}\right\}$

6. $x = 0;\ 3x - 7 = 0$
$3x = 7$
$\left\{0,\frac{7}{3}\right\}$

8. $2x - 7 = 0;\ x + 1 = 0$
$2x = 7$
$\left\{\frac{7}{2},-1\right\}$

10. $x + 5 = 0;\ 3x - 1 = 0$
$3x = 1$
$\left\{-5,\frac{1}{3}\right\}$

12. $3x - 4 = 0;\ 2x + 5 = 0$
$3x = 4 \qquad 2x = -5$
$\left\{\frac{4}{3},\frac{-5}{2}\right\}$

14. $x(x + 5) = 0$
$x = 0;\ x + 5 = 0$
$\{0,-5\}$

16. $3x^2 - 3x = 0$
$3x(x - 1) = 0$
$3x = 0;\ x - 1 = 0$
$\{0,1\}$

18. $(x + 2)(x - 2) = 0$
$x + 2 = 0;\ x - 2 = 0$
$\{-2,2\}$

20. $3(x^2 - 1) = 0$
$3(x + 1)(x - 1) = 0$
$x + 1 = 0;\ x - 1 = 0$
$\{-1,1\}$

22. $25x^2 - 4 = 0$
$(5x + 2)(5x - 2) = 0$
$5x + 2 = 0;\ 5x - 2 = 0$
$5x = -2; \qquad 5x = 2$
$x = \frac{-2}{5}; \qquad x = \frac{2}{5}$
$\left\{\frac{-2}{5},\frac{2}{5}\right\}$

24. $(3x + 5)(3x - 5) = 0$
$3x + 5 = 0;\ 3x - 5 = 0$
$3x = -5; \qquad 3x = 5$
$\left\{\frac{-5}{3},\frac{5}{3}\right\}$

26. $(x + 2)(x + 3) = 0$
$x + 2 = 0;\ x + 3 = 0$
$\{-2,-3\}$

28. $(x + 6)(x - 7) = 0$
$x + 6 = 0;\ x - 7 = 0$
$\{-6,7\}$

30. $12x^2 - 8x - 15 = 0$
$(2x - 3)(6x + 5) = 0$
$2x - 3 = 0; \quad 6x + 5 = 0$
$2x = 3; \quad 6x = -5$
$\left\{\frac{3}{2}, \frac{-5}{6}\right\}$

32. $2x^2 - 4x = x + 3$
$2x^2 - 5x - 3 = 0$
$(2x + 1)(x - 3) = 0$
$2x + 1 = 0; \quad x - 3 = 0$
$2x = -1$
$\left\{\frac{-1}{2}, 3\right\}$

34. $3x^2 + 2x = x^2 + 4x + 4$
$2x^2 - 2x - 4 = 0$
$2(x^2 - x - 2) = 0$
$2(x + 1)(x - 2) = 0$
$x + 1 = 0; \quad x - 2 = 0$
$\{-1, 2\}$

36. $x^2 + x = 4 - (x^2 + 4x + 4)$
$x^2 + x = 4 - x^2 - 4x - 4$
$2x^2 + 5x = 0$
$x(2x + 5) = 0$
$x = 0; \quad 2x + 5 = 0$
$\left\{0, \frac{-5}{2}\right\}$

38. $z^2 + 5z + 18 = 4 - 4z$
$z^2 + 9z + 14 = 0$
$(z + 2)(z + 7) = 0$
$z + 2 = 0; \quad z + 7 = 0$
$\{-2, -7\}$

40. $n^2 - n - 6 = 6$
$n^2 - n - 12 = 0$
$(n - 4)(n + 3) = 0$
$n - 4 = 0; \quad n + 3 = 0$
$\{-3, 4\}$

42. $2x^2 - x - 3 = 3$
$2x^2 - x - 6 = 0$
$(2x + 3)(x - 2) = 0$
$2x + 3 = 0; \quad x - 2 = 0$
$2x = -3; \quad x = 2$
$\left\{-\frac{3}{2}, 2\right\}$

44. $6y = y^2 + 2y + 1 + 3$
$0 = y^2 - 4y + 4$
$0 = (y - 2)(y - 2)$
$y - 2 = 0$
$\{2\}$

46. $2x - [x^2 - x - 6 + 8] = 0$
$2x - x^2 + x + 6 - 8 = 0$
$-x^2 + 3x - 2 = 0$
$-(x^2 - 3x + 2) = 0$
$-(x - 2)(x - 1) = 0$
$x - 2 = 0; \quad x - 1 = 0$
$\{1, 2\}$

48. $3[x^2 + 4x + 4 - 4x] = 15$
$3x^2 + 12 = 15$
$3x^2 - 3 = 0$
$3(x^2 - 1) = 0$
$3(x + 1)(x - 1) = 0$
$\{-1, 1\}$

50. $[x - (-4)](x - 3) = 0$
$(x + 4)(x - 3) = 0$
$x^2 + x - 12 = 0$

52. $x(x - 5) = 0$
$x^2 - 5x = 0$

54. $(x - 4)(x - 4) = 0$
$x^2 - 16 = 0$

56. $[x - (-a)][x - (-b)] = 0$
$[x + a][x + b] = 0$
$x^2 + ax + bx + ab = 0$

58. $[x - (a + b)][x + (a + b)] = 0$
$x - (a + b) = 0; \quad x + (a + b) = 0$
$x = a + b; \quad x = -(a + b)$

or

$x = -a - b$
$\{-a - b, a + b\}$

60. $(x + 2b)(x - 6b) = 0$
$x + 2b = 0; \quad x - 6b = 0$
$x = -2b; \quad x = 6b$
$\{-2b, 6b\}$

62. $(x + 2a)(x - b) = 0$
$x + 2a = 0; \quad x - b = 0$
$x = -2a; \quad x = b$
$\{-2a, b\}$

64. $(x - \frac{1}{2}a)(x - \frac{1}{2}b) = 0$
$x - \frac{1}{2}a = 0; \quad x - \frac{1}{2}b = 0$
$x = \frac{1}{2}a; \quad x = \frac{1}{2}b$
$\left\{\frac{1}{2}a, \frac{1}{2}b\right\}$

EXERCISE 4.5

2. The smaller negative number: x
The larger negative number: $x + 7$

$$x(x + 7) = 60$$
$$x^2 + 7x - 60 = 0$$
$$(x + 12)(x - 5) = 0; \quad \{-12, 5\}$$

Since it is required that the numbers be negative, 5 is rejected and $x = -12$ and $x + 7 = -5$. The required numbers are -12 and -5.

4. The smaller positive integer: x
The larger positive (consecutive) integer: $x + 1$

$$x^2 + (x + 1)^2 = 85$$
$$x^2 + x^2 + 2x + 1 = 85$$
$$2x^2 + 2x - 84 = 0$$
$$2(x^2 + x - 42) = 0$$
$$2(x + 7)(x - 6) = 0; \quad \{-7, 6\}$$

Since x is required to be positive, $x = 6$ and $x + 1 = 7$. The required integers are 6 and 7.

6. The smallest negative integer: x
The next (consecutive) negative integer: x + 1
The third (consecutive) negative integer: x + 2

$$x^2 + (x + 1)^2 + (x + 2)^2 = 434$$
$$x^2 + x^2 + 2x + 1 + x^2 + 4x + 4 = 434$$
$$3x^2 + 6x - 429 = 0$$
$$3(x^2 + 2x - 143) = 0$$
$$3(x + 13)(x - 11) = 0; \quad \{-13,11\}$$

Since x is required to be negative, $x = -13$, $x + 1 = -12$, and $x + 2 = -11$ and the required integers are -13, -12, and -11.

8. The width of the plate: x
The length of the plate: 2x + 2

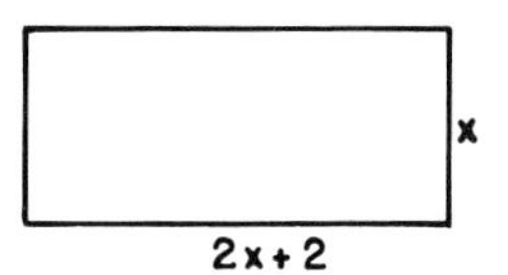

The area of a rectangle is equal to the product of its length and width.

$$x(2x + 2) = 40$$
$$2x^2 + 2x - 40 = 0$$
$$x^2 + x - 20 = 0$$
$$(x + 5)(x - 4) = 0; \quad \{-5,4\}$$

Since a dimension of a geometric figure cannot be negative, -5 is rejected. Then when $x = 4$, $2x + 2 = 10$. The dimensions are 4 centimeters by 10 centimeters.

10. The altitude: x
The base: x + 13

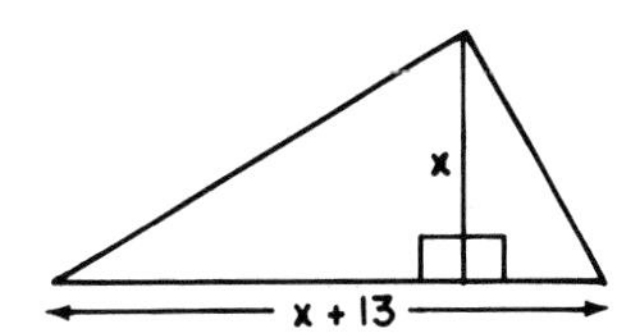

The area of a triangle is equal to one-half of the product of the base and altitude.

$$\frac{1}{2}(x)(x + 13) = 70$$
$$2\left[\frac{1}{2}x(x + 13)\right] = 2[70]$$
$$x(x + 13) = 140$$
$$x^2 + 13x - 140 = 0$$
$$(x + 20)(x - 7) = 0; \quad \{-20,7\}$$

Since -20 is not meaningful as a length, we have $x = 7$ and $x + 13 = 20$. The altitude and the base are 7 and 20 centimeters, respectively.

12. The length of one side: x
The length of the second side: $x + 7$

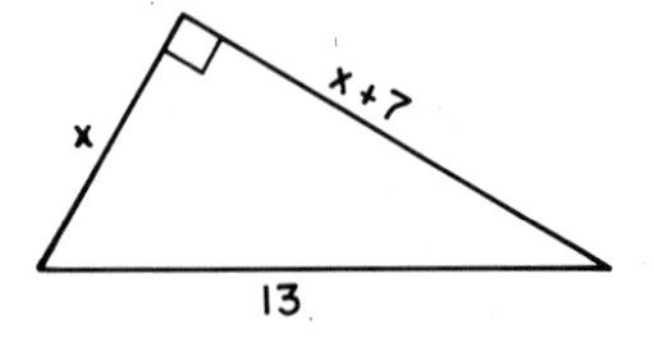

In a right triangle, the square of the hypotenuse is equal to the sum of the squares of the other two sides. Hence,

$$x^2 + (x + 7)^2 = 13^2$$
$$x^2 + x^2 + 14x + 49 = 169$$
$$2x^2 + 14x - 120 = 0$$
$$2(x^2 + 7x - 60) = 0$$
$$2(x + 12)(x - 5) = 0; \quad \{-12,5\}$$

Since a dimension of a geometric figure cannot be negative, $x = 5$ and $x + 7 = 12$. Hence the two shorter sides are 5 cm and 12 cm long.

14. Time it will take the ball to return to the ground: t
At the ground level, $h = 0$; hence,

$$0 = 64t - 16t^2$$
$$0 = 16t(4 - t)$$
$$t = 0; \quad t = 4$$

The ball will return to the ground after 4 seconds.

16. When $t = 2$, $s = 2^2 - 5(2) = -6$. It will have moved 6 units in a negative direction.
When $t = 3$, $s = 3^2 - 5(3) = -6$. It will have returned to where it was when $t = 2$.
It will be back to its starting point when $A = 0$.
Hence we solve

$$t^2 - 5t = 0$$
$$t(t - 5) = 0 \quad \{0,5\}$$

Thus, it will return to the starting point in 5 seconds.

18. The width of page: w
The length of page: $w + 2$

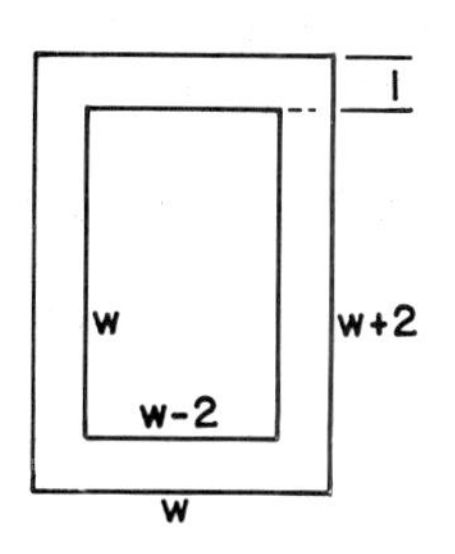

$$w(w - 2) = 35$$
$$w^2 - 2w - 35 = 0$$
$$(w - 7)(w + 5) = 0; \quad \{-5,7\}$$

Since w cannot be negative, the width of the page is 7 inches and the height is 9 inches.

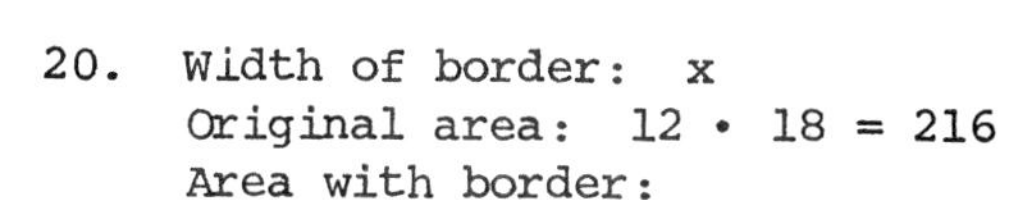

20. Width of border: x
Original area: $12 \cdot 18 = 216$
Area with border:

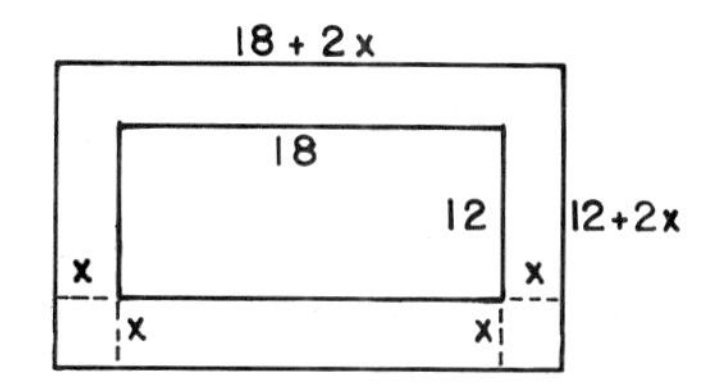

$$(12 + 2x)(18 + 2x)$$

$$(12 + 2x)(18 + 2x) - 216 = 216$$
$$216 + 60x + 4x^2 - 216 = 216$$
$$4x^2 + 60x - 216 = 0$$
$$x^2 + 15x - 54 = 0$$
$$(x + 18)(x - 3) = 0; \quad \{-18,3\}$$

Since -18 is not meaningful in this context, we have $x = 3$, the width of the border in meters.

22. Width of piece of metal: x
Length of piece of metal: $2x$
Width of "floor" of tray: $x - 2$
Length of "floor" of tray: $2x - 2$
Height (or depth) of tray: 1

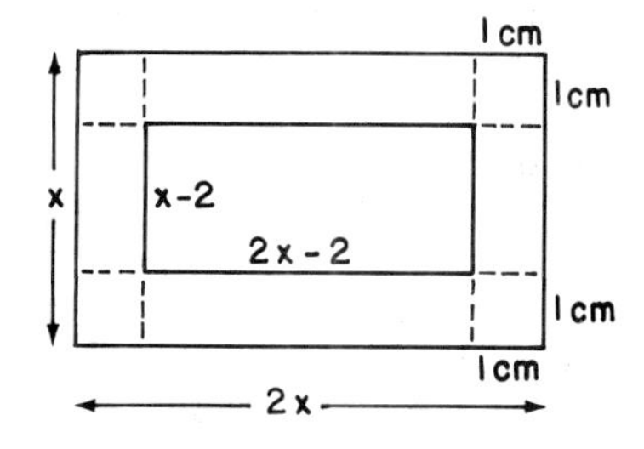

Since the volume (V) of the tray is given by

$$V = (\text{length}) \cdot (\text{width}) \cdot (\text{height})$$

We have

$$(2x - 2)(x - 2)(1) = 144$$
$$2x^2 - 6x + 4 = 144$$
$$2x^2 - 6x - 140 = 0$$
$$2(x^2 - 3x - 70) = 0$$
$$2(x - 10)(x + 7) = 0; \quad \{-7,10\}$$

Since x cannot be negative, $x = 10$.
Hence, in centimeters, the dimensions of the tray are:

length: $2x - 2 = 2(10) - 2 = 18$
width: $x - 2 = 10 - 2 = 8$
height: 1

24. Number of original seats: x

Income per seat: $\frac{1200}{x}$

$$\begin{pmatrix}\text{Number of}\\ \text{seats}\end{pmatrix} \cdot \begin{pmatrix}\text{Income}\\ \text{per seat}\end{pmatrix} = \begin{pmatrix}\text{Total}\\ \text{income}\end{pmatrix}$$

$$(x + 60)\left(\frac{1200}{x} - 1\right) = 1200$$

$$1200 - x + \frac{72,000}{x} - 60 = 1200$$

$$x^2 + 60x - 72,000 = 0$$

$$(x + 300)(x - 240) = 0, \text{ from which } x = -300 \text{ or } x = 240$$

The solution -300 is not meaningful. Hence, there are 240 seats.

5 Rational Expressions

EXERCISE 5.1

2. $\frac{-1}{3}$ 4. $\frac{3}{4}$ 6. $\frac{6}{7}$

8. $\frac{-4}{5}$ 10. $\frac{-x}{3y}$ $(y \neq 0)$ 12. $\frac{x}{2y}$ $(y \neq 0)$

14. $\frac{-(x + 3)}{x}$ or $\frac{-x - 3}{x}$ $(x \neq 0)$ 16. $\frac{-(y - x)}{y - 1}$ or $\frac{x - y}{y - 1}$ $(y \neq 1)$

18. $\frac{-3}{-(x - 2)} = \frac{3}{x - 2}$ 20. $\frac{-6}{-(y - x)} = \frac{6}{y - x}$ 22. $\frac{2x - 5}{-(y - 3)} = \frac{-(2x - 5)}{y - 3}$ or $\frac{5 - 2x}{y - 3}$

24. $\frac{x + 3}{-(x - y)} = \frac{-(x + 3)}{x - y} = \frac{-x - 3}{x - y}$ 26. $-\frac{x - 4}{x - 2y} = \frac{-(x - 4)}{-(2y - x)} = \frac{x - 4}{2y - x}$

28. $\frac{-a - 1}{2b - 3a} = \frac{-(a + 1)}{-(3a - 2b)} = \frac{a + 1}{3a - 2b}$

30. Let $x = 2$ and $y = 1$. Then

$$-\frac{2x - y}{3} = -\frac{2(2) - 1}{3} = -1 \quad \text{and} \quad \frac{-2x - y}{3} = \frac{-2(2) - 1}{3} = \frac{-5}{3}.$$

Therefore, $-\frac{2x - y}{3}$ is not equivalent to $\frac{-2x - y}{3}$.

32. $\frac{12a^4b^2}{4a^2b^4} = \frac{3a^2}{b^2}$ 34. $\frac{22a^2bc^3}{11a^4c^2} = \frac{2bc}{a^2}$ 36. $\frac{100mn}{-5m^2n^3} = \frac{-20}{mn^2}$

38. $\frac{xyz}{xy^2z^3} = \frac{1}{yz^2}$ 40. $\frac{abc}{-7abc^3} = \frac{-1}{7c^2}$ 42. $\frac{-15xy^3z}{-3y^2z^4} = \frac{5xy}{z^3}$

44. $\dfrac{\cancel{2}(y - 4)}{\cancel{8}_4} = \dfrac{y - 4}{4}$

46. $\dfrac{\cancel{5}(y - 2)}{\cancel{5}} = y - 2$

48. $\dfrac{\overset{-1}{\cancel{3x}}(x^2 - 2x + 1)}{\cancel{-3x}} = -(x^2 - 2x + 1)$
or
$-x^2 + 2x - 1$

50. $\dfrac{\cancel{(x + 3)}(x + 2)}{\cancel{x + 3}} = x + 2$

52. $\dfrac{3\cancel{(2x + 3)}}{5\cancel{(2x + 3)}} = \dfrac{3}{5}$

54. $\dfrac{(x - 4)\cancel{(x + 4)}}{2\cancel{(x + 4)}} = \dfrac{x - 4}{2}$

56. $\dfrac{4(x - 1)\cancel{(x + 1)}}{\underset{x + 1}{\cancel{(x + 1)^2}}} = \dfrac{4(x - 1)}{x + 1}$

58. $\dfrac{5\cancel{(y - 2)}(y + 2)}{2\cancel{(y - 2)}} = \dfrac{5(y + 2)}{2}$

60. $\dfrac{-2\cancel{(y - 2)}}{\cancel{(y - 2)}(y + 2)} = \dfrac{-2}{y + 2}$

62. $\dfrac{(2x - y)^2}{-(4x^2 - y^2)} = \dfrac{-\overset{(2x - y)}{\cancel{(2x - y)^2}}}{\cancel{(2x - y)}(2x + y)}$
$= \dfrac{-(2x - y)}{2x + y}$
or $\dfrac{-2x + y}{2x + y}$

64. $\dfrac{\cancel{(y - 3)}(y + 1)}{\cancel{(y - 3)}(y + 3)} = \dfrac{y + 1}{y + 3}$

66. $\dfrac{\cancel{(2x - 1)}(3x + 1)}{\cancel{(2x - 1)}(x + 5)} = \dfrac{3x + 1}{x + 5}$

68. $\dfrac{\cancel{(2x - 3y)}(2x + 3y)}{\cancel{(2x - 3y)}(x + 2y)} = \dfrac{2x + 3y}{x + 2y}$

70. $\dfrac{x(ax + 1) + a(ax + 1)}{3(x + a)}$
$= \dfrac{(ax + 1)\cancel{(x + a)}}{3\cancel{(x + a)}}$
$= \dfrac{ax + 1}{3}$

72. $\dfrac{2a(x - 2y) + b(x - 2y)}{(x - 2y)(x + 2y)}$
$= \dfrac{\cancel{(x - 2y)}(2a + b)}{\cancel{(x - 2y)}(x + 2y)}$
$= \dfrac{2a + b}{x + 2y}$

74. $$\frac{(2x)^3 - 1^3}{(2x-1)(2x+1)} = \frac{\cancel{(2x-1)}(4x^2+2x+1)}{\cancel{(2x-1)}(2x+1)} = \frac{4x^2+2x+1}{2x+1}$$

76. Let $x = 2$ and $y = 1$. Then

$$\frac{4x-y}{4} = \frac{4(2)-1}{4} = \frac{7}{4}$$

and

$$x - y = 2 - 1 = 1.$$

Therefore, $\frac{4x-y}{4}$ is not equivalent to $x - y$.

EXERCISE 5.2

2. $$\frac{15t^3}{3t^2} - \frac{12t^2}{3t^2} + \frac{5t}{3t^2} = 5t - 4 + \frac{5}{3t}$$

4. $$\frac{21n^4}{7n^2} + \frac{14n^2}{7n^2} - \frac{7}{7n^2} = 3n^2 + 2 - \frac{1}{n^2}$$

6. $$\frac{12x^3}{4x} - \frac{8x^2}{4x} + \frac{3x}{4x} = 3x^2 - 2x + \frac{3}{4}$$

8. $$\frac{9a^2b^2}{ab^2} + \frac{3ab^2}{ab^2} + \frac{4a^2b}{ab^2} = 9a + 3 + \frac{4a}{b}$$

10. $$\frac{36t^5}{-12t^2} + \frac{24t^3}{-12t^2} - \frac{12t}{-12t^2} = -3t^3 + (-2t) - \left(-\frac{1}{t}\right) = -3t^3 - 2t + \frac{1}{t}$$

12. $$\frac{15s^{10}}{3s^2} - \frac{21s^5}{3s^2} + \frac{6}{3s^2} = 5s^8 - 7s^3 + \frac{2}{s^2}$$

14.
$$\begin{array}{r|l} & \quad\quad\quad n + 7 \\ 2n - 1 & 2n^2 + 13n - 6 \\ & 2n^2 - n \\ & \quad\quad 14n - 6 \\ & \quad\quad 14n - 7 \\ & \quad\quad\quad\quad 1 \end{array}$$

$$\frac{2n^2 + 13n - 6}{2n - 1} = n + 7 + \frac{1}{2n-1}$$

16.
$$\begin{array}{r|l} & \quad\quad\quad x - 4 \\ 2x + 5 & 2x^2 - 3x - 15 \\ & 2x^2 + 5x \\ & \quad\quad -8x - 15 \\ & \quad\quad -8x - 20 \\ & \quad\quad\quad\quad 5 \end{array}$$

$$\frac{2x^2 - 3x - 15}{2x + 5} = x - 4 + \frac{5}{2x+5}$$

18.

$$\begin{array}{r|l} & 2x^2 - 5x + 3 \\ \hline x + 1 & 2x^3 - 3x^2 - 2x + 4 \\ & \underline{2x^3 + 2x^2} \\ & \quad - 5x^2 - 2x \\ & \quad \underline{- 5x^2 - 5x} \\ & \qquad\qquad 3x + 4 \\ & \qquad\qquad \underline{3x + 3} \\ & \qquad\qquad\qquad 1 \end{array}$$

$$\frac{2x^3 - 3x^2 - 2x + 4}{x + 1} = 2x^2 - 5x + 3 + \frac{1}{x + 1}$$

20.

$$\begin{array}{r|l} & 2b^3 + 8b^2 + 34b + 136 \\ \hline b - 4 & 2b^4 + 0b^3 + 2b^2 + 0b + 3 \\ & \underline{2b^4 - 8b^3} \\ & \quad 8b^3 + 2b^2 \\ & \quad \underline{8b^3 - 32b^2} \\ & \qquad 34b^2 + 0b \\ & \qquad \underline{34b^2 - 136b} \\ & \qquad\qquad 136b + 3 \\ & \qquad\qquad \underline{136b - 544} \\ & \qquad\qquad\qquad 547 \end{array}$$

$$\frac{3 + 2b^2 + 2b^4}{b - 4} = 2b^3 + 8b^2 + 34b + 136 + \frac{547}{b - 4}$$

22.

$$\begin{array}{r|l} & 5t^3 - 9t^2 + 2t - 3 \\ \hline 2t + 3 & 10t^4 - 3t^3 - 23t^2 + 0t + 7 \\ & \underline{10t^4 + 15t^3} \\ & \quad - 18t^3 - 23t^2 \\ & \quad \underline{- 18t^3 - 27t^2} \\ & \qquad 4t^2 + 0t \\ & \qquad \underline{4t^2 + 6t} \\ & \qquad\qquad - 6t + 7 \\ & \qquad\qquad \underline{- 6t - 9} \\ & \qquad\qquad\qquad 16 \end{array}$$

$$\frac{7 - 3t^3 - 23t^2 + 10t^4}{2t + 3} = 5t^3 - 9t^2 + 2t - 3 + \frac{16}{2t + 3}$$

24.

$$\begin{array}{r|l} & y^4 + y^3 + y^2 + y + 1 \\ \hline y - 1 & y^5 + 0y^4 + 0y^3 + 0y^2 + 0y + 1 \\ & \underline{y^5 - y^4} \\ & \quad y^4 + 0y^3 \\ & \quad \underline{y^4 - y^3} \\ & \qquad y^3 + 0y^2 \\ & \qquad \underline{y^3 - y^2} \\ & \qquad\quad y^2 + 0y \\ & \qquad\quad \underline{y^2 - y} \\ & \qquad\qquad y + 1 \\ & \qquad\qquad \underline{y - 1} \\ & \qquad\qquad\quad 2 \end{array}$$

$$\frac{y^5 + 1}{y - 1} = y^4 + y^3 + y^2 + y + 1 + \frac{2}{y - 1}$$

26.

$$\begin{array}{r|l} & 2y + 7 \\ \hline y^2 - y - 3 & 2y^3 + 5y^2 - 3y + 2 \\ & \underline{2y^3 - 2y^2 - 6y} \\ & \quad 7y^2 + 3y + 2 \\ & \quad \underline{7y^2 - 7y - 21} \\ & \qquad 10y + 23 \end{array}$$

$$\frac{2y^3 + 5y^2 - 3y + 2}{y^2 - y - 3} = 2y + 7 + \frac{10y + 23}{y^2 - y - 3}$$

28.

$$\begin{array}{r|l} & 2b^2 - 2b + 5 \\ \hline b^2 + b - 3 & 2b^4 + 0b^3 - 3b^2 + b + 2 \\ & \underline{2b^4 + 2b^3 - 6b^2} \\ & \quad - 2b^3 + 3b^2 + b \\ & \quad \underline{- 2b^3 - 2b^2 + 6b} \\ & \qquad 5b^2 - 5b + 2 \\ & \qquad \underline{5b^2 + 5b - 15} \\ & \qquad -10b + 17 \end{array}$$

$$\frac{2b^4 - 3b^2 + b + 2}{b^2 + b - 3} = 2b^2 - 2b + 5 + \frac{-10b + 17}{b^2 + b - 3}$$

30.
$$
\begin{array}{r|l}
 & r + 1 \\
\cline{2-2}
r^3 + 0r^2 + 2r + 3 & r^4 + r^3 - 2r^2 + r + 5 \\
 & \underline{r^4 + 0r^3 + 2r^2 + 3r} \\
 & r^3 - 4r^2 - 2r + 5 \\
 & \underline{r^3 + 0r^2 + 2r + 3} \\
 & - 4r^2 - 4r + 2
\end{array}
$$

$$\frac{r^4 + r^3 - 2r^2 + r + 5}{r^3 + 2r + 3} = r + 1 + \frac{-4r^2 - 4r + 2}{r^3 + 2r + 3}$$

32. $x^3 + 2x^2 + k$ has $x + 3$ as a factor if $(x^3 + 2x^2 + k) \div (x + 3)$ has a zero remainder.

$$
\begin{array}{r|l}
 & x^2 - x + 3 \\
\cline{2-2}
x + 3 & x^3 + 2x^2 + 0x + k \\
 & \underline{x^3 + 3x^2} \\
 & - x^2 + 0x \\
 & \underline{- x^2 - 3x} \\
 & 3x + k \\
 & \underline{3x + 9} \\
 & k - 9
\end{array}
$$

The remainder, $k - 9$, must equal zero. Hence, $k = 9$.

EXERCISE 5.3

2.
$$
\begin{array}{r|rrr}
-3 & 1 & 1 & -6 \\
 & & -3 & 6 \\
\hline
 & 1 & -2 & 0
\end{array};\quad a - 2 \ (a \neq -3)
$$

4.
$$
\begin{array}{r|rrr}
-3 & 1 & 6 & 9 \\
 & & -3 & -9 \\
\hline
 & 1 & 3 & 0
\end{array};\quad x + 3 \ (x \neq -3)
$$

6.
$$
\begin{array}{r|rrrrr}
3 & 1 & 0 & 2 & -3 & 5 \\
 & & 3 & 9 & 33 & 90 \\
\hline
 & 1 & 3 & 11 & 30 & 95
\end{array};\quad x^3 + 3x^2 + 11x + 30 + \frac{95}{x - 3} \ (x \neq 3)
$$

8.
$$
\begin{array}{r|rrrr}
-2 & 3 & 1 & 0 & -7 \\
 & & -6 & 10 & -20 \\
\hline
 & 3 & -5 & 10 & -27
\end{array};\quad 3x^2 - 5x + 10 + \frac{-27}{x + 2} \ (x \neq -2)
$$

10.
$$
\begin{array}{r|rrrrr}
4 & 3 & 0 & -1 & 0 & 1 \\
 & & 12 & 48 & 188 & 752 \\
\hline
 & 3 & 12 & 47 & 188 & 753
\end{array};\quad 3x^3 + 12x^2 + 47x + 188 + \frac{753}{x - 4} \ (x \neq 4)
$$

12. $$\begin{array}{r|rrrr} -3 & 1 & -7 & -1 & 3 \\ & & -3 & 30 & -87 \\ \hline & 1 & -10 & 29 & -84 \end{array} ;\; x^2 - 10x + 29 + \frac{-84}{x + 3} \quad (x \neq -3)$$

14. $$\begin{array}{r|rrrrr} -1 & 1 & 1 & 0 & -2 & 3 \\ & & -1 & 0 & 0 & 2 \\ \hline & 1 & 0 & 0 & -2 & 5 \end{array} ;\; x^3 - 2 + \frac{5}{x + 1} \quad (x \neq -1)$$

16. $$\begin{array}{r|rrrrrr} -2 & 1 & 0 & -2 & 1 & 0 & -1 \\ & & -2 & 4 & -4 & 6 & -12 \\ \hline & 1 & -2 & 2 & -3 & 6 & -13 \end{array} ;\; x^4 - 2x^3 + 2x^2 - 3x + 6 + \frac{-13}{x + 2} \quad (x \neq -2)$$

18. $$\begin{array}{r|rrrrrr} 2 & 1 & 0 & 0 & -2 & 0 & 1 \\ & & 2 & 4 & 8 & 12 & 24 \\ \hline & 1 & 2 & 4 & 6 & 12 & 25 \end{array} ;\; x^4 + 2x^3 + 4x^2 + 6x + 12 + \frac{25}{x - 2} \quad (x \neq 2)$$

20. $$\begin{array}{r|rrrrrrr} 2 & 1 & 0 & 0 & 3 & 0 & -2 & -1 \\ & & 2 & 4 & 8 & 22 & 44 & 84 \\ \hline & 1 & 2 & 4 & 11 & 22 & 42 & 83 \end{array}$$

$$x^5 + 2x^4 + 4x^3 + 11x^2 + 22x + 42 + \frac{83}{x - 2} \quad (x \neq 2)$$

22. $$\begin{array}{r|rrrrrr} -1 & 1 & 0 & 0 & 0 & 0 & 1 \\ & & -1 & 1 & -1 & 1 & -1 \\ \hline & 1 & -1 & 1 & -1 & 1 & 0 \end{array}$$

$$x^4 - x^3 + x^2 - x + 1 \quad (x \neq -1)$$

24. $$\begin{array}{r|rrrrrrr} -1 & 1 & 0 & 0 & 0 & 0 & 0 & 1 \\ & & -1 & 1 & -1 & 1 & -1 & 1 \\ \hline & 1 & -1 & 1 & -1 & 1 & -1 & 2 \end{array}$$

$$x^5 - x^4 + x^3 - x^2 + x - 1 + \frac{2}{x + 1} \quad (x \neq -1)$$

EXERCISE 5.4

2. $\frac{3 \cdot 2}{4 \cdot 2} = \frac{6}{8}$

4. $\frac{-12 \cdot 4}{5 \cdot 4} = \frac{-48}{20}$

6. $\frac{6 \cdot 7}{1 \cdot 7} = \frac{42}{7}$

8. $\frac{5 \cdot 7}{3y \cdot 7} = \frac{35}{21y}$

10. $\frac{-a(ab)}{b(ab)} = \frac{-a^2b}{ab^2}$

12. $\frac{x(xy^3)}{1(xy^3)} = \frac{x^2y^3}{xy^3}$

14. $\dfrac{5}{(2a + b)} = \dfrac{?}{(2a + b)(2a - b)}$; $\dfrac{5(2a - b)}{(2a + b)(2a - b)} = \dfrac{10a - 5b}{4a^2 - b^2}$

16. $\dfrac{5x}{(y + 3)} = \dfrac{?}{(y + 3)(y - 2)}$; $\dfrac{5x(y - 2)}{(y + 3)(y - 2)} = \dfrac{5xy - 10x}{y^2 + y - 6}$

18. $\dfrac{-3}{(a + 2)} = \dfrac{?}{(a + 2)(a + 1)}$; $\dfrac{-3(a + 1)}{(a + 2)(a + 1)} = \dfrac{-3a - 3}{a^2 + 3a + 2}$

In Exercises 20-42, each given number is written as a number or expression in prime factor form.

20. $3, 2^2, 5$; 3 and 5 occur as factors at most once and 2 occurs as a factor at most twice. Hence, LCM $= 3 \cdot 5 \cdot 2^2 = 60$.

22. $2^2, 3 \cdot 5, 2 \cdot 3^2$; in any one number, 2 and 3 occur as factors at most twice and 5 occurs at most once. Hence, LCM $= 2^2 \cdot 3^2 \cdot 5 = 180$.

24. $2 \cdot 2, 11, 2 \cdot 11$; in any one number, 2 occurs as a factor at most twice and 11 occurs at most once. Hence, LCM $= 2^2 \cdot 11 = 44$.

26. $2^2 \cdot 3 \cdot x \cdot y, 2^3 \cdot 3 \cdot x^3 \cdot y^2$; in any one expression, 2 and x occur as factors at most three times, y occurs at most twice, and 3 occurs at most once. Hence, LCM $= 2^3 \cdot 3 \cdot x^3 \cdot y^2 = 24x^3y^2$.

28. $7 \cdot x, 2^3 \cdot y, 2 \cdot 3 \cdot z$; in any one expression, 7, 3, x, y, and z occur as factors at most once and 2 occurs at most three times. Hence, LCM $= 2^3 \cdot 3 \cdot 7 \cdot x \cdot y \cdot z = 168xyz$.

30. $2 \cdot 3 \cdot (x + y)^2, 2^2 \cdot x \cdot y^2$; in any one expression, 3 and x occur as factors at most once and 2, $(x + y)$, and y occur at most twice. Hence, LCM $= 2^2 \cdot 3 \cdot x \cdot y^2 \cdot (x + y)^2 = 12xy^2(x + y)^2$.

32. $x + 2, (x + 2)(x - 2)$; in any one expression, $x + 2$ and $x - 2$ occur as factors at most once. Hence, LCM $= (x + 2)(x - 2)$.

34. $(x - 1)(x - 2), (x - 1)^2$; in any one expression, $x - 1$ occurs as a factor at most twice and $x - 2$ occurs once. Hence, LCM $= (x - 1)^2(x - 2)$.

36. $(x - 2)(x + 1), (x - 2)^2$; in any one expression, $x + 1$ occurs as a factor at most once and $x - 2$ occurs at most twice. Hence, LCM $= (x + 1)(x - 2)$.

38. $y(y + 2), (y + 2)^2$; in any one expression, y occurs as a factor at most once and $y + 2$ occurs at most twice. Hence, LCM $= y(y + 2)^2$.

40. $3(x - 1)(x + 1), (x - 1)^2, 2^2$; in any one expression 3 and $x + 1$ occur as factors at most once and 2 and $x - 1$ occur at most twice. Hence, LCM $= 3 \cdot 2^2 \cdot (x + 1) \cdot (x - 1)^2 = 12(x + 1)(x - 1)^2$.

42. y, $y(y + 1)(y - 1)$, $(y - 1)^3$; in any one expression, y and $y + 1$ occur as factors at most once and $y - 1$ occurs three times. Hence, LCM $= y(y + 1)(y - 1)^3$.

44. LCD $= 8xy$. $\frac{1}{4y} = \frac{1(2x)}{4y(2x)} = \frac{2x}{8xy}$; $\frac{3}{8xy}$

46. LCD $= 10x^2$. $\frac{1}{5x} = \frac{1(2x)}{5x(2x)} = \frac{2x}{10x^2}$; $\frac{3}{2x^2} = \frac{3(5)}{2x^2(5)} = \frac{15}{10x^2}$

48. $4x - 4 = 2^2(x - 1)$ and $4 = 2^2$. Hence, LCD $= 4(x - 1)$.
Then, $\frac{3}{4x - 4} = \frac{4}{4(x - 1)}$; $\frac{1}{2^2} = \frac{1(x - 1)}{2^2(x - 1)} = \frac{x - 1}{4(x - 1)}$

50. LCD $= (y + 4)(y - 1)$.
$$\frac{1}{y + 4} = \frac{1(y - 1)}{(y + 4)(y - 1)} = \frac{y - 1}{(y + 4)(y - 1)};$$
$$\frac{3}{y - 1} = \frac{3(y + 4)}{(y + 4)(y - 1)} = \frac{3y + 12}{(y + 4)(y - 1)}$$

52. $2y + 4 = 2(y + 2)$ and $4y - 6 = 2(2y - 3)$. LCD $= 2(y + 2)(2y - 3)$.
$$\frac{3}{2(y + 2)} = \frac{3(2y - 3)}{2(y + 2)(2y - 3)} = \frac{6y - 9}{2(y + 2)(2y - 3)}$$
$$\frac{5}{2(2y - 3)} = \frac{5(y + 2)}{2(2y - 3)(y + 2)} = \frac{5y + 10}{2(2y - 3)(y + 2)}$$

54. LCD $= (a - 1)^2$. $\frac{4}{(a - 1)^2}$; $\frac{2}{a - 1} = \frac{2(a - 1)}{(a - 1)(a - 1)} = \frac{2a - 2}{(a - 1)^2}$

56. LCD $= (a - 1)(a + 1)(a + 2)$.
$$\frac{3}{a^2 - 1} = \frac{3(a + 2)}{(a - 1)(a + 1)(a + 2)} = \frac{3a + 6}{(a - 1)(a + 1)(a + 2)};$$
$$\frac{1}{(a + 2)(a - 1)} = \frac{1(a + 1)}{(a + 2)(a - 1)(a + 1)} = \frac{a + 1}{(a - 1)(a + 1)(a + 2)}$$

58. LCD $= (y + 1)(y + 4)(y - 4)$.
$$\frac{y}{y^2 + 5y + 4} = \frac{y(y - 4)}{(y + 1)(y + 4)(y - 4)} = \frac{y^2 - 4y}{(y + 1)(y + 4)(y - 4)};$$
$$\frac{y}{y^2 - 16} = \frac{y(y + 1)}{(y + 4)(y - 4)(y + 1)} = \frac{y^2 + y}{(y + 1)(y + 4)(y - 4)}$$

60. LCD = $(3x - y)(3x + y)^2$.

$$\frac{x}{(3x + y)^2} = \frac{x(3x - y)}{(3x + y)^2(3x - y)} = \frac{3x^2 - xy}{(3x + y)^2(3x - y)}$$

$$\frac{2x}{(3x - y)(3x + y)} = \frac{2x(3x + y)}{(3x - y)(3x + y)^2} = \frac{6x^2 + 2xy}{(3x - y)(3x + y)^2}$$

62. Since $x^2 - 6xy + 9y^2 = (x - 3y)^2$ and $3x - 9y = 3(x - 3y)$, the LCD is $3(x - 3y)^2$.

$$\frac{2x}{(x - 3y)^2} = \frac{3 \cdot 2x}{3(x - 3y)^2} = \frac{6x}{3(x - 3y)^2}$$

$$\frac{4}{3(x - 3y)} = \frac{4(x - 3y)}{3(x - 3y)^2} = \frac{4x - 12y}{3(x - 3y)^2}$$

64. Since $x^3 + 8 = x^3 + 2^3 = (x + 2)(x^2 - 2x + 4)$, the LCD is $(x + 2)(x^2 - 2x + 4)$.

$$\frac{3}{(x + 2)(x^2 - 2x + 4)}$$

$$\frac{1}{x + 2} = \frac{1 \cdot (x^2 - 2x + 4)}{(x + 2)(x^2 - 2x + 4)} = \frac{x^2 - 2x + 4}{(x + 2)(x^2 - 2x + 4)}$$

66. Since $x^3 - y^3 = (x - y)(x^2 + xy + y^2)$, the LCD is $(x - y)(x^2 + xy + y^2)$.

$$\frac{y}{(x - y)(x^2 + xy + y^2)}$$

$$\frac{2x}{(x - y)} = \frac{2(x^2 + xy + y^2)}{(x - y)(x^2 + xy + y^2)} = \frac{2x^2 + 2xy + 2y^2}{(x - y)(x^2 + xy + y^2)}$$

68. Since $y - 2x = -(2x - y)$, the LCD is $(2x - y)^2$.

$$\frac{2y}{-(2x - y)} = \frac{-2y}{2x - y} = \frac{-2y(2x - y)}{(2x - y)^2} = \frac{-4xy + 2y^2}{(2x - y)^2}$$

70. Since $6x - 3y = 3(2x - y)$ and $y - 2x = -(2x - y)$, the LCD is $3(2x - y)$.

$$\frac{4}{3(2x - y)} \quad \text{and} \quad \frac{2}{-(2x - y)} = \frac{-2}{2x - y} = \frac{-2(3)}{3(2x - y)} = \frac{-6}{3(2x - y)}$$

EXERCISE 5.5

2. $\dfrac{y - 5}{7}$

4. $\dfrac{x}{3} - \dfrac{2y}{3} + \dfrac{z}{3} = \dfrac{x - 2y + z}{3}$

6. $\dfrac{y + 1 + y - 1}{b} = \dfrac{2y}{b}$

8. $\dfrac{2 - (b - 2) + b}{a - 3b} = \dfrac{2 - b + 2 + b}{a - 3b}$

$$= \frac{4}{a - 3b}$$

10. $\dfrac{(x + 4) + (2x - 3)}{x^2 - x + 2} \quad \dfrac{3x + 1}{x^2 - x + 2}$

12. LCD $= 2^2 \cdot 3 = 12.$

$$\frac{3y(3)}{4(3)} + \frac{(4)y}{(4) \cdot 3} = \frac{9y + 4y}{12} = \frac{13y}{12}$$

14. LCD $= 5 \cdot 2 = 10.$

$$\frac{(2)3x}{(2)5} - \frac{(5)x}{(5)2} = \frac{6x - 5x}{10} = \frac{x}{10}$$

16. LCD $= 2 \cdot 3 = 6.$

$$\frac{(3)y}{(3)2} + \frac{(2)2y}{(2)3} = \frac{3y + 4y}{6} = \frac{7y}{6}$$

18. LCD $= 2^2 \cdot 3 = 12.$

$$\frac{(3)3}{(3)4}\,y + \frac{(4)2}{(4)3}\,y = \left(\frac{9}{12} + \frac{8}{12}\right)y = \frac{17}{12}\,y$$

20. LCD $= 2^2 \cdot 3 = 12.$

$$\frac{(3)3}{(3)4}\,x - \frac{(2) \cdot 1}{(2)6}\,x = \left(\frac{9}{12} - \frac{2}{12}\right)x = \frac{7}{12}\,x$$

22. LCD $= 2^2 \cdot 3 = 12.$

$$\frac{(3)3}{(3)4}\,y + \frac{(4) \cdot 1}{(4)3}\,y - \frac{(2)5}{(2)6}\,y = \left(\frac{9}{12} + \frac{4}{12} - \frac{10}{12}\right)y$$

$$= \frac{3}{12}\,y = \frac{1}{4}\,y$$

24. LCD = xy

$$\frac{(x)(y-2)}{(x)y} + \frac{(y)(2x-3)}{(y)x} = \frac{x(y-2) + y(2x-3)}{xy}$$

$$= \frac{xy - 2x + 2xy - 3y}{xy}$$

$$= \frac{-2x + 3xy - 3y}{xy}$$

26. LCD = 2 • 3 • x • y = 6xy

$$\frac{(2y)2}{(2y)3x} - \frac{(3x) \cdot 1}{(3x)2y} = \frac{4y - 3x}{6xy}$$

28. LCD = (x - 1)(y + 2)

$$\frac{2(x-1)}{(x-1)(y+2)} - \frac{3(y+2)}{(x-1)(y+2)} = \frac{2(x-1) - 3(y+2)}{(x-1)(y+2)}$$

$$= \frac{2x - 2 - 3y - 6}{(x-1)(y+2)} = \frac{2x - 3y - 8}{(x-1)(y+2)}$$

30. LCD = (y - 1)(2y + 1)

$$\frac{2y(2y+1)}{(y-1)(2y+1)} + \frac{y(y-1)}{(2y+1)(y-1)} = \frac{2y(2y+1) + y(y-1)}{(y-1)(2y+1)}$$

$$= \frac{4y^2 + 2y + y^2 - y}{(y-1)(2y+1)}$$

$$= \frac{5y^2 + y}{(y-1)(2y+1)}$$

32. LCD = (3x + 1)(x - 2)

$$\frac{2x(x-2)}{(3x+1)(x-2)} - \frac{x(3x+1)}{(x-2)(3x+1)} = \frac{2x(x-2) - x(3x+1)}{(3x+1)(x-2)}$$

$$= \frac{2x^2 - 4x - 3x^2 - x}{(3x+1)(x-2)}$$

$$= \frac{-x^2 - 5x}{(3x+1)(x-2)}$$

34. LCD = $(y + 1)(y - 2)$

$$\frac{(y - 2)(y - 2)}{(y + 1)(y - 2)} + \frac{(y + 3)(y + 1)}{(y - 2)(y + 1)} = \frac{(y - 2)(y - 2) + (y + 3)(y + 1)}{(y + 1)(y - 2)}$$

$$= \frac{y^2 - 4y + 4 + y^2 + 4y + 3}{(y + 1)(y - 2)} = \frac{2y^2 + 7}{(y + 1)(y - 2)}$$

36. LCD = $(2x + 1)(x - 1)$

$$\frac{(x - 2)(x - 1)}{(2x + 1)(x - 1)} - \frac{(x + 1)(2x + 1)}{(x - 1)(2x + 1)} = \frac{(x - 2)(x - 1) - (x + 1)(2x + 1)}{(x - 1)(2x + 1)}$$

$$= \frac{x^2 - 3x + 2 - (x^2 + 3x + 1)}{(x - 1)(2x + 1)} = \frac{-6x + 1}{(x - 1)(2x + 1)}$$

38. LCD = $2^2 \cdot 5(y - 2) = 20(y - 2)$

$$\frac{5}{4(y - 2)} + \frac{3}{5(y - 2)} = \frac{(5)5}{(5)4(y - 2)} + \frac{(4)3}{(4)5(y - 2)}$$

$$= \frac{25}{20(y - 2)} + \frac{12}{20(y - 2)} = \frac{37}{20(y - 2)}$$

40. LCD = $2 \cdot 3(y + 2) = 6(y + 2)$

$$\frac{(2)2}{(2)3(y + 2)} - \frac{(3)3}{(3)2(y + 2)} = \frac{4}{6(y + 2)} - \frac{9}{6(y + 2)} = \frac{-5}{6(y + 2)}$$

42. LCD = $(y - 1)(y + 1)^2$

$$\frac{1 \cdot (y + 1)}{(y + 1)(y - 1)(y + 1)} + \frac{1 \cdot (y - 1)}{(y + 1)^2(y - 1)} = \frac{(y+1) + (y-1)}{(y - 1)(y + 1)^2} = \frac{2y}{(y - 1)(y + 1)^2}$$

44. LCD = $(x - 2)(x - 3)(x + 3)$

$$\frac{x(x + 3)}{(x - 3)(x - 2)(x + 3)} - \frac{(x - 1)(x - 2)}{(x + 3)(x - 3)(x - 2)} = \frac{x(x + 3) - (x - 1)(x - 2)}{(x - 3)(x - 2)(x + 3)}$$

$$= \frac{x^2 + 3x - [x^2 - 3x + 2]}{(x - 3)(x + 3)(x - 2)} = \frac{6x - 2}{(x - 3)(x + 3)(x - 2)}$$

46. LCD = $(y + 2)(y - 2)(y + 5)$

$$\frac{y(y + 2)}{(y + 5)(y - 2)(y + 2)} - \frac{(2y - 1)(y + 5)}{(y - 2)(y + 2)(y + 5)} = \frac{y(y + 2) - (2y - 1)(y + 5)}{(y + 5)(y - 2)(y + 2)}$$

$$= \frac{y^2 + 2y - [2y^2 + 9y - 5]}{(y + 5)(y - 2)(y + 2)} = \frac{-y^2 - 7y + 5}{(y + 5)(y - 2)(y + 2)}$$

48. LCD = $x(x + 2)(x - 3)$

$$\frac{(x + 1)(x - 3)}{x(x + 2)(x - 3)} - \frac{(x - 1)(x + 2)}{x(x - 3)(x + 2)} = \frac{(x + 1)(x - 3) - (x - 1)(x + 2)}{x(x + 2)(x - 3)}$$

$$= \frac{x^2 - 2x - 3 - [x^2 + x - 2]}{x(x + 2)(x - 3)} = \frac{-3x - 1}{x(x + 2)(x - 3)}$$

50. LCD = $(y - 1)(y - 3)^2$

$$\frac{(3y - 1)(y - 3)}{(y - 3)(y - 1)(y - 3)} - \frac{(y + 2)(y - 1)}{(y - 3)^2(y - 1)} = \frac{(3y - 1)(y - 3) - (y + 2)(y - 1)}{(y - 1)(y - 3)^2}$$

$$= \frac{3y^2 - 10y + 3 - [y^2 + y - 2]}{(y - 1)(y - 3)^2} = \frac{2y^2 - 11y + 5}{(y - 1)(y - 3)^2}$$

52. LCD = $(a + b)(a + 2b)(a - 2b)$

$$\frac{4(a + b)}{(a + 2b)(a - 2b)(a + b)} + \frac{2(a - 2b)}{(a + 2b)(a + b)(a - 2b)}$$

$$+ \frac{4(a + 2b)}{(a + b)(a - 2b)(a + 2b)}$$

$$= \frac{4a + 4b + 2a - 4b + 4a + 8b}{(a + b)(a + 2b)(a - 2b)} = \frac{10a + 8b}{(a + b)(a + 2b)(a - 2b)}$$

54. LCD = $(y + 1)(y + 4)(y - 5)$

$$\frac{(2y + 5)(y - 5)}{(y + 4)(y + 1)(y - 5)} + \frac{(y + 13)(y + 1)}{(y + 4)(y - 5)(y + 1)} + \frac{(y + 7)(y + 4)}{(y - 5)(y + 1)(y + 4)}$$

$$= \frac{(2y + 5)(y - 5) + (y + 13)(y + 1) + (y + 7)(y + 4)}{(y + 4)(y + 1)(y - 5)}$$

$$= \frac{2y^2 - 5y - 25 + y^2 + 14y + 13 + y^2 + 11y + 28}{(y + 4)(y + 1)(y - 5)}$$

$$= \frac{4y^2 + 20y + 16}{(y + 4)(y + 1)(y - 5)} = \frac{4\cancel{(y + 4)}\cancel{(y + 1)}}{\cancel{(y + 4)}\cancel{(y + 1)}(y - 5)}$$

$$= \frac{4}{y - 5}$$

56. LCD = $(y - 1)(y + 2)(y - 3)$

$$\frac{3y(y - 3)}{(y + 2)(y - 1)(y - 3)} - \frac{(3y - 4)(y - 1)}{(y - 3)(y + 2)(y - 1)} + \frac{(2y - 4)(y + 2)}{(y - 3)(y - 1)(y + 2)}$$

$$= \frac{3y(y - 3) - (3y - 4)(y - 1) + (2y - 4)(y + 2)}{(y + 2)(y - 1)(y - 3)}$$

$$= \frac{3y^2 - 9y - [3y^2 - 7y + 4] + 2y^2 - 8}{(y + 2)(y - 1)(y - 3)}$$

$$= \frac{2y^2 - 2y - 12}{(y + 2)(y - 1)(y - 3)} = \frac{2\cancel{(y - 3)}\cancel{(y + 2)}}{\cancel{(y + 2)}(y - 1)\cancel{(y - 3)}}$$

$$= \frac{2}{y - 1}$$

58. LCD = $(y + 1)(y - 1)$

$$\frac{y(y + 1)(y - 1)}{1 \cdot (y + 1)(y - 1)} - \frac{2}{(y + 1)(y - 1)} + \frac{3}{(y + 1)(y - 1)} = \frac{y(y + 1)(y - 1) - 2 + 3}{(y + 1)(y - 1)}$$

$$= \frac{y^3 - y + 1}{(y + 1)(y - 1)}$$

60. LCD = $(x + 2)(x - 1)$

$$\frac{x(x + 2)(x - 1)}{1 \cdot (x + 2)(x - 1)} + \frac{2x^2(x - 1)}{(x + 2)(x - 1)} - \frac{3x^2(x + 2)}{(x - 1)(x + 2)}$$

$$= \frac{x(x + 2)(x - 1) + 2x^2(x - 1) - 3x^2(x + 2)}{(x - 1)(x + 2)}$$

$$= \frac{x^3 + x^2 - 2x + 2x^3 - 2x^2 - 3x^3 - 6x^2}{(x - 1)(x + 2)}$$

$$= \frac{-7x^2 - 2x}{(x - 1)(x + 2)}$$

62. LCD = $(y - 2)(y + 2)^2$

$$\frac{1 \cdot (y - 2)(y + 2)^2}{1 \cdot (y - 2)(y + 2)^2} + \frac{(y + 1)(y + 2)}{(y - 2)(y + 2)(y + 2)} - \frac{(y - 1)(y - 2)}{(y + 2)^2(y - 2)}$$

$$= \frac{(y - 2)(y + 2)^2 + (y + 1)(y + 2) - (y - 1)(y - 2)}{(y - 2)(y + 2)^2}$$

$$= \frac{y^3 + 2y^2 - 4y - 8 + y^2 + 3y + 2 - [y^2 - 3y + 2]}{(y - 2)(y + 2)^2}$$

$$= \frac{y^3 + 2y^2 + 2y - 8}{(y - 2)(y + 2)^2}$$

64. a. LCD = $x - 1$

$$x + 3 + \frac{1}{x - 1} = \frac{(x + 3)(x - 1)}{1(x - 1)} + \frac{1}{x - 1} = \frac{(x + 3)(x - 1) + 1}{x - 1}$$

$$= \frac{x^2 + 2x - 3 + 1}{x - 1} = \frac{x^2 + 2x - 2}{x - 1}$$

b.

$$\begin{array}{r|l} & x + 3 \\ \hline x - 1 & x^2 + 2x - 2 \\ & \underline{x^2 - x} \\ & \quad 3x - 2 \\ & \quad \underline{3x - 3} \\ & \qquad + 1 \end{array}$$

; $x + 3 + \dfrac{1}{x - 1}$

66. a. $\dfrac{(x-3)(x-2)}{1 \cdot (x-2)} - \dfrac{3}{x-2} = \dfrac{(x-3)(x-2)-3}{x-2} = \dfrac{x^2-5x+6-3}{x-2}$

$$= \frac{x^2-5x+3}{x-2}$$

b.
$$
\begin{array}{r|l}
 & \quad\; x - 3 \\
\hline
x - 2 & x^2 - 5x + 3 \\
 & \underline{x^2 - 2x} \\
 & \quad\; -3x + 3 \\
 & \quad\; \underline{-3x + 6} \\
 & \qquad\quad -3
\end{array}
\; ; \quad x - 3 - \frac{3}{x-2}
$$

68. a. $\dfrac{(x^2+x-4)(x-1)}{1 \cdot (x-1)} - \dfrac{1}{x-1} = \dfrac{(x^2+x-4)(x-1)-1}{x-1}$

$$= \frac{x^3-5x+4-1}{x-1} = \frac{x^3-5x+3}{x-1}$$

b.
$$
\begin{array}{r|l}
 & \quad x^2 + x - 4 \\
\hline
x - 1 & x^3 + 0x^2 - 5x + 3 \\
 & \underline{x^3 - x^2} \\
 & \qquad x^2 - 5x \\
 & \qquad \underline{x^2 - x} \\
 & \qquad\quad -4x + 3 \\
 & \qquad\quad \underline{-4x + 4} \\
 & \qquad\qquad\quad -1
\end{array}
\; ; \quad x^2 + x - 4 - \frac{1}{x-1}
$$

70. LCD $= c^2s^2$; then $\dfrac{1}{c^2} + \dfrac{1}{s^2} = \dfrac{1(s^2)}{c^2(s^2)} + \dfrac{1(c^2)}{s^2(c^2)} = \dfrac{s^2+c^2}{c^2s^2}$;

substituting 1 for $s^2 + c^2$ gives $\dfrac{1}{c^2s^2}$.

72. LCD $= s^2$; then $\dfrac{c^2}{s^2} + 1 = \dfrac{c^2}{s^2} + \dfrac{1(s^2)}{s^2} = \dfrac{c^2+s^2}{s^2}$;

substituting 1 for $c^2 + s^2$ gives $\dfrac{1}{s^2}$.

74. LCD = s; then $s + \frac{c(c - s)}{s} = \frac{s(s)}{(s)} + \frac{c(c - s)}{s} = \frac{s^2 + c^2 - cs}{s}$;

substituting 1 for $s^2 + c^2$ gives $\frac{1 - cs}{s}$.

76. LCD = $c^2 - s^2$; $\frac{c^3 + c^2 s}{c^2 - s^2} - \frac{s(c^2 - s^2)}{c^2 - s^2} = \frac{c^3 + c^2 s - c^2 s + s^3}{c^2 - s^2} = \frac{c^3 + s^3}{c^2 - s^2}$

$$= \frac{\cancel{(c + s)}(c^2 - cs + s^2)}{\cancel{(c + s)}(c - s)}$$

$$= \frac{c^2 + s^2 - cs}{c - s};$$

substituting 1 for $c^2 + s^2$ gives $\frac{1 - cs}{c - s}$.

EXERCISE 5.6

2. $\frac{\cancel{4}}{\underset{5}{\cancel{15}}} \cdot \frac{\cancel{3}}{\underset{4}{\cancel{16}}} = \frac{1}{20}$

4. $\frac{7}{8} \cdot \frac{\overset{3}{\cancel{48}}}{\underset{4}{\cancel{64}}} = \frac{21}{32}$

6. $\frac{\cancel{3}}{\underset{2}{\cancel{10}}} \cdot \frac{\overset{4}{\cancel{16}}}{27} \cdot \frac{\overset{\cancel{5}}{\cancel{30}}}{\underset{\underset{2}{\cancel{6}}}{\cancel{36}}} = \frac{4}{27}$

8. $\frac{\overset{2a}{\cancel{4a^{\cancel{2}}}}}{\cancel{3}} \cdot \frac{\overset{2}{\cancel{6}}b}{\cancel{2a}} = 4ab$

10. $\frac{\overset{3t}{\cancel{21t^{\cancel{2}}}}}{\cancel{5s}} \cdot \frac{\overset{3s}{\cancel{15s^{\cancel{3}}}}}{\cancel{7st}} = 9st$

12. $\frac{\overset{2a}{\cancel{14a^{\cancel{3}}b}}}{\cancel{3b}} \cdot \frac{\overset{-2}{\cancel{-6}}}{\cancel{7a^{2}}} = -4a$

14. $\frac{\cancel{2}}{\cancel{3}}y \cdot \frac{\overset{3}{\cancel{9}}}{\underset{5}{\cancel{10}}}y^2 = \frac{3}{5}y^3$

16. $\frac{1}{\underset{2}{\cancel{4}}}x^3y \cdot \frac{\cancel{2}}{5}xy = \frac{1}{10}x^4y^2$

18. $-\frac{\cancel{3}}{\cancel{5}}x^2y \cdot \frac{\cancel{5}}{\underset{2}{\cancel{6}}}xy^2z = -\frac{1}{2}x^3y^3z$

20. $\frac{\overset{a}{\cancel{a^{2}}}}{\cancel{xy}} \cdot \frac{3\overset{x^2}{\cancel{x^{3}}}\cancel{y}}{4\cancel{a}} = \frac{3ax^2}{4}$

22. $\frac{\overset{\cancel{5}}{\cancel{10}}x}{\underset{\cancel{6}}{\cancel{12y}}} \cdot \frac{\cancel{3x^{2}z}}{\underset{\cancel{x}}{\cancel{5x^{3}z}}} \cdot \frac{\cancel{6y^{2}x}}{\cancel{3y}z} = \frac{x}{z}$

24. $\frac{\overset{x}{\cancel{15x^{2}y}}}{1} \cdot \frac{\cancel{3}}{\underset{3y}{\cancel{45xy^{2}}}} = \frac{x}{y}$

26. $\frac{\cancel{3y}}{2\cancel{y}\cancel{(2x-3y)}} \cdot \frac{\cancel{(2x-3y)}}{\underset{4}{\cancel{12}}x} = \frac{1}{8x}$

28. $\frac{(3x+5)\cancel{(3x-5)}}{2\cancel{(x-1)}} \cdot \frac{(x+1)\cancel{(x-1)}}{2\cancel{(3x-5)}} = \frac{(3x+5)(x+1)}{4}$

30. $\frac{(2x+3)\cancel{(2x+1)}}{\cancel{(2x-3)}(x-1)} \cdot \frac{3x\cancel{(2x-3)}}{\cancel{(1+2x)}(1-2x)} = \frac{3x(2x+3)}{(x-1)(1-2x)}$

32. $\frac{(x-2)\cancel{(x+1)}}{(x+3)\cancel{(x+1)}} \cdot \frac{(x+1)\cancel{(x-5)}}{\cancel{(x-5)}(x+2)} = \frac{(x-2)(x+1)}{(x+3)(x+2)}$

34. $\frac{(x-3)\cancel{(3x+2)}}{\cancel{(2x+1)}(x-1)} \cdot \frac{\cancel{(2x+1)}\cancel{(x-5)}}{\cancel{(3x+2)}\cancel{(x-5)}} = \frac{x-3}{x-1}$

36. $\frac{5x\cancel{(x-1)}}{3} \cdot \frac{\cancel{(x+1)}\cancel{(x-10)}}{4\cancel{(x-10)}} \cdot \frac{y^2}{-2\cancel{(x+1)}\cancel{(x-1)}} = \frac{-5xy^2}{24}$

38. $\frac{\cancel{3}}{4}y \cdot \frac{1}{\underset{2}{\cancel{6}}}y + \frac{3}{\cancel{4}}y \cdot \frac{\overset{2}{\cancel{8}}}{1} = \frac{1}{8}y^2 + 6y$

40. $(y - \frac{1}{3})(y - \frac{1}{3}) = y^2 - \frac{1}{3}y - \frac{1}{3}y + \frac{1}{9}$

$= y^2 - \frac{2}{3}y + \frac{1}{9}$

42. $(y + \frac{1}{4})(y + \frac{1}{4}) = y^2 + \frac{1}{4}y + \frac{1}{4}y + \frac{1}{16} = y^2 + \frac{1}{2}y + \frac{1}{16}$

44. $\frac{2}{\cancel{3}} \cdot \frac{\overset{5}{\cancel{15}}}{9} = \frac{10}{9}$

46. $\frac{\overset{3}{\cancel{9}}ab^3}{\cancel{x}} \cdot \frac{2\overset{x^2}{\cancel{x^3}}}{\cancel{3}} = 6ab^3x^2$

48. $\frac{\overset{8\,a}{\cancel{24}\cancel{a^3}}\cancel{b}}{1} \cdot \frac{7x}{\cancel{3}\cancel{a^2}\cancel{b}} = 56ax$

50. $\frac{3\cancel{(2y-9)}}{5\cancel{x}} \cdot \frac{\cancel{x}}{2\cancel{(2y-9)}} = \frac{3}{10}$

52. $\frac{\cancel{(a+5)}\cancel{(a-3)}}{\cancel{(a+5)}\cancel{(a-2)}} \cdot \frac{\cancel{(a-2)}(a-7)}{(a+3)\cancel{(a-3)}} = \frac{a-7}{a+3}$

54. $\frac{\cancel{(x+7)}\cancel{(x-1)}}{\cancel{(x+2)}\cancel{(x-1)}} \cdot \frac{(x-5)\cancel{(x+2)}}{(x-2)\cancel{(x+7)}} = \frac{x-5}{x-2}$

56. $\frac{\cancel{(3x+2)}\cancel{(3x-1)}}{\cancel{(4x-1)}\cancel{(3x+2)}} \cdot \frac{\cancel{(4x-1)}(2x+3)}{\cancel{(3x-1)}(3x-1)} = \frac{2x+3}{3x-1}$

58. $1 \cdot \frac{x-2}{x^2+3x+1} = \frac{x-2}{x^2+3x+1}$

60. $\frac{\cancel{(x-3)}(x+3)}{1} \cdot \frac{3x}{\cancel{(x-3)}(x-3)}$

$= \frac{3x(x+3)}{x-3} = \frac{3x^2+9x}{x-3}$

62. $\dfrac{\cancel{y}(2y + 1)}{3x} \cdot \dfrac{1}{2\cancel{y}} \quad \dfrac{2y + 1}{6x}$

64. $\dfrac{3}{2y - 1} - \dfrac{4\cancel{(y + 3)}}{(3y + 1)\cancel{(y + 4)}} \cdot \dfrac{\cancel{(2y + 1)}\cancel{(y + 4)}}{\cancel{(2y + 1)}\cancel{(y + 3)}}$

$= \dfrac{3}{2y - 1} - \dfrac{4}{3y + 1} = \dfrac{3[3y + 1]}{(2y - 1)[3y + 1]} - \dfrac{4[2y - 1]}{(3y + 1)[2y - 1]}$

$= \dfrac{3[3y + 1] - 4(2y - 1)}{(2y - 1)(3y + 1)} = \dfrac{9y + 3 - 8y + 4}{(2y - 1)(3y + 1)}$

$= \dfrac{y + 7}{(2y - 1)(3y + 1)}$

66. $\dfrac{2}{y - 4} + \dfrac{3\cancel{(2y - 1)}}{\cancel{(3y - 4)}(4y - 3)} \cdot \dfrac{\cancel{(3y - 4)}\cancel{(3y + 4)}}{\cancel{(3y + 4)}\cancel{(2y - 1)}}$

$= \dfrac{2}{y - 4} + \dfrac{3}{4y - 3} = \dfrac{2[4y - 3]}{(y - 4)[4y - 3]} + \dfrac{3[y - 4]}{(4y - 3)[y - 4]}$

$= \dfrac{8y - 6 + 3y - 12}{(4y - 3)(y - 4)} = \dfrac{11y - 18}{(4y - 3)(y - 4)}$

68. $\left(\dfrac{3[y + 4]}{(y - 3)[y + 4]} + \dfrac{4[y - 3]}{(y + 4)[y - 3]}\right)\left(\dfrac{1 \cdot y}{3 \cdot y} - \dfrac{1 \cdot 3}{y \cdot 3}\right)$

$= \left(\dfrac{3y + 12 + 4y - 12}{(y + 4)\cancel{(y - 3)}}\right)\left(\dfrac{\cancel{y - 3}}{3y}\right) = \left(\dfrac{7\cancel{y}}{y + 4}\right)\left(\dfrac{1}{3\cancel{y}}\right)$

$= \dfrac{7}{3(y + 4)}$

70. $\dfrac{\cancel{(2x - y)}(4x^2 + 2xy + y^2)}{\cancel{(x + y)}} \cdot \dfrac{\cancel{(x + y)}(x - y)}{\cancel{(2x - y)}} = (x - y)(4x^2 + 2xy + y^2)$

72. Since $2xy + 4x + 3y + 6 = 2x(y + 2) + 3(y + 2) = (2x + 3)(y + 2)$, we have

$\dfrac{\cancel{(2x + 3)}\cancel{(y + 2)}}{\cancel{(2x + 3)}} \cdot \dfrac{(y - 1)}{\cancel{(y + 2)}} = y - 1$

76. $Q(x) \div R(x) = \dfrac{x^2}{x^2 - 1} \div \dfrac{x^3}{(x - 1)^2} = \dfrac{\cancel{x^2}}{(x + 1)\cancel{(x - 1)}} \cdot \dfrac{\overset{x - 1}{\cancel{(x - 1)^2}}}{\underset{x}{\cancel{x^3}}} \quad \dfrac{x - 1}{x(x + 1)}$

78. $P(x) \cdot R(x) \div Q(x) = \frac{x}{x-1} \cdot \frac{x^3}{(x-1)^2} \div \frac{x^2}{x^2-1}$

$$= \frac{\cancel{x}}{\cancel{x-1}} \cdot \frac{\overset{x^2}{\cancel{x^3}}}{(x-1)^2} \cdot \frac{(\cancel{x-1})(x+1)}{\underset{x}{\cancel{x^2}}} = \frac{x^2(x+1)}{(x-1)^2}$$

EXERCISE 5.7

2. $\frac{4}{\cancel{5}} \cdot \frac{\cancel{5}}{7} = \frac{4}{7}$

4. $\frac{5}{\cancel{2}} \cdot \frac{\overset{2}{\cancel{4}}}{21} = \frac{10}{21}$

6. $\frac{5\cancel{x}}{6\cancel{y}} \cdot \frac{5\cancel{y}}{4\cancel{x}} = \frac{25}{24}$

8. $\frac{\cancel{3}a\cancel{b}}{\cancel{4}} \cdot \frac{\overset{2}{\cancel{8}}a^2}{\cancel{3b}} = 2a^3$

10. $4 + \frac{2}{3} = \frac{14}{3}$, hence

$$\frac{\frac{1}{3}}{\frac{14}{3}} = \frac{1}{3} \cdot \frac{3}{14} = \frac{1 \cdot 3}{14 \cdot 3} = \frac{1}{14}$$

or

$$\frac{\frac{1}{3}\left(\frac{3}{1}\right)}{\left(4 + \frac{2}{3}\right)\left(\frac{3}{1}\right)} = \frac{1}{12 + 2} = \frac{1}{14}$$

12. $\frac{\left(\frac{1}{2} + \frac{3}{4}\right)\left(\frac{4}{1}\right)}{\left(\frac{1}{2} - \frac{3}{4}\right)\left(\frac{4}{1}\right)} = \frac{\frac{1}{2} \cdot \frac{4}{1} + \frac{3}{4} \cdot \frac{4}{1}}{\frac{1}{2} \cdot \frac{4}{1} - \frac{3}{4} \cdot \frac{4}{1}}$

$$= \frac{2 + 3}{2 - 3} = -5$$

14. $\frac{\left(\frac{1}{4}\right)\frac{12}{1}}{\left(\frac{2}{3} + \frac{1}{2}\right)\frac{12}{1}} = \frac{3}{\frac{12}{1} \cdot \frac{2}{3} + \frac{12}{1} \cdot \frac{1}{2}}$

$$= \frac{3}{8 + 6}$$

$$= \frac{3}{14}$$

16. $\frac{\left(\frac{3}{4} + \frac{1}{3}\right)\frac{12}{1}}{\left(\frac{1}{2} + \frac{5}{6}\right)\frac{12}{1}} = \frac{\frac{3}{4} \cdot \frac{12}{1} + \frac{1}{3} \cdot \frac{12}{1}}{\frac{1}{2} \cdot \frac{12}{1} + \frac{5}{6} \cdot \frac{12}{1}}$

$$= \frac{9 + 4}{6 + 10}$$

$$= \frac{13}{16}$$

18. $$\frac{\left(1+\frac{2}{a}\right)\left(\frac{a^2}{1}\right)}{\left(1-\frac{4}{a^2}\right)\left(\frac{a^2}{1}\right)} = \frac{a^2+2a}{a^2-4}$$

$$= \frac{a(a+2)}{(a-2)(a+2)}$$

$$= \frac{a}{a-2}$$

20. $$\frac{\left(1+\frac{1}{x}\right)\left(\frac{x}{1}\right)}{\left(1-\frac{1}{x}\right)\left(\frac{x}{1}\right)} = \frac{x+1}{x-1}$$

22. $$\frac{(4)(x)}{\left(\frac{2}{x}+2\right)\left(\frac{x}{1}\right)} = \frac{4x}{2+2x} = \frac{2 \cdot 2x}{2(1+x)} = \frac{2x}{1+x}$$

24. $$\frac{\left(y+3\right)\left(\frac{y}{1}\right)}{\left(\frac{9}{y}-y\right)\left(\frac{y}{1}\right)} = \frac{y(y+3)}{9-y^2} = \frac{y(3+y)}{(3-y)(3+y)} = \frac{y}{3-y} \text{ or } \frac{-y}{y-3}$$

26. $$\frac{\left(\frac{1}{y-1}\right)\left[\frac{y^2(y-1)}{1}\right]}{\left(\frac{1}{y^2}+1\right)\left[\frac{y^2(y-1)}{1}\right]} = \frac{y^2}{\frac{1}{y^2}\cdot\frac{y^2(y-1)}{1}+y^2(y-1)}$$

$$= \frac{y^2}{y-1+y^3-y^2} \text{ or } \frac{y^2}{(y-1)(1+y^2)}$$

or

$$\frac{\frac{1}{y-1}}{\frac{1+y^2}{y^2}} = \frac{1}{y-1}\cdot\frac{y^2}{1+y^2} = \frac{y^2}{(y-1)(1+y^2)} \text{ or } \frac{y^2}{y-1+y^3-y^2}$$

28. $$\frac{\left(y+\frac{x}{y}\right)\left(\frac{xy}{1}\right)}{\left(x-\frac{y}{x}\right)\left(\frac{xy}{1}\right)} = \frac{xy^2+x^2}{x^2y-y^2}$$

or

$$\frac{\frac{y^2+x}{y}}{\frac{x^2-y}{x}} = \frac{y^2+x}{y}\cdot\frac{x}{x^2-y} = \frac{x(y^2+x)}{y(x^2-y)} \text{ or } \frac{xy^2+x^2}{x^2y-y^2}$$

30. $$\frac{\frac{4}{y} - \frac{1}{3}}{\frac{16}{y^2} - \frac{1}{9}} = \frac{\left(\frac{4}{y} - \frac{1}{3}\right)\frac{9y^2}{1}}{\left(\frac{16}{y^2} - \frac{1}{9}\right)\frac{9y^2}{1}} = \frac{\frac{4}{y} \cdot \frac{9y^2}{1} - \frac{1}{3} \cdot \frac{9y^2}{1}}{\frac{16}{y^2} \cdot \frac{9y^2}{1} - \frac{1}{9} \cdot \frac{9y^2}{1}}$$

$$= \frac{36y - 3y^2}{144 - y^2} = \frac{3y(12 - y)}{(12 - y)(12 + y)} = \frac{3y}{12 + y}$$

or

$$\frac{\frac{4}{y} - \frac{1}{3}}{\frac{16}{y^2} - \frac{1}{9}} = \frac{\frac{12 - y}{3y}}{\frac{144 - y^2}{9y^2}} = \frac{\cancel{12 - y}}{\cancel{3y}} \cdot \frac{\cancel{9y^2}\,3y}{(\cancel{12 - y})(12 + y)} = \frac{3y}{12 + y}$$

32. $$\frac{1 + \frac{1}{y^3}}{1 + \frac{1}{y}} = \frac{\left(1 + \frac{1}{y^3}\right)\left(\frac{y^3}{1}\right)}{\left(1 + \frac{1}{y}\right)\left(\frac{y^3}{1}\right)} = \frac{y^3 + 1}{y^3 + y^2}$$

$$= \frac{(y + 1)(y^2 - y + 1)}{y^2(y + 1)} = \frac{y^2 - y + 1}{y^2}$$

or

$$\frac{\frac{y^3 + 1}{y^3}}{\frac{y + 1}{y}} = \frac{y^3 + 1}{\cancel{y^3}\,y^2} \cdot \frac{\cancel{y}}{y + 1} = \frac{(y + 1)(y^2 - y + 1)}{y^2(y + 1)}$$

$$= \frac{y^2 - y + 1}{y^2}$$

34. $$\frac{y - 1 - \frac{4}{3y - 2}}{y - \frac{1}{3y - 2}} = \frac{\left(y - 1 - \frac{4}{3y - 2}\right)\frac{3y - 2}{1}}{\left(y - \frac{1}{3y - 2}\right)\frac{3y - 2}{1}} = \frac{(y - 1)(3y - 2) - 4}{y(3y - 2) - 1}$$

$$= \frac{3y^2 - 5y - 2}{3y^2 - 2y - 1} = \frac{(3y + 1)(y - 2)}{(3y + 1)(y - 1)}$$

$$= \frac{y - 2}{y - 1}$$

34. cont.

or

$$\frac{\frac{(y-1)(3y-2)-4}{3y-2}}{\frac{y(3y-2)-1}{3y-2}} = \frac{3y^2-5y+2-4}{3y-2} \cdot \frac{3y-2}{3y^2-2y-1}$$

$$= \frac{3y^2-5y-2}{3y^2-2y-1} = \frac{(3y+1)(y-2)}{(3y+1)(y-1)}$$

$$= \frac{y-2}{y-1}$$

36. First simplify $\dfrac{1}{\frac{a}{b}+2}$ and obtain $\dfrac{1 \cdot b}{\left(\frac{a}{b}+2\right)\frac{b}{1}} = \dfrac{b}{a+2b}$

Then simplify $\dfrac{3}{\frac{a}{2b}+1}$ and obtain $\dfrac{3(2b)}{\left(\frac{a}{2b}+1\right)\frac{2b}{1}} = \dfrac{6b}{a+2b}$

Substituting the above results into the original complex fraction, we have

$$\frac{1-\frac{1}{\frac{a}{b}+2}}{1+\frac{3}{\frac{a}{2b}+1}} = \frac{1-\frac{b}{a+2b}}{1+\frac{6b}{a+2b}} = \frac{\left(1-\frac{b}{a+2b}\right)\left(\frac{a+2b}{1}\right)}{\left(1+\frac{6b}{a+2b}\right)\left(\frac{a+2b}{1}\right)}$$

$$= \frac{a+2b-b}{a+2b+6b} = \frac{a+b}{a+8b}$$

38. $$\frac{\left(a+4-\frac{7}{a-2}\right)\left(\frac{a-2}{1}\right)}{\left(a-1+\frac{2}{a-2}\right)\left(\frac{a-2}{1}\right)} = \frac{(a+4)(a-2)-7}{(a-1)(a-2)+2} = \frac{a^2+2a-15}{a^2-3a+4}$$

40. $$\frac{\left(\frac{a}{bc}-\frac{b}{ac}+\frac{c}{ab}\right)\left(\frac{a^2b^2c^2}{1}\right)}{\left(\frac{1}{a^2b^2}-\frac{1}{a^2c^2}+\frac{1}{b^2c^2}\right)\left(\frac{a^2b^2c^2}{1}\right)} = \frac{a^3bc-ab^3c+abc^3}{c^2-b^2+a^2}$$

$$= \frac{abc(a^2-b^2+c^2)}{(a^2-b^2+c^2)} = abc$$

42. $$\frac{s(s + c)}{\frac{s}{s + c}(s + c)} - \frac{2sc}{s + c} = \frac{s(s + c)}{s} - \frac{2sc}{s + c}$$

$$= s + c - \frac{2sc}{s + c} = \frac{(s + c)(s + c)}{s + c} - \frac{2cs}{s + c}$$

$$= \frac{s^2 + 2sc + c^2 - 2sc}{s + c} = \frac{s^2 + c^2}{s + c} \text{ and}$$

substituting 1 for $s^2 + c^2$ gives $\frac{1}{s + c}$.

44. $$\frac{\left(\frac{1}{s^2} + \frac{1}{c^2}\right)(s^2c^2)}{\frac{1}{sc}(s^2c^2)} - \frac{c}{s} = \frac{c^2 + s^2}{sc} - \frac{c}{s} = \frac{c^2 + s^2}{sc} - \frac{c(c)}{s(c)}$$

$$= \frac{c^2 + s^2 - c^2}{sc} = \frac{s^2}{sc} = \frac{s}{c}$$

EXERCISE 5.8

2. $$(5)x + (5)4 = (5)\frac{2}{5}x - (5)3$$

$$5x + 20 = 2x - 15$$

$$5x + 20 + \boxed{(-2x) + (-20)} = 2x - 15 + \boxed{(-2x) + (-20)}$$

$$3x = -35; \left\{\frac{-35}{3}\right\}$$

4. $$(15)4 + (15)\frac{x}{5} = (15)\frac{5}{3}$$

$$60 + 3x = 25$$

$$60 + 3x + \boxed{(-60)} = 25 + \boxed{(-60)}$$

$$3x = -35; \left\{\frac{-35}{3}\right\}$$

6. $$(12)\frac{1}{4}x = (12)2 - (12)\frac{1}{3}x$$

$$3x = 24 - 4x$$

$$3x + \boxed{(4x)} = 24 - 4x + \boxed{(4x)}$$

$$7x = 24; \left\{\frac{24}{7}\right\}$$

8. $$(6)\frac{2x}{3} - (6)\frac{(2x + 5)}{6} = (6)\frac{1}{2}$$

$$4x - (2x + 5) = 3$$

$$4x - 2x - 5 = 3$$

$$2x - 5 + \boxed{5} = 3 + \boxed{5}$$

$$2x = 8; \{4\}$$

10. $(x + 3)(x - 3)\frac{x}{x - 3} + (x + 3)(x - 3)\frac{9}{x + 3} = (x + 3)(x - 3) \cdot 1$

$$x(x + 3) + 9(x - 3) = (x + 3)(x - 3)$$
$$x^2 + 3x + 9x - 27 = x^2 - 9$$
$$12x - 27 = -9$$
$$12x - 27 + \boxed{27} = -9 + \boxed{27}$$
$$12x = 18$$
$$x = \frac{18}{12} = \frac{3}{2}; \left\{\frac{3}{2}\right\}$$

12. $(x - 3)\frac{5}{(x - 3)} = (x - 3)\frac{x + 2}{x - 3} + (x - 3)3 \quad (x \neq 3)$

$$5 = x + 2 + 3x - 9$$
$$5 = 4x - 7$$
$$5 + \boxed{7} = 4x - 7 + \boxed{7}$$
$$12 = 4x$$
$$3 = x$$

The solution set is ∅, since 3 will not satisfy the original equation.

14. $\frac{y}{y + 2} - \frac{3}{y - 2} = \frac{y^2 + 8}{(y + 2)(y - 2)}$

$$(y + 2)(y - 2)\frac{y}{(y + 2)} - (y + 2)(y - 2)\frac{3}{(y - 2)}$$
$$= (y + 2)(y - 2)\frac{y^2 + 8}{(y + 2)(y - 2)} \quad (x \neq -2, 2)$$

$$y(y - 2) - 3(y + 2) = y^2 + 8$$
$$y^2 - 2y - 3y - 6 = y^2 + 8$$
$$y^2 - 5y - 6 + \boxed{(-y^2) + 6} = y^2 + 8 + \boxed{(-y^2) + 6}$$
$$-5y = 14; \left\{\frac{-14}{5}\right\}$$

16. $7(x - 2) = 5x$

$$7x - 14 = 5x$$
$$2x = 14$$
$$x = 7; \{7\}$$

18. $2 \cdot y = 1 \cdot (6 - y)$

$$2y = 6 - y$$
$$3y = 6$$
$$y = 2; \{2\}$$

20. $-3(y + 14) = 4(y - 7)$

$$-3y - 42 = 4y - 28$$
$$-7y = 14; \{-2\}$$

22. $30(r - 10) = 20r$

$$30r - 300 = 20r$$
$$10r = 300$$
$$r = 30; \{30\}$$

24. $3\left(2x - \frac{5}{3}\right) = 3\left(\frac{x^2}{3}\right)$

$6x - 5 = x^2$

$0 = x^2 - 6x + 5$

$0 = (x - 1)(x - 5)$

$x - 1 = 0;\ x - 5 = 0;\ \{1,5\}$

26. $4x\left(\frac{x}{4} - \frac{3}{4}\right) = 4x\left(\frac{1}{x}\right)$

$x^2 - 3x = 4$

$x^2 - 3x - 4 = 0$

$(x + 1)(x - 4) = 0$

$x + 1 = 0;\ x - 4 = 0;\ \{-1,4\}$

28. $[3x(3x + 1)]\frac{4}{3x} + [3x(3x + 1)]\frac{3}{(3x + 1)} + [3x(3x + 1)]2 = [3x(3x + 1)] \cdot 0$

$4(3x + 1) + (3x)3 + 6x(3x + 1) = 0$

$12x + 4 + 9x + 18x^2 + 6x = 0$

$18x^2 + 27x + 4 = 0$

$(3x + 4)(6x + 1) = 0$

$3x + 4 = 0;\ 6x + 1 = 0;\ \left\{\frac{-4}{3}, \frac{-1}{6}\right\}$

30. Substituting -40 for F,

$-40 = \frac{9}{5}C + 32$

$5(-40) = 5(\frac{9}{5}C) + 5(32)$

$-200 = 9C + 160$

$-360 = 9C$

$-40 = C$

The Celsius temperature is -40°.

32. $s = 165;\ hp = 468$

$468 = \frac{62.4N(165)}{33,000}$

$468 = \frac{10,296N}{33,000}$

$15,444,000 = 10,296N$

$1500 = N$

1500 cubic feet of water flowed.

34. $R_1 = 20;\ R_2 = 40;\ R_n = 10$

$\frac{1}{10} = \frac{1}{20} + \frac{1}{40} + \frac{1}{R_3}$

$(40R_3)\frac{1}{10} = (40R_3)\frac{1}{20} + (40R_3)\frac{1}{40} + (40R_3)\frac{1}{R_3}$

$4R_3 = 2R_3 + R_3 + 40$

$4R_3 = 3R_3 + 40$

$R_3 = 40$

The third resistor must be 40 ohms.

36. $F = 45;\ D_1 = 25;\ D_2 = 22$

$$45 = \frac{w}{2}\left(\frac{25 - 22}{25}\right)$$

$$45 = \frac{w}{2}\left(\frac{3}{25}\right)$$

$$45 = \frac{3}{50}w$$

$$\left(\frac{50}{3}\right)45 = \left(\frac{50}{3}\right)\frac{3}{50}w$$

$$w = 750$$

The weight is 750 kilograms.

38. $F = 98.6°;$

$$C = \frac{5(98.6) - 160}{9}$$

$$= \frac{493 - 160}{9}$$

$$= \frac{333}{9} = 37;$$

$$K = 37 + 273 = 310$$

Kelvin temperature of 310° corresponds to Fahrenheit temperature of 98.6°.

40. Number of consecutive natural numbers: n

$$\frac{n}{2}(n + 1) = 406$$

$$2\left[\frac{n}{2}(n + 1)\right] = 2[406]$$

$$n(n + 1) = 812$$

$$n^2 + n - 812 = 0$$

$$(n + 29)(n - 28) = 0; \quad \{-29,28\}$$

Since -29 is not a natural number, we take $n = 28$. The number of consecutive natural numbers is 28.

42.

$$(2)A = (\not{2})\frac{h}{\not{2}}(b + c)$$

$$2A = hb + hc$$

$$2A - hb = hc$$

$$\frac{2A - hb}{h} = c$$

44.

$$(2)S = (\not{2})\frac{n}{\not{2}}(a + s_n)$$

$$2S = na + ns_n$$

$$2S - na = ns_n$$

$$\frac{2S - na}{n} = s_n$$

46. $\frac{1}{R_n} - \frac{1}{R_1} - \frac{1}{R_2} = \frac{1}{R_3}$

$$\frac{1}{R_3} = \frac{R_1R_2 - R_nR_2 - R_nR_1}{R_nR_1R_2}$$

$$R_3(R_1R_2 - R_nR_2 - R_nR_1) = R_nR_1R_2$$

$$R_3 = \frac{R_nR_1R_2}{R_1R_2 - R_nR_2 - R_nR_1}$$

48. Substitute $\frac{5F - 160}{9}$ for C in $K = C + 273$ and obtain

$$K = \frac{5F - 160}{9} + 273$$

$$9K = 5F - 160 + 9(273)$$

$$9K = 5F - 160 + 2457$$

$$9K = 5F + 2297$$

$$9K - 2297 = 5F$$

$$\frac{9K - 2297}{5} = F$$

EXERCISE 5.9

2. The number: n

$$\frac{1}{2}n + 3n = \frac{35}{2}$$
$$n + 6n = 35$$
$$7n = 35$$
$$n = 5$$

The number is 5.

4. The smaller integer: n
The next integer: $n + 1$

$$2(n + 1) - \frac{1}{2}n = 14$$
$$4(n + 1) - n = 28$$
$$4n + 4 - n = 28$$
$$3n = 24$$
$$n = 8$$
$$n + 1 = 9$$

The integers are 8 and 9.

6. The number: n

$$n - 2\left(\frac{1}{n}\right) = \frac{17}{3}$$
$$\frac{3n}{1}\left(n - \frac{2}{n}\right) = \frac{17}{3}\left(\frac{3n}{1}\right)$$
$$3n^2 - 6 = 17n$$
$$3n^2 - 17n - 6 = 0$$
$$(3n + 1)(n - 6) = 0$$
$$n = -\frac{1}{3} \quad \text{or} \quad n = 6$$

The number is $-\frac{1}{3}$ or 6.

8. The denominator: x
The positive numerator: $x - 2$

$$\frac{x - 2}{x} + 3\left(\frac{x}{x - 2}\right) = \frac{28}{5}$$
$$\text{LCD} = 5x(x - 2)$$
$$5(x - 2)^2 + 15x^2 = 28x(x - 2)$$
$$5x^2 - 20x + 20 + 15x^2 = 28x^2 - 56x$$
$$-8x^2 + 36x + 20 = 0$$
$$-4(2x^2 - 9x - 5) = 0$$
$$-4(2x + 1)(x - 5) = 0$$
$$x = 5 \quad \text{or} \quad x = -\frac{1}{2}$$

Since the numerator must be positive, it is $5 - 2 = 3$ and the denominator is 5.

10. The selling price of the lot: x

$$\frac{3}{8}(\text{Profit}) = 600$$
$$\frac{3}{8}(x - 4000) = 600$$
$$3(x - 4000) = 4800$$
$$3x - 12000 = 4800$$
$$3x = 16800$$
$$x = 5600$$

The lot should sell for \$5600.

12. The inheritance: x

$$\left(\text{Inheritance}\right) - \left(\text{Expenditures}\right) = 10{,}000$$
$$x - \left(\frac{2}{3}x + \frac{1}{6}x\right) = 10{,}000$$
$$6x - 4x - x = 60{,}000$$
$$x = 60{,}000$$

His inheritance was \$60,000.

14. The denominator: x
The numerator: $x - 8$

$$\frac{x - 8}{x} = \frac{3}{5}$$
$$5(x - 8) = 3x$$
$$5x - 40 = 3x$$
$$2x = 40$$
$$x = 20$$

The denominator is 20.

16. Gallons required for trip of 459 miles: x

Gallons		Miles
$\frac{6}{x}$	=	$\frac{102}{459}$

$$102x = 6(459)$$
$$102x = 2754$$
$$x = 27$$

27 gallons are required for a trip of 459 miles.

18. Distance walked in 5 hours: x

Hours		Miles
$\frac{2}{5}$	=	$\frac{6.4}{x}$

$$2x = 5(6.4)$$
$$2x = 32$$
$$x = 16$$

He can walk 16 miles in 5 hours.

20. Number kilometers in 8 centimeters: x

Kilometers		Centimeters
$\frac{x}{10}$	=	$\frac{9}{3/4}$

$$\frac{3}{4}x = 9(10)$$
$$3x = 360$$
$$x = 120$$

8 centimeters represents 120 kilometers.

22. Length of time to type 10 pages.

Words		Minutes
$\frac{200}{240(10)}$	=	$\frac{5}{x}$

$$200x = (2400)(5)$$
$$x = 60$$

It would take 60 minutes.

24. Time the airplanes are in flight: t

	r	t	d = rt
airplane A	440	t	440t
airplane B	560	t	560t

$$\begin{bmatrix}\text{the distance}\\ \text{airplane A flies}\end{bmatrix} + \begin{bmatrix}\text{the distance}\\ \text{airplane B flies}\end{bmatrix} = 2500$$

$$440t + 560t = 2500$$

$$1000t = 2500$$
$$t = 2.5$$

The airplanes are in flight for 2.5 hours.

26. Distance boats have traveled when the second overtakes the first: d

	r	d	t = d/r
first boat	36	d	$\frac{d}{36}$
second boat	45	d	$\frac{d}{45}$

$$\begin{bmatrix}\text{time first}\\ \text{boat travels}\\ \text{in hours}\end{bmatrix} = \begin{bmatrix}\text{time second}\\ \text{boat travels}\\ \text{in hours}\end{bmatrix} + 1$$

$$\frac{d}{36} = \frac{d}{45} + 1$$

$$180\left(\frac{d}{36}\right) = 180\left(\frac{d}{45}\right) + 180(1)$$

$$5d = 4d + 180$$
$$d = 180$$

The boats will have traveled 180 nautical miles.

28. Rate of slow driver: r
Rate of fast driver: $2r$

	r	d	t = d/r
slow driver	r	400	400/r
fast driver	2r	400	400/2r

$$\begin{bmatrix}\text{time for}\\ \text{slow driver}\end{bmatrix} = \begin{bmatrix}\text{time for}\\ \text{fast driver}\end{bmatrix} + 5$$

$$\frac{400}{r} = \frac{\overset{200}{\cancel{400}}}{\cancel{2r}} + 5$$

$$400 = 200 + 5r$$
$$200 = 5r$$
$$40 = r$$

The drivers were going 40 and 80 miles per hour, respectively.

30. Speed of the current: x

	r	d	t = d/r
upstream	20 - x	90	90/20 - x
downstream	20 + x	75	75/20 + x

$$\begin{bmatrix}\text{time to go}\\ \text{upstream}\end{bmatrix} = \begin{bmatrix}\text{time to go}\\ \text{downstream}\end{bmatrix} + 3$$

$$\frac{90}{20 - x} = \frac{75}{20 + x} + 3; \text{ LCD} = 3(20 - x)(20 + x)$$

$$90(20 + x) = 75(20 - x) + (20 + x)(20 - x)$$
$$1800 + 90x = 1500 - 75x + 1200 - 3x^2$$
$$3x^2 + 165x - 900 = 0$$
$$3(x^2 + 55x - 300) = 0$$
$$3(x + 60)(x - 5) = 0$$
$$x = -60 \quad \text{or} \quad x = 5$$

Since the speed of the current must be positive, -60 is rejected and the 5 miles per hour is the speed of the current.

32. Rate in light traffic: r
Rate in heavy traffic: $r - 20$

	r	d	t = d/r
light traffic	r	40	$40/r$
heavy traffic	$r - 20$	10	$10/r - 20$

$$\begin{bmatrix}\text{time in}\\ \text{light traffic}\end{bmatrix} + \begin{bmatrix}\text{time in}\\ \text{heavy traffic}\end{bmatrix} = \frac{3}{2}$$

$$\frac{40}{r} + \frac{10}{r - 20} = \frac{3}{2}$$

$$2r(r - 20)\left[\frac{40}{r} + \frac{10}{r - 20}\right] = \left[\frac{3}{2}\right]2r(r - 20)$$

$$2(r - 20) \cdot 40 + 2r \cdot 10 = 3r(r - 20)$$

$$80r - 1600 + 20r = 3r^2 - 60r$$

$$-3r^2 + 160r - 1600 = 0$$

$$3r^2 - 160r + 1600 = 0$$

$$(3r - 40)(r - 40) = 0;\ r = \frac{40}{3} \quad \text{or} \quad r = 40$$

If $r = \frac{40}{3}$, $r - 20$ is negative. Hence, in miles per hour, the rate in light traffic is 40 and in heavy traffic 20.

34. Number of hours to fill the tank: t

$$\begin{bmatrix}\text{part of tank}\\ \text{filled in 1 hour}\end{bmatrix} \cdot \begin{bmatrix}\text{hours}\\ \text{together}\end{bmatrix} = \begin{bmatrix}\text{part of tank}\\ \text{filled by each pipe}\end{bmatrix}$$

Pipe 1	$\frac{1}{4}$	t	$\frac{1}{4}t$
Pipe 2	$\frac{1}{6}$	t	$\frac{1}{6}t$

$$\begin{bmatrix}\text{part of tank}\\ \text{filled by pipe 1}\end{bmatrix} - \begin{bmatrix}\text{part of tank}\\ \text{emptied by pipe 2}\end{bmatrix} = \begin{bmatrix}\text{entire}\\ \text{tank filled}\end{bmatrix}$$

$$\left(\frac{1}{4}\right)t \quad - \quad \left(\frac{1}{6}\right)t \quad = \quad 1$$

$$(12)\left(\frac{1}{4}\right)t - (12)\left(\frac{1}{6}\right)t = (12)1$$

$$3t - 2t = 12$$

$$t = 12$$

It will take 12 hours to fill the tank.

36. Number of hours to run copies with both machines working: t

$$\begin{bmatrix}\text{part of job}\\ \text{done in 1 hour}\end{bmatrix} \cdot \begin{bmatrix}\text{hours}\\ \text{together}\end{bmatrix} = \begin{bmatrix}\text{part of job}\\ \text{done by each machine}\end{bmatrix}$$

Machine 1	$\frac{1}{5}$	t	$\frac{1}{5}t$
Machine 2	$\frac{1}{4}$	t	$\frac{1}{4}t$

$$\begin{bmatrix}\text{part of job done}\\ \text{by machine 1}\end{bmatrix} + \begin{bmatrix}\text{part of job done}\\ \text{by machine 2}\end{bmatrix} = \begin{bmatrix}\text{entire}\\ \text{job done}\end{bmatrix}$$

$$\frac{1}{5}t \quad + \quad \frac{1}{4}t \quad = \quad 1$$

$$(20)\frac{1}{5}t + (20)\frac{1}{4}t = (20)1$$
$$4t + 5t = 20$$
$$9t = 20$$
$$t = \frac{20}{9}$$

It would take $\frac{20}{9}$ or $2\frac{2}{9}$ hours.

38. Number of hours for second tractor to do job alone: t

$$\begin{bmatrix}\text{part of job}\\ \text{done in 1 hour}\end{bmatrix} \cdot \begin{bmatrix}\text{hours}\\ \text{together}\end{bmatrix} = \begin{bmatrix}\text{part of job done}\\ \text{by each tractor}\end{bmatrix}$$

tractor 1	$\frac{1}{10}\left(\frac{5}{9}\right) = \frac{1}{18}$	13	$13\left(\frac{1}{18}\right) = \frac{13}{18}$
tractor 2	$\frac{1}{t}$	3	$3\left(\frac{1}{t}\right) = \frac{3}{t}$

$$\begin{bmatrix}\text{part of job done}\\ \text{by tractor 1}\end{bmatrix} + \begin{bmatrix}\text{part of job done}\\ \text{by tractor 2}\end{bmatrix} = \begin{bmatrix}\text{entire job}\\ \text{done}\end{bmatrix}$$

$$\frac{13}{18} \quad + \quad \frac{3}{t} \quad = \quad 1$$

$$\cancel{18}t\left(\frac{13}{\cancel{18}}\right) + 18\cancel{t}\left(\frac{3}{\cancel{t}}\right) = 18t \cdot (1)$$
$$13t + 54 = 18t$$
$$54 = 5t$$
$$10.8 = t$$

It would take tractor 2 10.8 hours to do the job alone.

6 Exponents, Roots, and Radicals

EXERCISE 6.1

2. $y^{1+4} = y^5$

4. $b^{5+4} = b^9$

6. $b^{3\cdot 4} = b^{12}$

8. $y^{4\cdot 2} = y^8$

10. $x^{2\cdot 2}y^{3\cdot 2} = x^4y^6$

12. $a^{2\cdot 2}b^{3\cdot 2}c^{1\cdot 2} = a^4b^6c^2$

14. $(-1)^2 3^{1\cdot 2}x^{2\cdot 2}(-5x) = 3^2x^4(-5x)$
$= -45x^5$

16. $a^{2\cdot 3}b^{2\cdot 3}(-1)^3a^{1\cdot 3}b^{2\cdot 3}$
$= a^6b^6(-1)a^3b^6$
$= (-1)a^{6+3}b^{6+6} = -a^9b^{12}$

18. $\dfrac{1}{y^{6-2}} = \dfrac{1}{y^4}$

20. $x^{4-2}y^{3-1} = x^2y^2$

22. $\dfrac{y^{2\cdot 2}}{z^{3\cdot 2}} = \dfrac{y^4}{z^6}$

24. $\dfrac{3^{1\cdot 2}y^{2\cdot 2}}{x^{1\cdot 2}} = \dfrac{3^2y^4}{x^2} = \dfrac{9y^4}{x^2}$

26. $\dfrac{(-1)^4x^{2\cdot 4}}{2^{1\cdot 4}y^{1\cdot 4}} = \dfrac{x^8}{16y^4}$

28. $\dfrac{5^2x^2}{3^3x^6} = \dfrac{25}{27x^4}$

30. $\dfrac{(-1)^2x^2y^4}{x^6y^3} = \dfrac{y}{x^4}$

32. $\dfrac{(-1)^2x^2(-1)^4x^8}{x^6} = \dfrac{x^{10}}{x^6} = x^4$

34. $\dfrac{x^4z^2(-1)^3 2^3}{2^2 \cdot x^6z^3} = \dfrac{-2}{x^2z}$

36. $\dfrac{y^2 \cdot (-1)^3 3^3}{x^2 \cdot 4^3x^3y^3} = \dfrac{-27}{64x^5y}$

38. $\left[\dfrac{a^{12}b^4c^4 \cdot x^4y^2z^2}{x^8y^4 \cdot a^2b^4c^6}\right]^2 = \left[\dfrac{a^{10}z^2}{c^2x^4y^2}\right]^2 = \dfrac{a^{20}z^4}{c^4x^8y^4}$

40. $\dfrac{m^6n^4p^2 \cdot r^3s^3 \cdot (-1)^2m^2n^2p^2}{r^4s^2 \cdot m^3n^6p^6 \cdot r^2s^2} = \dfrac{m^8n^6p^4r^3s^3}{m^3n^6p^6r^6s^4} = \dfrac{m^5}{p^2r^3s}$

42. $\dfrac{2^2x^4y^2 \cdot 2^3z^6 \cdot (-1)^2 4^2x^2z^2}{3^2 \cdot z^2 \cdot 3^3x^3y^6 \cdot 3^2y^2} = \dfrac{2^5 4^2 x^6 y^2 z^8}{3^7 x^3 y^8 z^2} = \dfrac{512x^3z^6}{2187y^6}$

44. $(x^2 + y^2)^3 = (x^2 + y^2)(x^2 + y^2)(x^2 + y^2)$

$= x^6 + 3x^4y^2 + 3x^2y^4 + y^6$

46. $x^{2n+n-(n+1)} = x^{3n-n-1}$

$= x^{2n-1}$

48. $\left(\dfrac{y^5}{y}\right)^{2n} = (y^4)^{2n} = y^{8n}$

50. $\dfrac{y^{n(n+1)}}{y^n} = y^{(n^2+n)-n} = y^{n^2}$

EXERCISE 6.2

2. $\dfrac{1}{3^2} = \dfrac{1}{9}$

4. $3 \cdot \dfrac{1}{4^{-2}} = 3 \cdot 4^2 = 3 \cdot 16 = 48$

6. $(-3)^2 = 9$

8. $\dfrac{1}{3^{-2}} = 3^2 = 9$

10. $\dfrac{1^{-2}}{3^{-2}} = \dfrac{1}{3^{-2}} = 3^2 = 9$

12. $3^{-3} \cdot \dfrac{1}{6^{-2}} = \dfrac{1}{3^3} \cdot 6^2 = \dfrac{36}{27} = \dfrac{4}{3}$

14. $\dfrac{1}{5} + 1 = \dfrac{6}{5}$

16. $\dfrac{1}{8^2} - 1 = \dfrac{1}{64} - 1 = \dfrac{-63}{64}$

18. $x^3 \cdot \dfrac{1}{y^{-2}} = x^3y^2$

20. $\dfrac{1}{(xy^3)^2} = \dfrac{1}{x^2y^6}$

22. $\dfrac{2^3x^3}{y^6} = \dfrac{8x^3}{y^6}$

24. $\dfrac{3^2x^2}{y^4} \cdot \dfrac{2^2y^6}{x^2} = \dfrac{36x^2y^6}{y^4x^2} = 36y^2$

26. $x^{3-(-2)}$

$= x^{3+2} = x^5$

28. 1

30. $x^{-3} \cdot \dfrac{1}{y^{-2}} = \dfrac{1}{x^3} \cdot y^2 = \dfrac{y^2}{x^3}$

32. $\left(\dfrac{x^{-1}y^{3}}{2y^{-5}}\right)^{-2} = \left(\dfrac{x^{-1}y^{8}}{2}\right)^{-2}$
$= \dfrac{x^{2}y^{-16}}{2^{-2}} = x^{2}y^{-16} \cdot \dfrac{1}{2^{-2}}$
$= x^{2} \cdot \dfrac{1}{y^{16}} \cdot 2^{2} = \dfrac{4x^{2}}{y^{16}}$

34. $\left(\dfrac{x^{-1}y}{xy^{-1}z}\right)^{-1} = \left(\dfrac{x^{-2}y^{2}}{z}\right)^{-1}$
$= \dfrac{x^{2}y^{-2}}{z^{-1}} = x^{2}y^{-2} \cdot \dfrac{1}{z^{-1}}$
$= x^{2} \cdot \dfrac{1}{y^{2}} \cdot z = \dfrac{x^{2}z}{y^{2}}$

36. $\dfrac{2^{-1}y}{x^{-2}} \cdot \dfrac{y^{2}}{x} = \dfrac{2^{-1}y^{3}}{x^{-1}}$
$= 2^{-1}y^{3} \cdot \dfrac{1}{x^{-1}}$
$= \dfrac{1}{2}y^{3} \cdot x = \dfrac{xy^{3}}{2}$

38. $\dfrac{2^{2}y^{4}x^{2}}{3^{2}z^{2}} \cdot \dfrac{2^{-2}x^{-4}}{9^{-2}z^{-2}} = \dfrac{2^{0}x^{-2}y^{4}}{9 \cdot 9^{-2}z^{0}}$
$= \dfrac{x^{-2}y^{4}}{9^{-1}} = x^{-2}y^{4} \cdot \dfrac{1}{9^{-1}}$
$= \dfrac{1}{x^{2}}y^{4} \cdot 9 = \dfrac{9y^{4}}{x^{2}}$

40. $\dfrac{1}{x} - \dfrac{1}{y^{3}} = \dfrac{1 \cdot y^{3}}{x \cdot y^{3}} - \dfrac{1 \cdot x}{y^{3} \cdot x}$
$= \dfrac{y^{3} - x}{xy^{3}}$

42. $\dfrac{\frac{1}{x}}{\frac{1}{y}} + \dfrac{y}{x} = \dfrac{(xy)\frac{1}{x}}{(xy)\frac{1}{y}} + \dfrac{y}{x} = \dfrac{y}{x} + \dfrac{y}{x} = \dfrac{2y}{x}$

44. $\dfrac{1}{(x + y)^{3}}$

46. $\dfrac{x}{y} + \dfrac{y}{x} = \dfrac{x \cdot x}{y \cdot x} + \dfrac{y \cdot y}{x \cdot y} = \dfrac{x^{2} + y^{2}}{xy}$

48. $\dfrac{(y)\left(x + \frac{1}{y}\right)}{(y)\left(\frac{1}{y}\right)} = \dfrac{xy + 1}{1} = xy + 1$

50. $\dfrac{\frac{1}{x^{2}} - \frac{1}{y^{2}}}{\frac{1}{xy}} = \dfrac{x^{2}y^{2}\left(\frac{1}{x^{2}} - \frac{1}{y^{2}}\right)}{x^{2}y^{2}\left(\frac{1}{xy}\right)} = \dfrac{y^{2} - x^{2}}{xy}$

52. Let x = 1 and y = 2. Then

$$(x + y)^{-2} = (1 + 2)^{-2} = \frac{1}{3^{2}} = \frac{1}{9}$$

and

$$\frac{1}{x^{2}} + \frac{1}{y^{2}} = \frac{1}{1^{2}} + \frac{1}{2^{2}} = 1 + \frac{1}{4} = \frac{5}{4}.$$

Therefore, $(x + y)^{-2}$ is not equivalent to $\dfrac{1}{x^{2}} + \dfrac{1}{y^{2}}$.

54. $x^{-n+n+1} = x$

56. $x^{n-(2n-1)}y^{n+1-n} = x^{-n+1}y$

58. $[x^{n-1-(-2)}y^{n-(-n)}]^2$

$= [x^{n+1}y^{2n}]^2 = x^{2n+2}y^{4n}$

60. $[a^{2n-(n-1)}b^{n-1-1}]^2$

$= [a^{n+1}b^{n-2}]^2 = a^{2n+2}b^{2n-4}$

EXERCISE 6.3

2. 3.476×10^3

4. 6.8742×10^4

6. 4.81×10^5

8. 6.3×10^{-3}

10. 5.23×10^{-4}

12. 6×10^{-4}

14. 4,800

16. 83,100

18. 0.80

20. 0.00431

22. 143,800

24. 6.210

26. $\frac{1}{4} \times \frac{1}{10^4} = 0.25 \times 10^{-4}$

$= 0.000025$

28. $\frac{1}{5} \times \frac{1}{10^{-3}} = 0.2 \times 10^3 = 200$

30. $\frac{5}{8} \times \frac{1}{10^2} = 0.625 \times 10^{-2}$

$= 0.00625$

32. $\frac{10^{3-7+2}}{10^{-2+4}} = \frac{10^{-2}}{10^2} = 10^{-4}$ or 0.0001

34. $\frac{(4 \times 6) \times (10^3 \times 10^{-2})}{3 \times 10^{-7}}$

$= \frac{24 \times 10}{3 \times 10^{-7}}$

$= 8 \times 10^{1-(-7)} = 8 \times 10^8$

or 800,000,000

36. $\frac{(3^3 \times 10^3) \times (2 \times 10^{-1})}{2 \times 10^{-2}}$

$= \frac{27 \times \overset{1}{\cancel{2}} \times 10^2}{\underset{1}{\cancel{2}} \times 10^{-2}}$

$= 27 \times 10^{2-(-2)}$

$= 27 \times 10^4$ or 270,000

38. $\frac{(8^2 \times 10^8) \times (3^3 \times 10^3)}{6^2 \times 10^{-4}} = \frac{\overset{16}{\cancel{64}} \times \overset{3}{\cancel{27}} \times 10^{11}}{\underset{1}{\cancel{4}} \times \underset{1}{\cancel{9}} \times 10^{-4}}$

$= 48 \times 10^{11-(-4)} = 48 \times 10^{15}$

40. $29.4 = 2.94 \times 10^1$; hence there are 3 significant digits.

42. $0.00474 = 4.72 \times 10^{-3}$; hence there are 3 significant digits.

44. $0.600 = 6.00 \times 10^{-1}$; hence there are 3 significant digits.

46. $0.02004 = 2.004 \times 10^{-2}$; hence there are 4 significant digits.

48. $$\frac{(65 \times 10^{-3}) \times (22 \times 10^{-1}) \times (5 \times 10)}{(13 \times 10^{-1}) \times (11 \times 10^{-3}) \times (5 \times 10^{-2})} = \frac{\overset{5}{\cancel{65}} \times \overset{2}{\cancel{22}} \times \overset{1}{\cancel{5}} \times 10^{-3}}{\underset{1}{\cancel{13}} \times \underset{1}{\cancel{11}} \times \underset{1}{\cancel{5}} \times 10^{-6}}$$

$$= 10 \times 10^{-3-(-6)}$$

$$= 10 \times 10^3 = 10^4 \quad \text{or} \quad 10{,}000$$

50. $$\frac{(54 \times 10^{-4}) \times (5 \times 10^{-2}) \times (3 \times 10^2)}{(15 \times 10^{-4}) \times (27 \times 10^{-2}) \times (8 \times 10)} = \frac{\overset{2}{\cancel{54}} \times \overset{1}{\cancel{15}} \times 10^{-4}}{\underset{1}{\cancel{27}} \times \underset{1}{\cancel{15}} \times 8 \times 10^{-5}}$$

$$= \frac{2}{8} \times 10^{-4-(-5)}$$

$$= 0.25 \times 10 = 2.5$$

52. $$\frac{27 \times 10^{-4} \times 4 \times 10^{-3} \times 65 \times 10}{26 \times 10 \times 10^{-4} \times 9 \times 10^{-3}} = \frac{\overset{3}{\cancel{27}} \times \overset{5}{\cancel{65}} \times 4 \times \overset{1}{\cancel{10^{-6}}}}{\underset{1}{\cancel{9}} \times \underset{2}{\cancel{26}} \times \underset{1}{\cancel{10^{-6}}}}$$

$$= \frac{60}{2} = 30$$

54. In 1 hour there are 3600 seconds; in a day there are 24 hours; and in a year there are 365 days. Therefore, the number of miles in a light-year is given by

$$186{,}000 \times 3600 \times 24 \times 365$$

$$= 1.86 \times 10^5 \times 3.6 \times 10^3 \times 2.4 \times 10 \times 3.65 \times 10^2$$

$$= (1.86 \times 3.6 \times 2.4 \times 3.65) \times 10^{11}$$

$$= 5.865696 \times 10^{12}$$

56. 4.5×10^{-7}

58. 6.45×10^6

EXERCISE 6.4

2. 5

4. 3

6. -3

8. 9

10. $(125^{1/3})^2 = (5)^2 = 25$

12. $[(-64)^{1/3}]^2 = [-4]^2 = 16$

14. $\dfrac{1}{8^{1/3}} = \dfrac{1}{2}$

16. $(27^{1/3})^{-2} = (3)^{-2} = \dfrac{1}{3^2} = \dfrac{1}{9}$

18. $(32^{1/5})^3 = 2^3 = 8$

20. $(8^{1/3})^{-2} = 2^{-2} = \dfrac{1}{2^2} = \dfrac{1}{4}$

22. $y^{(1/2)+(3/2)} = y^{4/2} = y^2$

24. $x^{(3/4)-(1/4)} = x^{2/4} = x^{1/2}$

26. $b^{6\cdot(2/3)} = b^{12/3} = b^4$

28. $y^{(-2/3)+(5/3)} = y^{3/3} = y$

30. $a^{(1/2)\cdot 6}b^{(1/3)\cdot 6}$

$= a^{6/2}b^{6/3} = a^3b^2$

32. $\dfrac{a^{(1/2)\cdot 2}}{a^{2\cdot 2}} = \dfrac{a}{a^4} = \dfrac{1}{a^3}$

34. $x^{(1/4)\cdot 8}y^{(1/2)\cdot 8} = x^2y^4$

36. $\dfrac{a^{(-1/2)\cdot 6}}{b^{(1/3)\cdot 6}} = \dfrac{a^{-3}}{b^2} = a^{-3} \cdot \dfrac{1}{b^2} = \dfrac{1}{a^3} \cdot \dfrac{1}{b^2} = \dfrac{1}{a^3b^2}$

38. $\dfrac{x^{1/2}y^{3/2}z^{-2}}{x^{-3/2}y^{1/2}} = x^{(1/2)-(-3/2)}y^{(3/2)-(1/2)} \cdot \dfrac{1}{z^2}$

$= \dfrac{x^{(1/2)+(3/2)}y^{(3/2)-(1/2)}}{z^2} \quad \dfrac{x^2y}{z^2}$

40. $x^{(1/5)+(2/5)} + x^{(1/5)+(4/5)} = x^{3/5} + x$

42. $x^{(3/8)+(1/4)} - x^{(3/8)+(1/2)} = x^{(3/8)+(2/8)} - x^{(3/8)+(4/8)} = x^{5/8} - x^{7/8}$

44. $y^{(-1/4)+(3/4)} + y^{(-1/4)+(5/4)} = y^{1/2} + y$

46. $a^{(-2/7)+(9/7)} + a^{(-2/7)+(2/7)}$

$= a^{7/7} + a^0 = a + 1$

48. $x^{5/6+(-5)/6} + x^{5/6+1/6}$

$= x^0 + x^{6/6} = 1 + x$

50. $x^{(2/3)+(4/3)} - x^{(1/3)+(2/3)} + x^{(-1/3)+(2/3)} = x^2 - x + x^{1/3}$

52. $(2x^{1/2})(x^{1/2}) + x^{1/2} - 2x^{1/2} - 1 = 2x - x^{1/2} - 1$

54. $(x^{1/2} + 3)(x^{1/2} + 3) = x + 6x^{1/2} + 9$

56. $x^{(1/2)+(1/2)} - x^{1+(1/2)} + 2x^{1+(1/2)} - 2x^2 = x + x^{3/2} - 2x^2$

58. Let a = 9 and b = 16. Then

$$(a + b)^{-1/2} = \frac{1}{(a + b)^{1/2}} = \frac{1}{(9 + 16)^{1/2}} = \frac{1}{25^{1/2}} = \frac{1}{5};$$

$$a^{-1/2} + b^{-1/2} = \frac{1}{a^{1/2}} + \frac{1}{b^{1/2}} = \frac{1}{9^{1/2}} + \frac{1}{16^{1/2}} = \frac{1}{3} + \frac{1}{4} = \frac{7}{12}.$$

Hence, $(a + b)^{-1/2}$ is not equivalent to $a^{-1/2} + b^{-1/2}$.

60. $a^{2\cdot(n/2)}b^{2n(2/n)} = a^n b^4$

62. $\frac{a^{n/2}}{b^{1/2}} \cdot \frac{b^{3/2}}{a^{3n}} = \frac{b^{(3/2)-(1/2)}}{a^{3n-(n/2)}} = \frac{b}{a^{5n/2}}$

64. $\left(\frac{m^a}{n^{2a}}\right)^{1/a} = \frac{m^{a\cdot 1/a}}{n^{2a\cdot 1/a}} = \frac{m}{n^2}$

66. $(x^{n+1-1}y^{n+2-2})^{1/n} = (x^n y^n)^{1/n} = x^{n\cdot 1/n}y^{n\cdot 1/n} = xy$

68. Since $\left(\frac{1}{16}\right)^{1/4} = \frac{1}{2}$ and $\left(\frac{1}{16}\right)^{1/2} = \frac{1}{4}$, $\left(\frac{1}{16}\right)^{1/4} > \left(\frac{1}{16}\right)^{1/2}$.

$a^{1/n} > a^{1/m}$ when $n > m$ and $0 < a < 1$.

EXERCISE 6.5

2. $5^{1/2}$

4. $(4y)^{1/3}$

6. $x^{1/2} - 2y^{1/3}$

8. $x^{1/2}(xy)^{1/3} = x^{1/2}x^{1/3}y^{1/3}$
$= x^{5/6}y^{1/3}$

10. $\sqrt{7}$

12. $3\sqrt[4]{x}$

14. $\sqrt[3]{y + 2}$

16. $2\sqrt[5]{x + y}$

18. 6

20. 7

22. -3

24. -5

26. $\sqrt[4]{y^3}$

28. $5\sqrt[3]{y^2}$

30. $\sqrt[3]{(x - 2y)^2}$

32. $\sqrt[3]{x^{-1}}$

34. $\sqrt[7]{y^{-2}}$

36. $4\sqrt{x^{-3}}$

38. $y^{3/2}$

40. $(xy^2)^{1/3} = x^{1/3}y^{2/3}$

42. $(xy)^{3/2} = x^{3/2}y^{3/2}$

44. $\dfrac{2}{y^{1/3}}$

46. 5

48. -2

50. y

52. a^3

54. $3y^3$

56. $-a^4b^5$

58. $\frac{3}{4}ab^2$

60. $\frac{2}{3}ab^2$

62. $-2xy^2$

64. $-3a^2b^3$

66. $2^{2/6} = 2^{1/3} = \sqrt[3]{2}$

68. $5^{2/8} = 5^{1/4} = \sqrt[4]{5}$

70. $\sqrt[10]{25} = 2^{5/10} = 2^{1/2}$
$= \sqrt{2}$

72. $y^{3/9} = y^{1/3} = \sqrt[3]{y}$

74.

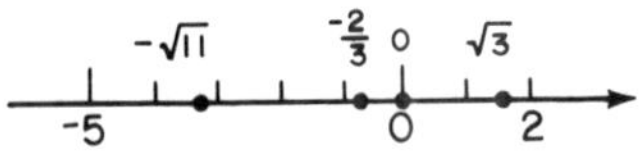

76.

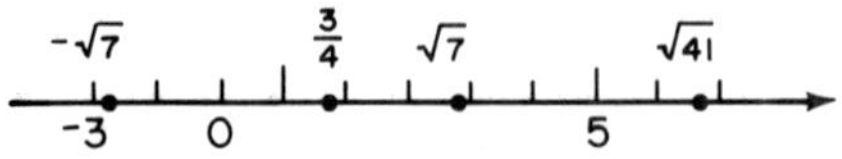

78. $3|x|y^2$ (Note that $|y^2| = y^2$.)

80. $\sqrt{(2x - 1)^2} = |2x - 1|$

82. $\sqrt{(x^2 + y^2)^2} = x^2 + y^2$ (Note that $|x^2 + y^2| = x^2 + y^2$.)

84. Let a = 0. Then $\sqrt{(a - 1)^2} = \sqrt{(0 - 1)^2} = \sqrt{1} = 1$ and $a - 1 = 0 - 1 = -1$. Hence $\sqrt{(a - 1)^2}$ is not equivalent to $a - 1$.

EXERCISE 6.6

2. $\sqrt{5^2}\ \sqrt{2} = 5\sqrt{2}$

4. $\sqrt{6^2} \cdot \sqrt{2} = 6\sqrt{2}$

6. $\sqrt{4^2}\ \sqrt{3} = 4\sqrt{3}$

8. $\sqrt{(y^3)^2} = y^3$

10. $\sqrt{y^2}\ \sqrt{y^2}\ \sqrt{y^2}\ \sqrt{y^2}\ \sqrt{y^2}\ \sqrt{y}$

$= y \cdot y \cdot y \cdot y \cdot y\sqrt{y}$

$= y^5\sqrt{y}$

12. $\sqrt{2^2}\ \sqrt{y^2}\ \sqrt{y^2}\ \sqrt{y} = 2 \cdot y \cdot y\sqrt{y}$

$= 2y^2\sqrt{y}$

14. $\sqrt{3^2}\ \sqrt{2}\ \sqrt{y^2}\ \sqrt{y^2}\ \sqrt{y^2}\ \sqrt{y^2}$

$= 3y \cdot y \cdot y \cdot y\sqrt{2}$

$= 3y^4\sqrt{2}$

16. $\sqrt[3]{y^3}\ \sqrt[3]{y} = y\sqrt[3]{y}$

18. $\sqrt[5]{a^5}\ \sqrt[5]{a^5}\ \sqrt[5]{a^2}\ \sqrt[5]{b^5}\ \sqrt[5]{b^5}\ \sqrt[5]{b^5} = a \cdot a \cdot \sqrt[5]{a^2} \cdot b \cdot b \cdot b = a^2b^3\ \sqrt[5]{a^2}$

20. $\sqrt[6]{m^6}\ \sqrt[6]{n^6}\ \sqrt[6]{m^2n} = mn\sqrt[6]{m^2n}$

22. $\sqrt[7]{4^7}\ \sqrt[7]{x^7}\ \sqrt[7]{y^7}\ \sqrt[7]{z^7}\ \sqrt[7]{z^7}\ \sqrt[7]{4x^2y^3} = 4xyz^2\ \sqrt[7]{4x^2y^3}$

24, $\sqrt{3}\ \sqrt{3^3} = \sqrt{3^4} = \sqrt{3^2}\ \sqrt{3^2} = 3 \cdot 3 = 9$

26. $\sqrt{a^2b^2} = \sqrt{a^2}\sqrt{b^2} = ab$

28. $\sqrt[4]{3}\ \sqrt[4]{3^3} = \sqrt[4]{3^4} = 3$

30. $\sqrt{5 \cdot (5 \cdot 2)^3} = \sqrt{5^4 \cdot 2^3}$

$= \sqrt{5^2}\ \sqrt{5^2}\ \sqrt{2^2}\ \sqrt{2}$

$= 5 \cdot 5 \cdot 2 \cdot \sqrt{2}$

$= 50\sqrt{2}$

32. $\sqrt{8 \cdot 10^5} = \sqrt{2^3 \cdot (2 \cdot 5)^5} = \sqrt{2^8 \cdot 5^5}$

$= \sqrt{2^2}\ \sqrt{2^2}\ \sqrt{2^2}\ \sqrt{2^2}\ \sqrt{5^2}\ \sqrt{5^2}\ \sqrt{5}$

$= 2 \cdot 2 \cdot 2 \cdot 2 \cdot 5 \cdot 5\ \sqrt{5} = 400\sqrt{5}$

or

$\sqrt{80 \cdot 10^4} = \sqrt{10^2}\ \sqrt{10^2}\ \sqrt{80}$

$= 100\ \sqrt{4^2}\ \sqrt{5} = 100 \cdot 4 \cdot \sqrt{5} = 400\sqrt{5}$

34. $\dfrac{\sqrt{2}}{\sqrt{3}} = \dfrac{\sqrt{2}\sqrt{3}}{\sqrt{3}\sqrt{3}} = \dfrac{\sqrt{6}}{3}$

36. $\dfrac{-\sqrt{3}\sqrt{7}}{\sqrt{7}\sqrt{7}} = \dfrac{-\sqrt{21}}{7}$

38. $\dfrac{-\sqrt{y}\sqrt{3}}{\sqrt{3}\sqrt{3}} = \dfrac{-\sqrt{3y}}{3}$

40. $\dfrac{\sqrt{2a}\sqrt{b}}{\sqrt{b}\sqrt{b}} = \dfrac{\sqrt{2ab}}{b}$

42. $\dfrac{-x\sqrt{2y}}{\sqrt{2y}\sqrt{2y}} = \dfrac{-x\sqrt{2y}}{2y}$

44. $\dfrac{\sqrt{y}}{\sqrt{6x}} \quad \dfrac{\sqrt{y}\sqrt{6x}}{\sqrt{6x}\sqrt{6x}} = \dfrac{\sqrt{6xy}}{6x}$

46. $\dfrac{1\sqrt[4]{y}}{\sqrt[4]{y^3}\sqrt[4]{y}} = \dfrac{\sqrt[4]{y}}{\sqrt[4]{y^4}} = \dfrac{\sqrt[4]{y}}{y}$

48. $\dfrac{\sqrt[4]{2}\ \sqrt[4]{3^3x^3}}{\sqrt[4]{3x}\ \sqrt[4]{3^3x^3}} = \dfrac{\sqrt[4]{2 \cdot 27x^3}}{\sqrt[4]{3^4x^4}} = \dfrac{\sqrt[4]{54x^3}}{3x}$

50. $\dfrac{\sqrt[4]{x}\ \sqrt[4]{2y}}{\sqrt[4]{(2y)^3}\ \sqrt[4]{2y}} = \dfrac{\sqrt[4]{2xy}}{\sqrt[4]{(2y)^4}} = \dfrac{\sqrt[4]{2xy}}{2y}$

52. $\dfrac{\sqrt[5]{2}\ \sqrt[5]{(3y)^3}}{\sqrt[5]{(3y)^2}\ \sqrt[5]{(3y)^3}} = \dfrac{\sqrt[5]{2 \cdot 27y^3}}{\sqrt[5]{(3y)^5}} = \dfrac{\sqrt[5]{54y^3}}{3y}$

54. $\sqrt{\dfrac{x^2y^3}{y}} = \sqrt{x^2y^2} = xy$

56. $\sqrt{\dfrac{45x^3y^3}{5y}} = \sqrt{9x^3y^2} = \sqrt{3^2x^3y^2}$

$= 3xy\sqrt{x}$

58. $\sqrt[3]{\dfrac{16r^4}{4t^3}} = \sqrt[3]{\dfrac{4r^3r}{t^3}} = \dfrac{r\sqrt[3]{4r}}{t}$

60. $\sqrt[5]{\dfrac{x^2y^3}{xy^2}} = \sqrt[5]{xy}$

62. $\dfrac{\sqrt{2}\sqrt{2}}{3\sqrt{2}} = \dfrac{2}{3\sqrt{2}}$

64. $\dfrac{\sqrt{xy}\sqrt{xy}}{x\sqrt{xy}} = \dfrac{xy}{x\sqrt{xy}} = \dfrac{y}{\sqrt{xy}}$

66. Let $a = 9$ and $b = 16$. Then

$$\sqrt{a + b} = \sqrt{9 + 16} = 5 \quad \text{and} \quad \sqrt{a} + \sqrt{b} = \sqrt{9} + \sqrt{16} = 7.$$

Therefore, $\sqrt{a + b}$ is not equivalent to $\sqrt{a} + \sqrt{b}$.

68. $\sqrt{2^2(x + 2)^2 \cdot 3(x + 2)}$

$= 2(x + 2)\sqrt{3(x + 2)}$

70. $\sqrt{y^2 \cdot y^2 \cdot y \cdot (x - 1)^2 \cdot (x - 1)}$

$= y^2(x - 1)\sqrt{y(x - 1)}$

72. $\sqrt{\dfrac{(x + 2)^2(x+2)^2(x + 2)}{x^2 \cdot xy}}$

$= \dfrac{(x + 2)^2}{x}\sqrt{\dfrac{(x + 2)xy}{xy \cdot xy}}$

$= \dfrac{(x + 2)^2}{x^2y}\sqrt{xy(x + 2)}$

74. $\sqrt[3]{y^3(2y - 1)} = y\sqrt[3]{2y - 1}$

76. $\sqrt[3]{\dfrac{(y + 1)^3(y + 1)}{x^2y}} = (y + 1)\sqrt[3]{\dfrac{y + 1}{x^2y}}$

$= (y + 1)\sqrt[3]{\dfrac{(y + 1)xy^2}{x^2y \cdot xy^2}}$

$= (y + 1)\sqrt[3]{\dfrac{xy(y + 1)}{x^3y^3}}$

$= \dfrac{y + 1}{xy}\sqrt[3]{xy^2 + xy}$

78. $\sqrt{\dfrac{\overset{(y + 2)}{\cancel{4}\cancel{(y + 2)^2}}}{\cancel{4}y \cdot y^2\cancel{(y + 2)}}} = \dfrac{1}{y}\sqrt{\dfrac{y + 2}{y}}$

$= \dfrac{1}{y}\sqrt{\dfrac{y(y + 2)}{y \cdot y}} = \dfrac{1}{y^2}\sqrt{y^2 + 2y}$

EXERCISE 6.7

2. $2\sqrt{2}$

4. $\sqrt{25}\sqrt{3} + 2\sqrt{3} = 5\sqrt{3} + 2\sqrt{3} = 7\sqrt{3}$

6. $\sqrt{4}\sqrt{2y} - \sqrt{9}\sqrt{2y} = 2\sqrt{2y} - 3\sqrt{2y} = -\sqrt{2y}$

8. $2\sqrt{4y^2}\sqrt{2z} + 3\sqrt{16y^2}\sqrt{2z} = 2(2y)\sqrt{2z} + 3(4y)\sqrt{2z}$

$= 4y\sqrt{2z} + 12y\sqrt{2z} = 16y\sqrt{2z}$

10. $\sqrt{3b} - 2\sqrt{4}\sqrt{3b} + 3\sqrt{16}\sqrt{3b} = \sqrt{3b} - 2(2)\sqrt{3b} + 3(4)\sqrt{3b}$

$= \sqrt{3b} - 4\sqrt{3b} + 12\sqrt{3b} = 9\sqrt{3b}$

12. $\sqrt[3]{27}\sqrt[3]{2} + 2\sqrt[3]{64}\sqrt[3]{2} = 3\sqrt[3]{2} + 2(4)\sqrt[3]{2} = 3\sqrt[3]{2} + 8\sqrt[3]{2} = 11\sqrt[3]{2}$

14. $10 - 5\sqrt{7}$

16. $\sqrt{36} - \sqrt{45} = 6 - \sqrt{9}\sqrt{5} = 6 - 3\sqrt{5}$

18. $2 + \sqrt{2} - 2\sqrt{2} - 2 = -\sqrt{2}$

20. $(2)^2 - (\sqrt{x})^2 = 4 - x$

22. $2(3) + \sqrt{15} - 2\sqrt{15} - 5$

$= 1 - \sqrt{15}$

24. $(\sqrt{2} - 2\sqrt{3})(\sqrt{2} - 2\sqrt{3})$

$= 2 - 2\sqrt{6} - 2\sqrt{6} + 4(3)$

$= 14 - 4\sqrt{6}$

26. $5 + 5 \cdot 2\sqrt{2} = 5(1 + 2\sqrt{2})$

28. $5\sqrt{5} - 5 = 5(\sqrt{5} - 1)$

30. $3 + \sqrt{9}\sqrt{2x} = 3 + 3\sqrt{2x}$
$= 3(1 + \sqrt{2x})$

32. $\sqrt{4}\sqrt{3} - (2\sqrt{3})\sqrt{2} = (2\sqrt{3}) - (2\sqrt{3})\sqrt{2}$
$= 2\sqrt{3}(1 - \sqrt{2})$

34. $\dfrac{2 \cdot 3 + 2\sqrt{5}}{2} = \dfrac{2(3 + \sqrt{5})}{2}$
$= 3 + \sqrt{5}$

36. $\dfrac{8 - 2\sqrt{4}\sqrt{3}}{4} = \dfrac{8 - 4\sqrt{3}}{4} = \dfrac{4(2 - \sqrt{3})}{4}$
$= 2 - \sqrt{3}$

38. $\dfrac{xy - x\sqrt{y^2}\sqrt{x}}{xy} = \dfrac{xy - xy\sqrt{x}}{xy}$
$= \dfrac{xy(1 - \sqrt{x})}{xy}$
$= 1 - \sqrt{x}$

40. $\dfrac{\sqrt{x} - y\sqrt{x^2}\sqrt{x}}{\sqrt{x}} = \dfrac{\sqrt{x} - xy\sqrt{x}}{\sqrt{x}}$
$= \dfrac{(1 - xy)\sqrt{x}}{\sqrt{x}} = 1 - xy$

42. $\dfrac{1(2 + \sqrt{2})}{(2 - \sqrt{2})(2 + \sqrt{2})} = \dfrac{2 + \sqrt{2}}{4 - 2}$
$= \dfrac{2 + \sqrt{2}}{2}$

44. $\dfrac{2(4 + \sqrt{5})}{(4 - \sqrt{5})(4 + \sqrt{5})} = \dfrac{2(4 + \sqrt{5})}{16 - 5}$
$= \dfrac{2(4 + \sqrt{5})}{11}$
or $\dfrac{8 + 2\sqrt{5}}{11}$

46. $\dfrac{y(\sqrt{3} + y)}{(\sqrt{3} - y)(\sqrt{3} + y)} = \dfrac{y(\sqrt{3} + y)}{3 - y^2}$
or $\dfrac{y\sqrt{3} + y^2}{3 - y^2}$

48. $\dfrac{(\sqrt{x} + \sqrt{y})(\sqrt{x} + \sqrt{y})}{(\sqrt{x} - \sqrt{y})(\sqrt{x} + \sqrt{y})} = \dfrac{x + 2\sqrt{xy} + y}{x - y}$

50. $\dfrac{(\sqrt{3} + \sqrt{2})(\sqrt{3} - \sqrt{2})}{\sqrt{3}(\sqrt{3} - \sqrt{2})} = \dfrac{3 - 2}{\sqrt{3}(\sqrt{3} - \sqrt{2})} = \dfrac{1}{\sqrt{3}(\sqrt{3} - \sqrt{2})}$ or $\dfrac{1}{3 - \sqrt{6}}$

52. $\dfrac{(4 - \sqrt{2y})(4 + \sqrt{2y})}{2(4 + \sqrt{2y})} = \dfrac{16 - 2y}{2(4 + \sqrt{2y})} = \dfrac{2(8 - y)}{2(4 + \sqrt{2y})} = \dfrac{8 - y}{4 + \sqrt{2y}}$

54. $\dfrac{(2\sqrt{x} + \sqrt{y})(2\sqrt{x} - \sqrt{y})}{\sqrt{xy}(2\sqrt{x} - \sqrt{y})} = \dfrac{4x - y}{\sqrt{xy}(2\sqrt{x} - \sqrt{y})}$ or $\dfrac{4x - y}{2x\sqrt{y} - y\sqrt{x}}$

56. $$\frac{\sqrt{x^2-2}}{1} - \frac{x^2+1}{\sqrt{x^2-2}} = \frac{\sqrt{x^2-2}\sqrt{x^2-2}}{1\sqrt{x^2-2}} - \frac{x^2+1}{\sqrt{x^2-2}}$$

$$= \frac{x^2-2-(x^2+1)}{\sqrt{x^2-2}} = \frac{-3}{\sqrt{x^2-2}}$$

$$= \frac{-3\sqrt{x^2-2}}{\sqrt{x^2-2}\sqrt{x^2-2}} = \frac{-3\sqrt{x^2-2}}{x^2-2}$$

58. $$\frac{x}{\sqrt{x^2-1}} + \frac{\sqrt{x^2-1}}{x} = \frac{x \cdot x}{x\sqrt{x^2-1}} + \frac{\sqrt{x^2-1}\sqrt{x^2-1}}{x\sqrt{x^2-1}} = \frac{x^2+x^2-1}{x\sqrt{x^2-1}} = \frac{2x^2-1}{x\sqrt{x^2-1}}$$

$$= \frac{(2x^2-1)\sqrt{x^2-1}}{x\sqrt{x^2-1}\sqrt{x^2-1}} = \frac{(2x^2-1)\sqrt{x^2-1}}{x(x^2-1)}$$

60. $$\frac{(1-\sqrt{x+1})(1+\sqrt{x+1})}{\sqrt{x+1}(1+\sqrt{x+1})} = \frac{1-(x+1)}{\sqrt{x+1}+x+1} = \frac{-x}{\sqrt{x+1}+x+1}$$

62. $$\frac{(\sqrt{x-1}+\sqrt{x})(\sqrt{x-1}-\sqrt{x})}{(\sqrt{x-1}+\sqrt{x})(\sqrt{x-1}-\sqrt{x})} = \frac{x-1-x}{x-1-2\sqrt{x}\sqrt{x-1}+x}$$

$$= \frac{-1}{2x-2\sqrt{x(x-1)}-1}$$

EXERCISE 6.8

2. $\sqrt{-1}\sqrt{9} = 3i$

4. $\sqrt{-1 \cdot 25 \cdot 2} = \sqrt{-1}\sqrt{25}\sqrt{2}$
$= i \cdot 5 \cdot \sqrt{2}$
$= 5i\sqrt{2}$

6. $4\sqrt{-1 \cdot 9 \cdot 2} = 4\sqrt{-1}\sqrt{9}\sqrt{2}$
$= 4 \cdot i \cdot 3\sqrt{2}$
$= 12i\sqrt{2}$

8. $2\sqrt{-1 \cdot 4 \cdot 10} = 2 \cdot \sqrt{-1} \cdot 2 \cdot \sqrt{10}$
$= 4i\sqrt{10}$

10. $7\sqrt{-1 \cdot 81} = 7 \cdot \sqrt{-1} \cdot \sqrt{81}$
$= 7 \cdot i \cdot 9$
$= 63i$

12. $-3\sqrt{-1 \cdot 25 \cdot 3} = -3 \cdot \sqrt{-1} \cdot \sqrt{25} \cdot \sqrt{3}$
$= -3 \cdot i \cdot 5 \cdot \sqrt{3}$
$= -15i\sqrt{3}$

14. $5 - 3i$

16. $5\sqrt{-1 \cdot 4 \cdot 3} - 1 = 5\sqrt{-1}\sqrt{4}\sqrt{3} - 1$
$= 5i \cdot 2\sqrt{3} - 1$
$= 10i\sqrt{3} - 1$

18. $\sqrt{4 \cdot 5} - \sqrt{-1 \cdot 4 \cdot 5}$
$= \sqrt{4}\sqrt{5} = \sqrt{-1}\sqrt{4}\sqrt{5}$
$= 2\sqrt{5} - i \cdot 2 \cdot \sqrt{5}$
$= 2\sqrt{5} - 2i\sqrt{5}$

20. $(2 + 3) + [-1 + (-2)]i = 5 + (-3)i$
$= 5 - 3i$

22. $(2 + i) + (-4 + 2i)$
$= [2 + (-4)] + (1 + 2)i$
$= -2 + 3i$

24. $(2 - 6i) + (-3 + 0i)$
$= [2 + (-3)] + (-6 + 0)i$
$= -1 - 6i$

26. $4 - 5i - 12i + 15i^2$
$= 4 - 17i + 15(-1)$
$= -11 - 17i$

28. $-6 - 2i + 9i + 3i^2$
$= -6 + 7i + 3(-1)$
$= -9 + 7i$

30. $-14 - 6i - 21i - 9i^2$
$= -14 - 27i - 9(-1)$
$= -5 - 27i$

32. $(2 + 3i)(2 + 3i)$
$= 4 + 6i + 6i + 9i^2$
$= 4 + 12i + 9(-1)$
$= -5 + 12i$

34. $1 - 4i^2 = 1 - 4(-1)$
$= 5$

36. $\dfrac{-2 \cdot i}{5i \cdot i} = \dfrac{-2i}{5i^2} = \dfrac{-2i}{5(-1)} = \dfrac{2}{5}i$

38. $\dfrac{(4 + 2i)i}{3i \cdot i} = \dfrac{4i + 2i^2}{3i^2}$
$= \dfrac{4i + 2(-1)}{3(-1)}$
$= \dfrac{-2 + 4i}{-3}$
$= \dfrac{2}{3} - \dfrac{4}{3}i$

40. $\dfrac{-3(2 - i)}{(2 + i)(2 - i)} = \dfrac{-6 + 3i}{4 - i^2}$
$= \dfrac{-6 + 3i}{4 - (-1)}$
$= \dfrac{-6 + 3i}{5}$
$= -\dfrac{6}{5} + \dfrac{3}{5}i$

42. $\dfrac{(3 - i)(1 - i)}{(1 + i)(1 - i)} = \dfrac{3 - 4i + i^2}{1 - i^2}$

$= \dfrac{3 - 4i + (-1)}{1 - (-1)}$

$= \dfrac{2 - 4i}{2}$

$= 1 - 2i$

44. $\dfrac{(6 + i)(2 + 5i)}{(2 - 5i)(2 + 5i)} = \dfrac{12 + 32i + 5i^2}{4 - 25i^2}$

$= \dfrac{12 + 32i + 5(-1)}{4 - 25(-1)}$

$= \dfrac{7 + 32i}{29}$

$= \dfrac{7}{29} + \dfrac{32}{29}i$

46. $\dfrac{(-4 - 3i)(2 - 7i)}{(2 + 7i)(2 - 7i)}$

$= \dfrac{-8 + 22i + 21i^2}{4 - 49i^2}$

$= \dfrac{-8 + 22i + 21(-1)}{4 - 49(-1)}$

$= \dfrac{-29 + 22i}{53} = -\dfrac{29}{53} + \dfrac{22}{53}i$

48. $i\sqrt{9}(3 + i\sqrt{16}) = 3i(3 + 4i)$
$= 9i + 12i^2$
$= 9i - 12$
$= -12 + 9i$

50. $(4 - i\sqrt{2})(3 + i\sqrt{2})$

$= 12 + 4i\sqrt{2} - 3i\sqrt{2} - 2i^2$

$= 12 + i\sqrt{2} + 2 = 14 + i\sqrt{2}$

52. $\dfrac{-i}{i\sqrt{25}} = \dfrac{(-1) \cdot i}{(5i) \cdot i} = \dfrac{-i}{5i^2}$

$= \dfrac{-i}{-5} = \dfrac{1}{5}i$

54. $\dfrac{1 + i\sqrt{2}}{3 - i\sqrt{2}} = \dfrac{(1 + i\sqrt{2})(3 + i\sqrt{2})}{(3 - i\sqrt{2})(3 + i\sqrt{2})} = \dfrac{3 + 4i\sqrt{2} + 2i^2}{9 - 2i^2}$

$= \dfrac{3 + 4i\sqrt{2} - 2}{9 + 2} = \dfrac{1 + 4i\sqrt{2}}{11}$

$= \dfrac{1}{11} + \dfrac{4\sqrt{2}}{11}i$

56. For $\sqrt{x + 3}$ to be real,

$x + 3 \geq 0$ or $x \geq -3$.

For $\sqrt{x + 3}$ to be imaginary,

$x + 3 < 0$ or $x < -3$.

58. Assuming that properties of exponents that hold for real number bases also hold for bases that are imaginary numbers:

a. $i^{-1} = \frac{1}{i} = \frac{1 \cdot i}{i \cdot i} = \frac{i}{-1} = -i$

b. $i^{-2} = \frac{1}{i^2} = \frac{1}{-1} = -1$

c. $i^{-3} = (i^{-1})(i^{-2}) = (-i)(-1) = i$

d. $i^{-6} = (i^{-4})(i^{-2}) = (1)(-1) = -1$

60. $$\begin{aligned} 2(2 - i)^2 - (2 - i) + 2 &= 2(4 - 4i + i^2) - 2 + i + 2 \\ &= 8 - 8i + 2i^2 - 2 + i + 2 \\ &= 8 - 8i - 2 - 2 + i + 2 \\ &= 6 - 7i \end{aligned}$$

7 Nonlinear Equations and Inequalities

EXERCISE 7.1

2. $x = 4;\ x = -4$
$\{-4,4\}$

4. $x^2 = \frac{9}{4}$

$x = \frac{3}{2};\ x = \frac{-3}{2}$

$\left\{\frac{-3}{2},\frac{3}{2}\right\}$

6. $x^2 = 5$

$x = \sqrt{5};\ x = -\sqrt{5}$

$\{-5\sqrt{5},\sqrt{5}\}$

8. $3x^2 = -9$

$x^2 = -3$

$x = i\sqrt{3};\ x = -i\sqrt{3}$

$\{-i\sqrt{3},\ i\sqrt{3}\}$

10. $5\left(\frac{3x^2}{5}\right) = 5(3)$

$3x^2 = 15$

$x^2 = 5$

$x = \sqrt{5};\ x = -\sqrt{5}$

$\{-\sqrt{5},\sqrt{5}\}$

12. $2\left(\frac{9x^2}{2}\right) = 2(-50)$

$9x^2 = -100$

$x^2 = \frac{-100}{9}$

$x = \frac{10}{3}i;\quad x = \frac{-10}{3}i$

$\left\{\frac{10}{3},\frac{-10}{3}\right\}$

14. $x + 3 = 2;\ x + 3 = -2$
$x = -1;\quad x = -5$
$\{-5,-1\}$

16. $3x + 1 = 5;\ 3x + 1 = -5$
$3x = 4;\quad 3x = -6$
$\left\{-2,\frac{4}{3}\right\}$

18. $x - 5 = \sqrt{-7};\ x - 5 = -\sqrt{-7}$

$x = 5 + i\sqrt{7};\ x = 5 - i\sqrt{7}$

$\{5 - i\sqrt{7},\ 5 + i\sqrt{7}\}$

20. $x - \frac{2}{3} = \frac{\sqrt{5}}{3};\ x - \frac{2}{3} = \frac{-\sqrt{5}}{3}$

$x = \frac{2}{3} + \frac{\sqrt{5}}{3};\ x = \frac{2}{3} - \frac{\sqrt{5}}{3}$

$\left\{\frac{2 - \sqrt{5}}{3},\ \frac{2 + \sqrt{5}}{3}\right\}$

22. $x + \frac{1}{2} = \frac{1}{4}$; $x + \frac{1}{2} = -\frac{1}{4}$

$x = -\frac{1}{2} + \frac{1}{4}$; $x = -\frac{1}{2} - \frac{1}{4}$

$= -\frac{1}{4}$ $\qquad = -\frac{3}{4}$

$\left\{-\frac{3}{4}, \frac{-1}{4}\right\}$

24. $3x + 4 = 4$; $3x + 4 = -4$

$3x = 0$; $\qquad 3x = -8$

$x = 0$ $\qquad x = \frac{-8}{3}$

$\left\{\frac{-8}{3}, 0\right\}$

26. $2x + 3 = 6$; $2x + 3 = -6$

$2x = 3$; $\qquad 2x = -9$

$\left\{\frac{-9}{2}, \frac{3}{2}\right\}$

28. $5x + 3 = \sqrt{-7}$; $5x + 3 = -\sqrt{-7}$

$5x = i\sqrt{7} - 3$; $5x = -i\sqrt{7} - 3$

$\left\{\frac{i\sqrt{7} - 3}{5}, \frac{-i\sqrt{7} - 3}{5}\right\}$

30. $5x - 12 = \sqrt{-24}$; $5x - 12 = -\sqrt{-24}$

$5x = 2i\sqrt{6} + 12$; $5x = -2i\sqrt{6} + 12$

$\left\{\frac{2i\sqrt{6} + 12}{5}, \frac{-2i\sqrt{6} + 12}{5}\right\}$

32. $x^2 - x = 6$

$x^2 - x + \frac{1}{4} = 6 + \frac{1}{4}$

$\left(x - \frac{1}{2}\right)^2 = \frac{25}{4}$

$x - \frac{1}{2} = \frac{5}{2}$; $x - \frac{1}{2} = \frac{-5}{2}$

$x = 3$; $\qquad x = -2$

$\{-2, 3\}$

34. $x^2 + 4x = -4$

$x^2 + 4x + 4 = -4 + 4$

$(x + 2)^2 = 0$

$x + 2 = 0$

$\{-2\}$

36. $x^2 - x = 20$

$x^2 - x + \frac{1}{4} = 20 + \frac{1}{4}$

$\left(x - \frac{1}{2}\right)^2 = \frac{81}{4}$

$x - \frac{1}{2} = \frac{9}{2}$; $x - \frac{1}{2} = \frac{-9}{2}$

$x = 5$ $\qquad x = -4$

$\{-4, 5\}$

38. $x^2 + 3x = 1$

$x^2 + 3x + \frac{9}{4} = 1 + \frac{9}{4}$

$\left(x + \frac{3}{2}\right)^2 = \frac{13}{4}$

$x + \frac{3}{2} = \frac{\sqrt{13}}{2}$; $x + \frac{3}{2} = \frac{-\sqrt{13}}{2}$

$x = \frac{-3}{2} + \frac{\sqrt{13}}{2}$; $x = \frac{-3}{2} - \frac{\sqrt{13}}{2}$

$\left\{\frac{-3 + \sqrt{13}}{2}; \frac{-3 - \sqrt{13}}{2}\right\}$

40.
$$x^2 + 5x = 5$$
$$x^2 + 5x + \frac{25}{4} = 5 + \frac{25}{4}$$
$$\left(x + \frac{5}{2}\right)^2 = \frac{45}{4}$$
$$x + \frac{5}{2} = \frac{\sqrt{45}}{2};\ x + \frac{5}{2} = \frac{-\sqrt{45}}{2}$$
$$x = -\frac{5}{2} + \frac{3\sqrt{5}}{2};\ x = -\frac{5}{2} - \frac{3\sqrt{5}}{2}$$
$$\left\{\frac{-5 + 3\sqrt{5}}{2};\ \frac{-5 - 3\sqrt{5}}{2}\right\}$$

42.
$$x^2 + \frac{1}{3}x = \frac{4}{3}$$
$$x^2 + \frac{1}{3}x + \frac{1}{36} = \frac{4}{3} + \frac{1}{36}$$
$$\left(x + \frac{1}{6}\right)^2 = \frac{49}{36}$$
$$x + \frac{1}{6} = \frac{7}{6};\ x + \frac{1}{6} = \frac{-7}{6}$$
$$x = 1;\quad x = \frac{-8}{6} = \frac{-4}{3}$$
$$\left\{1, \frac{-4}{3}\right\}$$

44.
$$x^2 - \frac{3}{4} = \frac{1}{2}x$$
$$x^2 - \frac{1}{2}x = \frac{3}{4}$$
$$x^2 - \frac{1}{2}x + \frac{1}{16} = \frac{3}{4} + \frac{1}{16}$$
$$\left(x - \frac{1}{4}\right)^2 = \frac{13}{16}$$
$$x - \frac{1}{4} = \frac{\sqrt{13}}{4};\ x - \frac{1}{4} = \frac{-\sqrt{13}}{4}$$
$$x = \frac{1}{4} + \frac{\sqrt{13}}{4};\ x - \frac{1}{4} - \frac{\sqrt{13}}{4}$$
$$\left\{\frac{1 + \sqrt{13}}{4};\ \frac{1 - \sqrt{13}}{4}\right\}$$

46.
$$x^2 - 5x = -8$$
$$x^2 - 5x + \frac{25}{4} = -8 + \frac{25}{4}$$
$$\left(x - \frac{5}{2}\right)^2 = \frac{-7}{4}$$
$$x - \frac{5}{2} = \frac{i\sqrt{7}}{2};\ x - \frac{5}{2} = \frac{-i\sqrt{7}}{2}$$
$$x = \frac{5}{2} + \frac{i\sqrt{7}}{2};\ x = \frac{5}{2} - \frac{i\sqrt{7}}{2}$$
$$\left\{\frac{5}{2} - \frac{\sqrt{7}}{2}i,\ \frac{5}{2} + \frac{\sqrt{7}}{2}i\right\}$$

48.
$$x^2 + \frac{1}{3}x = \frac{-4}{3}$$
$$x^2 + \frac{1}{3}x + \frac{1}{36} = \frac{-4}{3} + \frac{1}{36}$$
$$\left(x + \frac{1}{6}\right)^2 = \frac{-47}{36}$$
$$x + \frac{1}{6} = \frac{i\sqrt{47}}{6};\ x + \frac{1}{6} = \frac{-i\sqrt{47}}{6}$$
$$x = -\frac{1}{6} + \frac{i\sqrt{47}}{6};\ x = -\frac{1}{6} - \frac{i\sqrt{47}}{6}$$
$$\left\{-\frac{1}{6} - \frac{\sqrt{47}}{6}i,\ -\frac{1}{6} + \frac{\sqrt{47}}{6}i\right\}$$

50.
$$x^2 = 2a$$
$$x = \sqrt{2a};\ x = -\sqrt{2a}$$
$$\{\sqrt{2a}, -\sqrt{2a}\}$$

52. $\frac{c}{1}\left(\frac{bx^2}{c} - a\right) = \left(\frac{c}{1}\right)0$

$bx^2 - ac = 0$

$bx^2 = ac$

$x^2 = \frac{ac}{b}$

$x = \sqrt{\frac{ac}{b}} \quad x = -\sqrt{\frac{ac}{b}}$

$= \frac{\sqrt{abc}}{b}; \quad = -\frac{\sqrt{abc}}{b}$

$\left\{\frac{\sqrt{abc}}{b}, \frac{-\sqrt{abc}}{b}\right\}$

54. $(x + a)^2 = 36$

$x + a = 6; \; x + a = -6$

$x = -a + 6; \; x = -a - 6$

$\{-a + 6, -a - 6\}$

56. $ax - b = 5; \; ax - b = -5$

$ax = b + 5; \; ax = b - 5$

$\left\{\frac{b + 5}{a}, \frac{b - 5}{a}\right\}$

58. $r^2 + S = h$

$r^2 = h - S$

$r = \sqrt{h - S}; \; r = -\sqrt{h - S}$

$\{\sqrt{h - S}, -\sqrt{h - S}\}$

60. $2(s) = 2\left(\frac{1}{2}gt^2\right) + 2c$

$2s = gt^2 + 2c$

$gt^2 + 2c = 2s$

$t^2 = \frac{2s - 2c}{g}$

$t = \sqrt{\frac{2s - 2c}{g}}; \; t = -\sqrt{\frac{2s - 2c}{g}}$

$= \frac{\sqrt{2gs - 2gc}}{g}; \quad = \frac{-\sqrt{2gs - 2gc}}{g}$

$\left\{\frac{\sqrt{2gs - 2gc}}{g}, \frac{-\sqrt{2gs - 2gc}}{g}\right\}$

62. $4y^2 = 36 - 9x^2$

$y^2 = \frac{1}{4}(36 - 9x^2) = \frac{1}{4} \cdot 9(4 - x^2)$

$y = \frac{3}{2}\sqrt{4 - x^2}; \; y = -\frac{3}{2}\sqrt{4 - x^2}$

$\left\{\frac{3}{2}\sqrt{4 - x^2}, -\frac{3}{2}\sqrt{4 - x^2}\right\}$

64. $25y^2 = 9x^2$

$y^2 = \frac{9}{25}x^2$

$y = \frac{3}{5}x; \; y = -\frac{3}{5}x$

$\left\{\frac{3}{5}x, -\frac{3}{5}x\right\}$

EXERCISE 7.2

2. $a = 1,\ b = -4,\ c = 4$

$$x = \frac{-(-4) \pm \sqrt{(-4)^2 - 4(1)(4)}}{2(1)} = \frac{4 \pm 0}{2};$$

$\{2\}$

4. $y^2 - 5y - 6 = 0;\ a = 1,\ b = -5,\ c = -6$

$$y = \frac{-(-5) \pm \sqrt{(-5)^2 - 4(1)(-6)}}{2(1)} = \frac{5 \pm \sqrt{49}}{2}$$

$$y = \frac{5 + 7}{2},\ y = \frac{5 - 7}{2};\quad \{6,-1\}$$

6. $2z^2 - 7z + 6 = 0;\ a = 2,\ b = -7,\ c = 6$

$$z = \frac{-(-7) \pm \sqrt{(-7)^2 - 4(2)(6)}}{2(2)} = \frac{7 \pm \sqrt{1}}{4}$$

$$z = \frac{7 + 1}{4},\ z = \frac{7 - 1}{4};\quad \left\{2, \frac{3}{2}\right\}$$

8. $2x^2 - x + 1 = 0;\ a = 2,\ b = -1,\ c = 1$

$$x = \frac{-(-1) \pm \sqrt{(-1)^2 - 4(2)(1)}}{2(2)} = \frac{1 \pm \sqrt{-7}}{4};\quad \left\{\frac{1 + i\sqrt{7}}{4}, \frac{1 - i\sqrt{7}}{4}\right\}$$

10. $6z^2 - 13z - 5 = 0;\ a = 6,\ b = -13,\ c = -5$

$$z = \frac{-(-13) \pm \sqrt{(-13)^2 - 4(6)(-5)}}{2(6)} = \frac{13 \pm \sqrt{289}}{12}$$

$$z = \frac{13 + 17}{12};\quad z = \frac{13 - 17}{12}$$

$$z = \frac{30}{12};\quad z = \frac{-4}{12};\quad \left\{\frac{5}{2}, \frac{-1}{3}\right\}$$

12. $a = 1,\ b = 3,\ c = 0$

$$y = \frac{-3 \pm \sqrt{3^2 - 4(1)(0)}}{2(1)} = \frac{-3 \pm 3}{2}$$

$$y = \frac{-3 + 3}{2};\quad y = \frac{-3 - 3}{2};\quad \{0,-3\}$$

14. $a = 2,\ b = 0,\ c = 1$

$$z = \frac{0 \pm \sqrt{0^2 - 4(2)(1)}}{2(2)} = \frac{\pm\sqrt{-8}}{4} = \frac{\pm 2i\sqrt{2}}{4};\quad \left\{\frac{i\sqrt{2}}{2}, \frac{-i\sqrt{2}}{2}\right\}$$

16. $x^2 + 2x + 5 = 0;\ a = 1,\ b = 2,\ c = 5$

$$x = \frac{-2 \pm \sqrt{2^2 - 4(1)(5)}}{2(1)} = \frac{-2 \pm \sqrt{-16}}{2} = \frac{-2 \pm 4i}{2}$$

$$= -1 \pm 2i;\quad \{-1 + 2i, -1 - 2i\}$$

18. $2z(z - 2) = 3$

$2z^2 - 4z - 3 = 0;\ a = 2,\ b = -4,\ c = -3$

$$z = \frac{-(-4) \pm \sqrt{(-4)^2 - 4(2)(-3)}}{2(2)} = \frac{4 \pm \sqrt{40}}{4}$$

$$= \frac{4 \pm 2\sqrt{10}}{4} = \frac{2(2 \pm \sqrt{10})}{2 \cdot 2} = \frac{2 \pm \sqrt{10}}{2}$$

$$\left\{\frac{2 + \sqrt{10}}{2}, \frac{2 - \sqrt{10}}{2}\right\}$$

20. $2x(x - 1) = x + 1$

$2x^2 - 2x = x + 1$

$2x^2 - 3x - 1 = 0;\ a = 2,\ b = -3,\ c = -1$

$$x = \frac{-(-3) \pm \sqrt{(-3)^2 - 4(2)(-1)}}{2(2)} = \frac{3 \pm \sqrt{17}}{4}$$

$$\left\{\frac{3 + \sqrt{17}}{4}, \frac{3 - \sqrt{17}}{4}\right\}$$

22. $3z^2 + z + 2 = 0;\ a = 3,\ b = 1,\ c = 2$

$$z = \frac{-1 \pm \sqrt{1^2 - 4(3)(2)}}{2(3)} = \frac{-1 \pm \sqrt{-23}}{6} = \frac{-1 \pm i\sqrt{23}}{6}$$

$$\left\{\frac{-1}{6} + \frac{\sqrt{23}}{6}i, \frac{-1}{6} - \frac{\sqrt{23}}{6}i\right\}$$

24. $y(y - 1) = -1$

$y^2 - y + 1 = 0;\ a = 1,\ b = -1,\ c = 1$

$$y = \frac{-(-1) \pm \sqrt{(-1)^2 - 4(1)(1)}}{2(1)} = \frac{1 \pm \sqrt{-3}}{2} = \frac{1 \pm i\sqrt{3}}{2}$$

$$\left\{\frac{1}{2} + \frac{\sqrt{3}}{2}i, \frac{1}{2} - \frac{\sqrt{3}}{2}i\right\}$$

26. $a = 1,\ b = -2,\ c = -3$

$b^2 - 4ac = 4 - (-12) = 16$

$b^2 - 4ac > 0$; therefore, the solutions are real and unequal.

28. $a = 2,\ b = 3,\ c = 7$

$b^2 - 4ac = 9 - 56 = -47$

$b^2 - 4ac < 0$; therefore, the solutions are imaginary and unequal.

30. $a = 4,\ b = -12,\ c = 9$

$b^2 - 4ac = 144 - 144 = 0$

$b^2 - 4ac = 0$; there is one real solution.

32. $a = 2,\ b = -k,\ c = 3$

$$x = \frac{-(-k) \pm \sqrt{(-k)^2 - 4(2)(3)}}{2(2)} = \frac{k \pm \sqrt{k^2 - 24}}{4}$$

34. $x^2 + 2x + (k + 3) = 0;\ a = 1,\ b = 2,\ c = k + 3$

$$x = \frac{-2 \pm \sqrt{2^2 - 4(1)(k + 3)}}{2(1)} = \frac{-2 \pm \sqrt{-4k - 8}}{2}$$

$$= \frac{-2 \pm 2\sqrt{-k - 2}}{2} = -1 \pm \sqrt{-k - 2}$$

36. $a = 2,\ b = -3,\ c = 2y$

$$x = \frac{-(-3) \pm \sqrt{(-3)^2 - 4(2)(2y)}}{2(2)} = \frac{3 \pm \sqrt{9 - 16y}}{4}$$

38. $x^2 + (-3y)x + (y^2 - 3) = 0;\ a = 1,\ b = -3y,\ c = y^2 - 3$

$$x = \frac{-(-3y) \pm \sqrt{(-3y)^2 - 4(1)(y^2 - 3)}}{2(1)} = \frac{3y \pm \sqrt{5y^2 + 12}}{2}$$

40. $y^2 + (-3x)y + (x^2 - 3) = 0;\ a = 1,\ b = -3x,\ c = x^2 - 3$

$$y = \frac{-(-3x) \pm \sqrt{(-3x)^2 - 4(1)(x^2 - 3)}}{2(1)} = \frac{3x \pm \sqrt{5x^2 + 12}}{2}$$

EXERCISE 7.3

2. $\sqrt{x} = 5$

$x = 25$

Check: $\sqrt{25} = 5$? Yes, $\{25\}$

4. $y - 3 = 25$

$y = 28$

Check: $\sqrt{28 - 3} = 5$? Yes, $\{28\}$

6. $(2z - 3)^2 = 7z - 3$

$4z^2 - 12z + 9 = 7z - 3$

$4z^2 - 19z + 12 = 0$

$(4z - 3)(z - 4) = 0$

$z = \frac{3}{4}, \quad z = 4$

Check: $2\left(\frac{3}{4}\right) - 3 = \sqrt{7\left(\frac{3}{4}\right) - 3}$? No.

$2(4) - 3 = \sqrt{7(4) - 3}$? Yes.

{4}

8. $(4x + 5)^2 = 3x + 4$

$16x^2 + 40x + 25 = 3x + 4$

$16x^2 + 37x + 21 = 0$

$(16x + 21)(x + 1) = 0$, from which

$x = \frac{-21}{16}$ or $x = -1$.

Check: $4\left(\frac{-21}{16}\right) + 5 = \sqrt{3\left(\frac{-21}{16}\right) + 4}$?

$\frac{-21}{4} + \frac{20}{4} = \sqrt{\frac{-63}{16} + \frac{64}{16}}$?

$-\frac{1}{4} = \sqrt{\frac{1}{16}}$? No.

$4(-1) + 5 = \sqrt{3(-1) + 4}$?

$1 = \sqrt{1}$? Yes.

{-1}

10. $16(x - 4) = x^2$

$16x - 64 = x^2$

$-x^2 + 16x - 64 = 0$

$x^2 - 16x + 64 = 0$

$(x - 8)(x - 8) = 0$

$x = 8$

Check: $4\sqrt{8 - 4} = 8$?

$4\sqrt{4} = 8$? Yes.

{8}

12. $4y + 1 = 6y - 3$

$-2y + 1 = -3$

$-2y = -4$

$y = 2$

Check: $\sqrt{4(2) + 1} = \sqrt{6(2) - 3}$?

$\sqrt{9} = \sqrt{9}$? Yes.

{2}

14. $x(x - 5) = 36$

$x^2 - 5x - 36 = 0$

$(x - 9)(x + 4) = 0$

$x = 9; \ x = -4$

Check: $\sqrt{9}\sqrt{9 - 5} = 6$? Yes.

$\sqrt{-4}\sqrt{-4 - 5} = 6$? No.

{9}

16. $x = (-4)^3$

$x = -64$

Check: $\sqrt[3]{-64} = -4$? Yes.

{-64}

18. $x - 1 = 81$

$x = 82$

Check: $\sqrt[4]{82 - 1} = 3?$ Yes.

$\{82\}$

20. $\sqrt{1 + 16y} = 5 - 4\sqrt{y}$

$1 + 16y = (5 - 4\sqrt{y})^2$

$1 + 16y = 25 - 40\sqrt{y} + 16y$

$-24 = -40\sqrt{y}$

$\sqrt{y} = \frac{24}{40} = \frac{3}{5}; \quad y = \frac{9}{25}$

Check: $4\sqrt{\frac{9}{25}} + \sqrt{1 + 16\left(\frac{9}{25}\right)} = 5?$ Yes.

$\left\{\frac{9}{25}\right\}$

22. $4x + 17 = (4 - \sqrt{x + 1})^2$

$4x + 17 = 16 - 8\sqrt{x + 1} + (x + 1)$

$3x = -8\sqrt{x + 1}$

$9x^2 = 64(x + 1)$

$9x^2 - 64x - 64 = 0$

$(9x + 8)(x - 8) = 0$

$x = \frac{-8}{9}; \quad x = 8$

Check: $\sqrt{4\left(\frac{-8}{9}\right) + 17} = 4 - \sqrt{\frac{-8}{9} + 1}?$ Yes.

$\sqrt{4(8) + 17} = 4 - \sqrt{8 + 1}?$ No.

$\left\{\frac{-8}{9}\right\}$

24. $(y + 7)^{1/2} = 3 - (y + 4)^{1/2}$

$y + 7 = 9 - 6(y + 4)^{1/2} + (y + 4)$

$-6 = -6(y + 4)^{1/2}$

$(y + 4)^{1/2} = 1$

$y + 4 = 1$

$y = -3$

Check: $(-3 + 7)^{1/2} + (-3 + 4)^{1/2} = 3?$ Yes.

$\{-3\}$

26. $(z + 5)^{1/2} = 4 - (z - 3)^{1/2}$

$$z + 5 = 16 - 8(z - 3)^{1/2} + (z - 3)$$

$$-8 = -8(z - 3)^{1/2}$$

$$(z - 3)^{1/2} = 1$$

$$z - 3 = 1; \; z = 4$$

Check: $(4 - 3)^{1/2} + (4 + 5)^{1/2} = 4$? Yes.

$\{4\}$

28. $t^2 = \dfrac{2v}{g}$

$$g = \frac{2v}{t^2}$$

30. $P^2 = \pi^2\left(\dfrac{\ell}{g}\right)$

$$gP^2 = \pi^2\ell$$

$$g = \frac{\pi^2\ell}{P^2}$$

32. $(q - 1)^2 = 4\left(\dfrac{r^2 - 1}{3}\right)$

$$[q^2 - 2q + 1](3) = \left[4\left(\frac{r^2 - 1}{3}\right)\right](3)$$

$$3q^2 - 6q + 3 = 4r^2 - 4$$

$$3q^2 - 6q + 7 = 4r^2$$

$$r^2 = \frac{3q^2 - 6q + 7}{4}; \quad r = \frac{\pm\sqrt{3q^2 - 6q + 7}}{2}$$

34. $C\sqrt{D - E^2} = B - A$

$$C^2(D - E^2) = (B - A)^2$$

$$D - E^2 = \left(\frac{B - A}{C}\right)^2$$

$$E^2 = D - \left(\frac{B - A}{C}\right)^2$$

$$E = \pm\sqrt{D - \left(\frac{B - A}{C}\right)^2}$$

36. $v \ge 0$ and $g > 0$

or

$v < 0$ and $g < 0$

38. Longer leg of triangle: x

Hypotenuse: $x + 1$

Shorter leg of triangle:

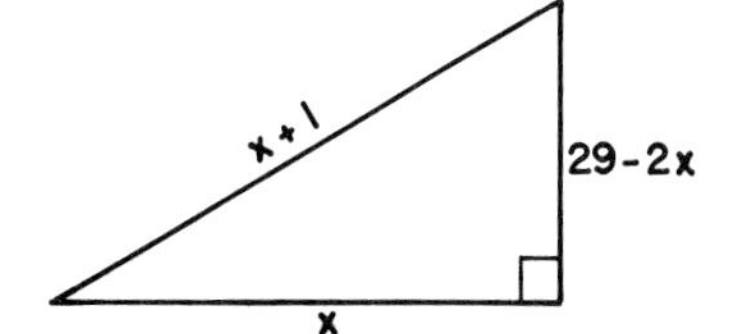

$$30 - (x + x + 1) = 29 - 2x$$

$$(x + 1)^2 = x^2 + (29 - 2x)^2$$

$$x^2 + 2x + 1 = x^2 + 841 - 116x + 4x^2$$

$$4x^2 - 118x + 840 = 0$$

$$2x^2 - 59x + 420 = 0$$

$(2x - 35)(x - 12) = 0$, from which

$x = \frac{35}{2}$ or $x = 12$. $\frac{35}{2}$ is not meaningful because it would make the shorter leg of the triangle negative. Hence, the hypotenuse is $x + 1$, or 13.

40. Length of segment AB: x

Referring to the figure, $BC = 12 - AB = 12 - x$.

Also, from $AB + BD = 14$, $BD = 14 - AB = 14 - x$.

$$\sqrt{(BC)^2 + (CD)^2} = BD$$

$$\sqrt{(12 - x)^2 + 4^2} = 14 - x$$

$$(12 - x)^2 + 16 = (14 - x)^2$$

$$144 - 24x + x^2 + 16 = 196 - 28x + x^2$$

$$-24x + 160 = 196 - 28x$$

$$4x = 36$$

$$x = 9$$

The length of segment AB is 9 centimeters.

EXERCISE 7.4

2. Let $u = x^2$; $u^2 = x^4$;

$u^2 - 13u + 36 = 0$

$(u - 9)(u - 4) = 0$

$u = 9$; $u = 4$

Since $u = x^2$,

$x^2 = 9$; $x^2 = 4$

$\{-2,2,-3,3\}$

4. Let $u = y^2$; $u^2 = y^4$;

$u^2 - 6u + 5 = 0$

$(u - 5)(u - 1) = 0$

$u = 5$; $u = 1$

Since $u = y^2$,

$y^2 = 5$; $y^2 = 1$

$\{-\sqrt{5},\sqrt{5},-1,1\}$

6. Let $u = x^2$, $u^2 = x^4$;

$u^2 - 2u - 24 = 0$

$(u - 6)(u + 4) = 0$

$u = 6$; $u = -4$

$x^2 = 6$; $x^2 = -4$

$\{-\sqrt{6},\sqrt{6},-2i,2i\}$

8. Let $u = z^2$, $u^2 = z^4$;

$u^2 + 7u + 10 = 0$

$(u + 5)(u + 2) = 0$

$u = -5$; $u = -2$

$z^2 = -5$; $z^2 = -2$

$\{-i\sqrt{5},i\sqrt{5},-i\sqrt{2},i\sqrt{2}\}$

10. $u = \sqrt{x}$, $u^2 = x$;

$u^2 + 3u - 10 = 0$

$(u + 5)(u - 2) = 0$

$u = -5$; $u = 2$

Replace u by $\sqrt{x}$ and solve for x. Since $\sqrt{x}$ cannot be negative only 2 need be considered.

$\sqrt{x} = 2$

$x = 4$

Check: Does $4 + 3\sqrt{4} - 10 = 0$? Yes.

$\{4\}$

12. Let $u = \sqrt{y}$, $u^2 = y$

$u^2 - 2u + 1 = 0$

$(u - 1)(u - 1) = 0$

$u = 1$

$\sqrt{y} = 1$

$y = 1$

Check: Does $1 - 2\sqrt{1} + 1 = 0$? Yes.

$\{1\}$

14. Let $u = \sqrt{y^2 - 5}$, $u^2 = y^2 - 5$

$u^2 - 5u + 6 = 0$

$(u - 3)(u - 2) = 0$

$u = 3$ $\qquad$ $u = 2$

$\sqrt{y^2 - 5} = 3;$ $\qquad$ $\sqrt{y^2 - 5} = 2$

$y^2 - 5 = 9;$ $\qquad$ $y^2 - 5 = 4$

$y^2 = 14;$ $\qquad$ $y^2 = 9$

$y = \pm\sqrt{14};$ $\qquad$ $y = \pm 3$

Check: Does $[(\pm\sqrt{14})^2 - 5] - 5\sqrt{(\pm\sqrt{14})^2 - 5} + 6 = 0?$

$[14 - 5] \quad - 5\sqrt{14 - 5} \quad + 6 = 0?$

$9 \quad - 5\sqrt{9} \quad + 6 = 0?$ Yes.

Does $[(\pm 3)^2 - 5] - 5\sqrt{(\pm 3)^2 - 5} + 6 = 0?$

$[9 - 5] \quad - 5[\sqrt{9 - 5}] \quad + 6 = 0?$ Yes.

$\{-\sqrt{14}, \sqrt{14}, -3, 3\}$

16. Let $u = \sqrt{x^2 - 4}$; $u^2 = x^2 - 4$

$u^2 + 4u + 3 = 0$

$(u + 3)(u + 1) = 0$

$u = -3;$ $\quad u = -1$

Since $u = \sqrt{x^2 - 4}$, we have

$\sqrt{x^2 - 4} = -3;$ $\quad \sqrt{x^2 - 4} = -1.$

The equations cannot be true for any value of x because $\sqrt{x^2 - 4}$ cannot be negative. Hence the solution set is $\emptyset$.

18. Let $z^{1/3} = u$, $z^{2/3} = u^2$

then substitute for $z^{1/3}$ and $z^{2/3}$.

$u^2 - 2u - 35 = 0$

$(u - 7)(u + 5) = 0$

$u = 7$ or $u = -5$

Hence,

$z^{1/3} = 7$ or $z^{1/3} = -5$

$x = 343$ or $z = -125.$

Check: $(-125)^{2/3} - 2(-125)^{1/3} = 35?$

$25 - 2(-5) = 35?$ Yes.

$(345)^{2/3} - 2(343)^{1/3} = 35?$

$49 - 2(7) = 35?$ Yes.

$\{343, -125\}$

20. Let $y^{1/3} = u$, $y^{2/3} = u^2$;
then substitute for $y^{1/3}$ and $y^{2/3}$.

$$2u^2 + 5u = 3$$
$$2u^2 + 5u - 3 = 0$$
$$(2u - 1)(u + 3) = 0$$
$$u = \frac{1}{2} \quad \text{or} \quad u = -3$$

Hence,

$$y^{1/3} = \frac{1}{2} \quad \text{or} \quad y^{1/3} = -3$$
$$(y^{1/3})^3 = y = \left(\frac{1}{2}\right)^3 \quad \text{or} \quad (y^{1/3})^3 = y = (-3)^3$$
$$y = \frac{1}{8} \quad \text{or} \quad y = -27.$$

Check: $2\left(\frac{1}{8}\right)^{2/3} + 5\left(\frac{1}{8}\right)^{1/3} = 3?$

$2\left(\frac{1}{4}\right) + 5\left(\frac{1}{2}\right) = 3?$

$\frac{1}{2} + \frac{5}{2} = 3?$ Yes

$2(-27)^{2/3} + 5(-27)^{1/3} = 3?$

$2(9) + 5(-3) = 3?$ Yes

$\left\{-27, \frac{1}{8}\right\}$

22. Let $z^{1/2} = u$, $z = u^2$;
then substitute u^2 for
z and u for $z^{1/2}$.

$$u^2 + u = 72$$
$$u^2 + u - 72 = 0$$
$$(u + 9)(u - 8) = 0$$
$$u = -9 \quad \text{or} \quad u = 8$$

Hence,
$z^{1/2} = 8$ or $z = 64$;
$z^{1/2}$ cannot equal -9.

Check: $64 + 64^{1/2} = 72?$ Yes

$\{64\}$

24. Let $x^{1/4} = u$, $x^{1/2} = u^2$;
then substitute u^2 for
$x^{1/2}$ and u for $x^{1/4}$.

$$8u^2 + 7u = 1$$
$$8u^2 + 7u - 1 = 0$$
$$(8u - 1)(u + 1) = 0$$
$$8u - 1 = 0 \quad \text{or} \quad u + 1 = 0$$
$$u = \frac{1}{8} \quad \text{or} \quad u = -1$$

Hence,
$x^{1/4} = \frac{1}{8}$; $\quad x = \left(\frac{1}{8}\right)^4 = \frac{1}{4096}$;
$x^{1/4}$ cannot equal -1.

Check: $8\left(\frac{1}{4096}\right)^{1/2} + 7\left(\frac{1}{4096}\right)^{1/4} = 1?$
Yes

$\left\{\frac{1}{4096}\right\}$

26. Let $z^{-1} = u$, $z^{-2} = u^2$;

$$u^2 + 9u - 10 = 0$$
$$(u + 10)(u - 1) = 0$$
$$u = -10 \quad \text{or} \quad u = 1$$

Hence,

$$z^{-1} = -10 \quad \text{or} \quad z^{-1} = 1$$
$$\frac{1}{z} = -10 \quad \text{or} \quad \frac{1}{z} = 1$$
$$z = \frac{-1}{10} \quad \text{or} \quad z = 1.$$

Check: $\left(\frac{-1}{10}\right)^{-2} + 9\left(\frac{-1}{10}\right)^{-1} - 10 = 0?$ Yes

$1^{-2} + 9(1)^{-1} - 10 = 0?$ Yes

$\left\{\frac{-1}{10}, 1\right\}$

28. Let $(x - 2)^{1/4} = u$, $(x - 2)^{1/2} = u^2$; then substitute.

$$u^2 - 11u + 18 = 0$$
$$(u - 9)(u - 2) = 0$$
$$u = 9 \quad \text{or} \quad u = 2$$

Hence,

$$(x - 2)^{1/4} = 9 \quad \text{or} \quad (x - 2)^{1/4} = 2$$
$$x - 2 = 9^4 \quad \text{or} \quad x - 2 = 2^4$$
$$x - 2 = 6561 \quad \text{or} \quad x - 2 = 16$$
$$x = 6563 \quad \text{or} \quad x = 18.$$

Check: $(6563 - 2)^{1/2} - 11(6563 - 2)^{1/4} + 18 = 0?$ Yes

$(18 - 2)^{1/2} - 11(18 - 2)^{1/4} + 18 = 0?$ Yes

$\{6563, 18\}$

30. a. $\sqrt{x} = 6 - x$

$$x = (6 - x)^2$$
$$x = 36 - 12x + x^2$$
$$0 = x^2 - 13x + 36$$
$$0 = (x - 9)(x - 4)$$
$$x = 9 \quad \text{or} \quad x = 4$$

Check:

$9 + \sqrt{9} - 6 = 0?$ No

$4 + \sqrt{4} - 6 = 0?$ Yes

$\{4\}$

b. Let $\sqrt{x} = u$, $x = u^2$;

$$u^2 + u - 6 = 0$$
$$(u + 3)(u - 2) = 0$$
$$u = -3 \quad \text{or} \quad u = 2$$

Hence,

$\sqrt{x} = 2$ or $x = 4$;

$\sqrt{x}$ cannot equal -3.

Check: See part a.

$\{4\}$

32. a. $2\sqrt{y} = 15 - y$

$4y = 225 - 30y + y^2$

$0 = y^2 - 34y + 225$

$0 = (y - 25)(y - 9)$

$y = 25$ or $y = 9$

Check:

$25 + 2\sqrt{25} \stackrel{?}{=} 15$? No

$9 + 2\sqrt{9} \stackrel{?}{=} 15$? Yes

$\{9\}$

b. Let $\sqrt{y} = u$, $y = u^2$;

$u^2 + 2u = 15$

$u^2 + 2u - 15 = 0$

$(u + 5)(u - 3) = 0$

$u = -5$ or $u = 3$

Hence,

$\sqrt{y} = 3$ or $y = 9$

$\sqrt{y}$ cannot equal -5.

Check: See part a.

$\{9\}$

EXERCISE 7.5

2. The critical numbers are -2 and 3 because $(x - 3)(x + 2) = 0$ for these values. We check the intervals shown on the number line below using arbitrary numbers in each interval, say, -3, 0, and 4 for the variable in $(x - 3)(x + 2) > 0$.

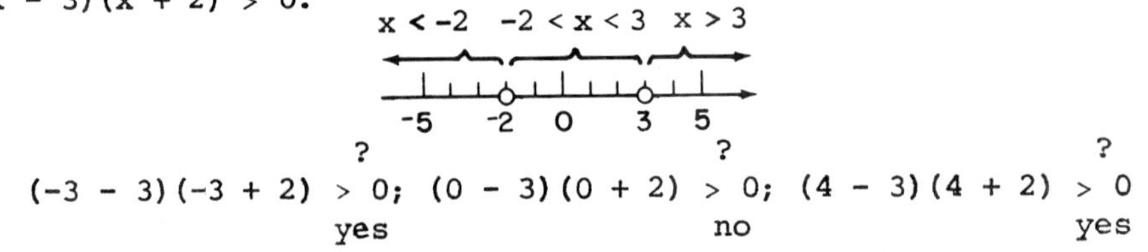

$(-3 - 3)(-3 + 2) \stackrel{?}{>} 0$; yes $(0 - 3)(0 + 2) \stackrel{?}{>} 0$; no $(4 - 3)(4 + 2) \stackrel{?}{>} 0$ yes

The solution set is $(-\infty,-2) \cup (3,+\infty)$.

The graph is:

4. The critical numbers are -2 and -5 because $(x + 2)(x + 5) = 0$ when x has these values.

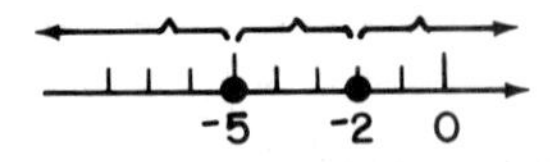

A number from each of the three intervals shown on the number line, say -6, -3, and 0, is substituted into the given inequality.

$(-6 + 2)(-6 + 5) \stackrel{?}{\le} 0$; no $(-3 + 2)(-3 + 5) \stackrel{?}{\le} 0$; yes

$(0 + 2)(0 + 5) \stackrel{?}{\le} 0$ no

4. cont'd.

Hence, the solution set is $(-5,-2)$.

The graph is:

6. The critical numbers are -4 and 0. We check the intervals shown on the number line below using arbitrary numbers in each interval, say, -5, -1, and 1, for the variable in $x(x + 4) < 0$.

$$-5(-5 + 4) \overset{?}{<} 0; \quad -1(-1 + 4) \overset{?}{<} 0; \quad 1(1 + 4) \overset{?}{<} 0$$

no yes no

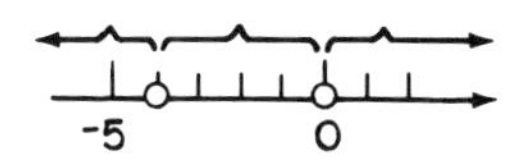

The solution set is $(-4,0)$.

The graph is:

8. The critical numbers are -3 and 0. We check the intervals shown on the number line below using arbitrary numbers in each interval, say, -4, -1, and 1, for the variable in $x(x + 3) \geq 0$.

$$-4(-4 + 3) \overset{?}{\geq} 0; \quad -1(-1 + 3) \overset{?}{\geq} 0; \quad 1(1 + 3) \overset{?}{\geq} 0$$

yes no yes

The solution set is $(-\infty,-3] \cup [0,+\infty)$

The graph is:

10. The critical numbers are -1 and 6 because $x^2 - 5x - 6 = (x + 1)(x - 6) = 0$ for these values. We check the intervals shown on the number line below using arbitrary numbers in each interval, say, -2, 0, and 7, for the variable in $x^2 - 5x - 6 \geq 0$.

$$(-2)^2 - 5(-2) - 6 \overset{?}{\geq} 0; \quad 0^2 - 5(0) - 6 \overset{?}{\geq} 0; \quad 7^2 - 5(7) - 6 \overset{?}{\geq} 0$$

yes no yes

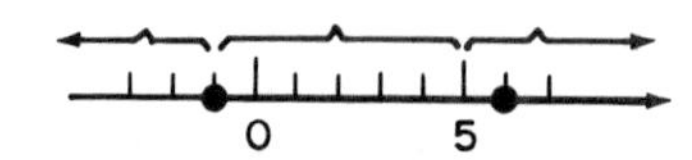

The solution set is $(-\infty,-1] \cup [6,+\infty)$.

The graph is:

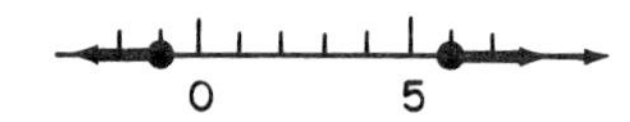

12. The critical numbers are -4 and 3 because $x^2 + 4x - 12 = (x + 4)(x - 3) = 0$ for these values. We check the intervals shown on the number line below using arbitrary numbers in each interval, say, -5, 0, and 4, for the variable in $x^2 + 4x - 12 < 0$.

$$(-5)^2 + 4(-5) - 12 \overset{?}{<} 0; \quad 0^2 + 4(0) - 12 \overset{?}{<} 0; \quad 4^2 + 4(4) - 12 \overset{?}{<} 0$$

no yes no

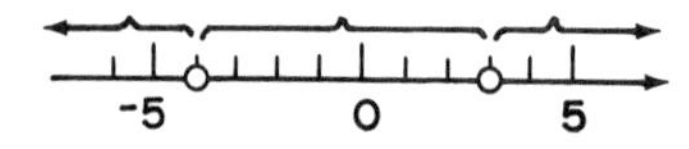

The solution set is $(-4,3)$

The graph is:

14. The critical numbers are -2 and 5, because $y(y - 3) - 10 = y^2 - 3y - 10 = (y + 2)(y - 5) = 0$ for these values. We check the intervals shown on the number line below using arbitrary numbers in each interval, say, -3, 0, and 6, for the variable in $y(y - 3) < 10$.

$$-3(-3 - 3) \overset{?}{<} 10; \quad 0(0 - 3) \overset{?}{<} 10; \quad 5(5 - 3) \overset{?}{<} 10$$

no yes no

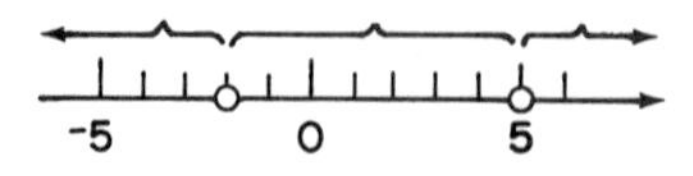

The solution set is $(-2,5)$.

14. cont'd.

The graph is:

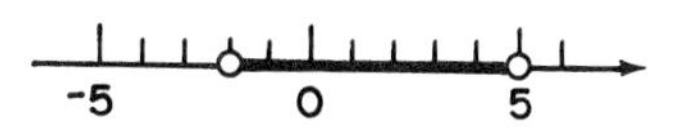

16. The critical numbers are -6 and 3, because $y(y + 3) - 18 = y^2 + 3y - 18 = (y + 6)(y - 3) = 0$ for these values. We check the intervals shown on the number line below using arbitrary numbers in each interval, say, -7, 0, and 4, for the variable in $y(y + 3) \geq 18$.

$$-7(-7 + 3) \overset{?}{\geq} 18; \quad 0(0 + 3) \overset{?}{\geq} 18; \quad 4(4 + 3) \overset{?}{\geq} 18$$

yes no yes

The solution set is $(-\infty,-6] \cup [3,+\infty)$.

The graph is:

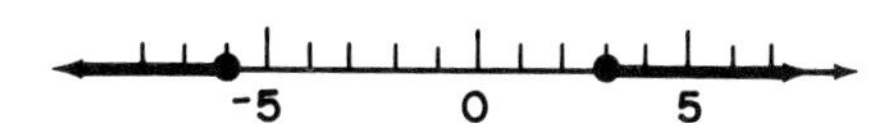

18. The critical numbers are $-\sqrt{7}$ and $\sqrt{7}$ because $x^2 - 7 = (x - \sqrt{7})(x + \sqrt{7}) = 0$ for these values. We check the intervals shown on the number line below using arbitrary numbers in each interval, say, -3, 0, and 3 for the variable in $x^2 \leq 7$. Note: $(\sqrt{7} \approx 2.6)$

$$(-3)^2 \overset{?}{\leq} 7; \quad 0^2 \overset{?}{\leq} 7; \quad 3^2 \overset{?}{\leq} 7$$

no yes no

The solution set is $[-\sqrt{7},\sqrt{7}]$.

The graph is:

20. Because $4x^2 + 1$ is greater than zero for all real-number replacements for x, the solution set is $\emptyset$.
The graph has no points.

22. The critical numbers are 1 and -4 because x - 1 and x + 4 are 0 when x is 1 and -4, respectively. The critical numbers are graphed with open dots because these numbers do not satisfy the given inequality.

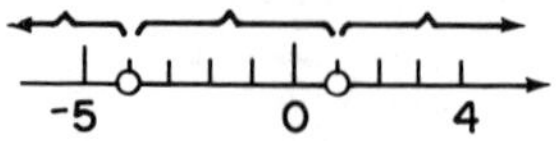

A number from each of the three intervals shown on the number line, say -5, 0, and 2, is substituted into the given inequality.

$$\frac{-5-1}{-5+4} \overset{?}{>} 0; \qquad \frac{0-1}{0+4} \overset{?}{>} 0; \qquad \frac{2-1}{2+4} \overset{?}{>} 0$$

yes; no; yes

Hence, the solution set is $\{x|x < -4\} \cup \{x|x > 1\}$ or $(-\infty,-4) \cup (1,+\infty)$.

The graph is:

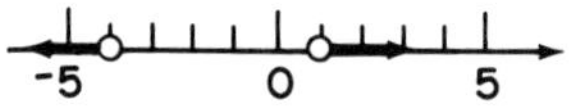

24. $\frac{x+2}{x-2} - 6 \geq 0; \quad \frac{x+2-6(x-2)}{x-2} \geq 0; \quad \frac{-5x+14}{x-2} \geq 0$

The critical numbers are $\frac{14}{5}$ and 2 because $\frac{-5x+14}{x-2} = 0$ when $x = \frac{14}{5}$ and because $\frac{-5x-14}{x-2}$ is undefined when x = 2. $\frac{14}{5}$ is graphed with a closed dot because it satisfies the given inequality, and 2 is graphed with an open dot because it doesn't.

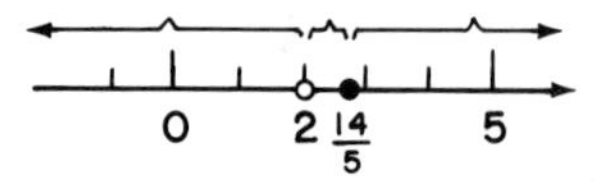

A number from each of the three intervals shown on the number line, say 0, $\frac{12}{5}$, and 3, is substituted into the given inequality.

$$\frac{0+2}{0-2} \overset{?}{\geq} 6; \qquad \frac{\frac{12}{5}+\frac{10}{5}}{\frac{12}{5}-\frac{10}{5}} \overset{?}{\geq} 6; \qquad \frac{3+2}{3-2} \overset{?}{\geq} 6$$

no; yes; no

Hence, the solution set is $\left\{x|2 < x \leq \frac{14}{5}\right\}$ or $\left(2, \frac{14}{5}\right]$

The graph is:

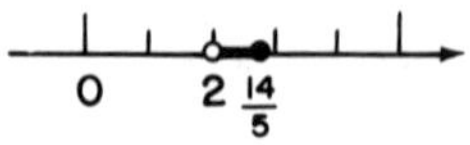

26. $\frac{x}{x-1} - 5 \geq 0;\quad \frac{x - 5(x-1)}{x-1} \geq 0;\quad \frac{-4x+5}{x-1} \geq 0$

The critical numbers are $\frac{5}{4}$ and 1 because $\frac{-4x+5}{x-1} = 0$ when $x = \frac{5}{4}$ and because $\frac{-4x+5}{x-1}$ is undefined when $x = 1$. $\frac{5}{4}$ is graphed with a closed dot because it satisfies the given inequality, and 1 is graphed with an open dot because it does not.

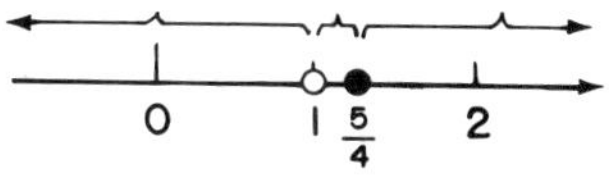

0 1 $\frac{5}{4}$ 2

A number from each of the three intervals shown on the number line, say 0, $\frac{9}{8}$, and 2, is substituted into the given inequality.

$$\frac{0}{0-1} \overset{?}{\geq} 5;\ \text{no}\qquad \frac{\frac{9}{8}}{\frac{9}{8}-1} = 9 \overset{?}{\geq} 5;\ \text{yes}\qquad \frac{2}{2-1} \overset{?}{\geq} 5\ \text{no}$$

Hence, the solution set is $\left\{x \middle| 1 < x \leq \frac{5}{4}\right\}$ or $\left(1, \frac{5}{4}\right]$.

The graph is:

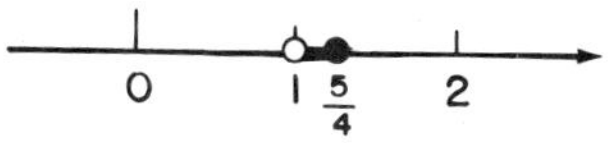

0 1 $\frac{5}{4}$ 2

28. $\frac{x+1}{x-1} - 1 < 0;\quad \frac{x+1-1(x-1)}{x-1} < 0;\quad \frac{2}{x-1} < 0.$

Because $\frac{2}{x-1}$ is undefined when $x = 1$, 1 is a critical number. No value of x can make $\frac{2}{x-1}$ equal zero; so 1 is the only critical number. 1 is graphed with an open dot because it does not satisfy the inequality.

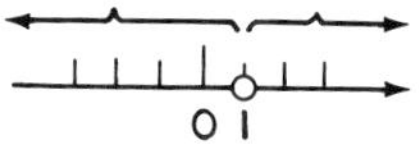

0 1

A number from each of the two intervals shown on the number line, say 0 and 2, is substituted into the given inequality.

$$\frac{0+1}{0-1} \overset{?}{<} 1;\ \text{yes}\qquad \frac{2+1}{2-1} \overset{?}{<} 1\ \text{no}$$

Hence, the solution set is $\{x | x < 1\}$ or $(-\infty, 1)$.

The graph is:

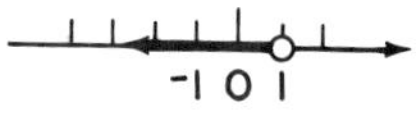

-1 0 1

30. $\dfrac{3}{x - 1} - \dfrac{1}{x} \geq 0; \quad \dfrac{3x - (x - 1)}{x(x - 1)} \geq 0; \quad \dfrac{2x + 1}{x(x - 1)} \geq 0$

The critical numbers are $\dfrac{-1}{2}$, 0, and 1 because $\dfrac{2x - 1}{x(x - 1)} = 0$ when $x = \dfrac{-1}{2}$ and because $\dfrac{2x + 1}{x(x - 1)}$ is undefined when x equals either 0 or 1. $\dfrac{-1}{2}$ is graphed with a closed dot because it satisfies the given inequality, and 0 and 1 are graphed with open dots because they don't.

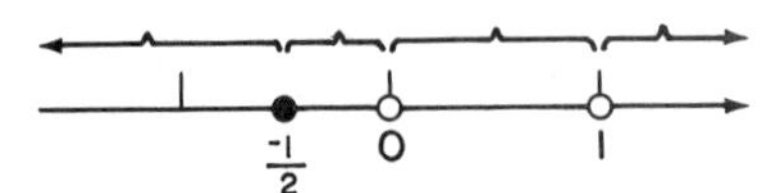

A number from each of the four intervals shown on the number line is substituted into the given inequality to determine the solution set:

$\left\{x \middle| \dfrac{-1}{2} \leq x < 0\right\} \cup \{x|x > 1\}$ or $\left[-\dfrac{1}{2}, 0\right) \cup (1, +\infty)$.

The graph is:

32. $\dfrac{3}{4x + 1} - \dfrac{2}{x - 5} > 0; \quad \dfrac{3(x - 5) - 2(4x + 1)}{(4x + 1)(x - 5)} > 0; \quad \dfrac{-5x - 17}{(4x + 1)(x - 5)} > 0$

The critical numbers are $\dfrac{-17}{5}$, $\dfrac{-1}{4}$, and 5 because $\dfrac{-5x - 17}{(4x + 1)(x - 5)} = 0$ when $x = \dfrac{-17}{5}$ and because $\dfrac{-5x - 17}{(4x + 1)(x - 5)}$ is undefined when x equals either $\dfrac{-1}{4}$ or 5. All of these critical numbers are graphed with open dots.

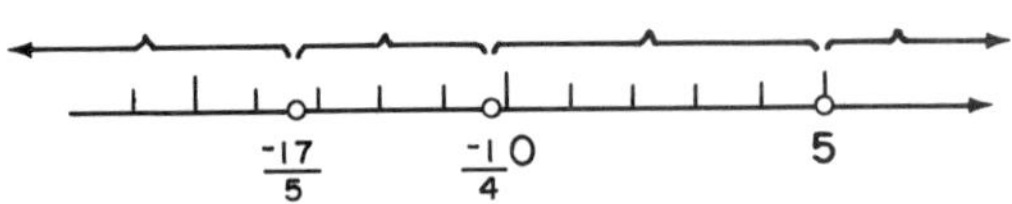

A number from each of the four intervals shown on the number line is substituted into the given inequality to determine the solution set:

$\left\{x \middle| x < \dfrac{-17}{5}\right\} \cup \left\{x \middle| -\dfrac{1}{4} < x < 5\right\}$ or $\left(-\infty, \dfrac{-17}{5}\right) \cup \left(-\dfrac{1}{4}, 5\right)$.

The graph is:

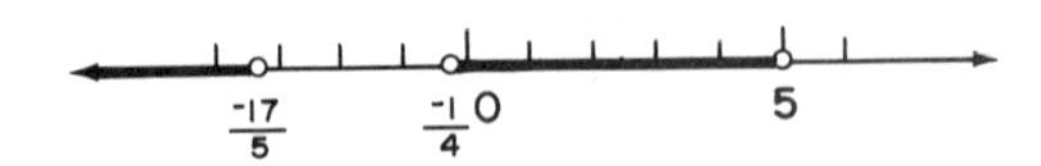

34. The given inequality is equivalent to

$$-2 < y^2 - 3 \quad \text{and} \quad y^2 - 3 < 13,$$

which can be written equivalently as

$$0 < y^2 - 1 \quad \text{and} \quad y^2 - 16 < 0.$$

The critical numbers are those values of y that are solutions of $y^2 - 1 = 0$ or $y^2 - 16 = 0$. These are 1, -1, 4, -4, which are graphed with open dots since they do not satisfy the given inequality.

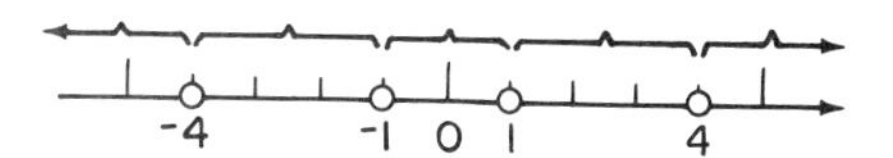

A number from each of the five intervals shown on the number line is substituted into the given inequality to determine the solution set: $\{y \mid -4 < y < -1\} \cup \{y \mid 1 < y < 4\}$ or $(-4,-1) \cup (1,4)$.

The graph is:

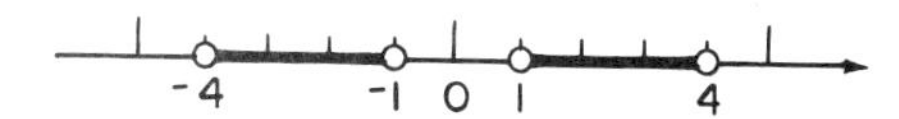

36. We must solve the inequality

$$40 < 56t - 16t^2 < 48$$

which can be written equivalently as

$$40 < 56t - 16t^2 \quad \text{and} \quad 56t - 16t^2 < 48$$

The critical numbers are the solutions of:

$$16t^2 - 56t + 40 = 0 \quad \text{or} \quad -16t^2 + 56t - 48 = 0$$
$$8(2t - 5)(t - 1) = 0 \quad \text{or} \quad -8(2t - 3)(t - 2) = 0$$

Hence the critical numbers are 1, $\frac{5}{2}$, $\frac{3}{2}$, and 2 which are graphed to determine the intervals shown on the number line below.

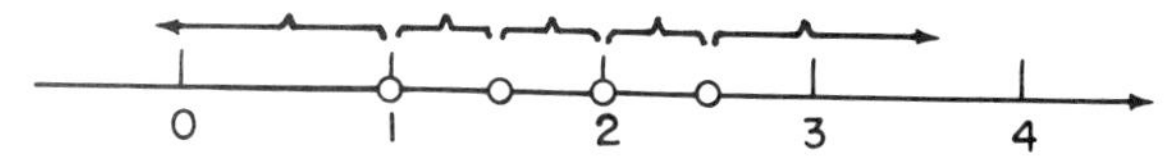

An arbitrary number from each interval, say, 0, $\frac{5}{4}$, $\frac{7}{4}$, $\frac{9}{4}$, and 3 is substituted into

$$40 < 56t - 16t^2 < 48.$$

36. cont'd.

$$40 \overset{?}{<} 56(0) - 16(0)^2 \overset{?}{<} 48; \text{ no}$$

$$40 \overset{?}{<} 56\left(\frac{5}{4}\right) - 16\left(\frac{5}{4}\right)^2 \overset{?}{<} 48; \text{ yes}$$

$$40 \overset{?}{<} 56\left(\frac{7}{4}\right) - 16\left(\frac{7}{4}\right)^2 \overset{?}{<} 48; \text{ no}$$

$$40 \overset{?}{<} 56\left(\frac{9}{4}\right) - 16\left(\frac{9}{4}\right)^2 \overset{?}{<} 48; \text{ yes}$$

$$40 \overset{?}{<} 56(3) - 16(3)^2 \overset{?}{<} 48; \text{ no}$$

The graph of the solution set is

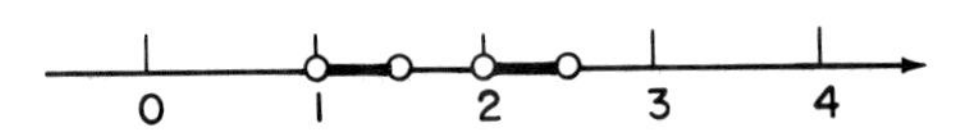

and the solution set is $(1,1.5) \cup (2,2.5)$.

Hence, the height of the ball is between 40 feet and 48 feet when the time elapsed is between 1 and 1.5 seconds and again when the time elapsed is between 2 and 2.5 seconds.

8 Equations and Inequalities in Two Variables

EXERCISE 8.1

2. a. $y = 6 - 2(0) = 6$; $(0,6)$
   b. $0 = 6 - 2x$; $(3,0)$
   c. $y = 6 - 2(-1)$
   $= 6 + 2 = 8$; $(-1,8)$

4. Solve for y: $y = \frac{5 - x}{2}$

   a. $y = \frac{5 - (0)}{2} = \frac{5}{2}$; $\left(0,\frac{5}{2}\right)$

   b. $y = \frac{5 - (5)}{2} = 0$; $(5,0)$

   c. $y = \frac{5 - (-3)}{2} = 4$; $(-3,4)$

6. For $x = -2$,
   $y = 2(-2) + 6 = 2$;

   for $x = 0$,
   $y = 2(0) + 6 = 6$;

   for $x = 2$,
   $y = 2(2) + 6 = 10$.

   $\{(-2,2),(0,6),(2,10)\}$

8. For $x = 0$,

   $$y = \frac{4(0)}{(0)^2 - 1} = \frac{0}{-1} = 0;$$

   for $x = 2$,

   $$y = \frac{4(2)}{(2)^2 - 1} = \frac{8}{4 - 1} = \frac{8}{3};$$

   for $x = 4$,

   $$y = \frac{4(4)}{(4)^2 - 1} = \frac{16}{16 - 1} = \frac{16}{15}.$$

   $\left\{(0,0), \left(2,\frac{8}{3}\right), \left(4,\frac{16}{15}\right)\right\}$

10. For $x = 0$,

$$y = \frac{1}{2}\sqrt{4 - (0)^2}$$

$$= \frac{1}{2}\sqrt{4} = 1;$$

for $x = 1$,

$$y = \frac{1}{2}\sqrt{4 - (1)^2}$$

$$= \frac{1}{2}\sqrt{3};$$

for $x = 2$,

$$y = \frac{1}{2}\sqrt{4 - (2)^2} = \frac{1}{2}\sqrt{4 - 4} = 0.$$

$\{(0,1), (1,\frac{1}{2}\sqrt{3}), (2,0)\}$

12. $y = 4x - 2$;

for $x = -2$,
$y = 4(-2) - 2 = -8 - 2 = -10$

for $x = -4$,
$y = 4(-4) - 2 = -16 - 2 = -18.$

14. $3x - xy = 6$

$xy = 3x - 6$

$$y = \frac{3x - 6}{x};$$

for $x = 1$,

$$y = \frac{3(1) - 6}{(1)};$$

$$= 3 - 6 = -3;$$

for $x = 3$,

$$y = \frac{3(3) - 6}{(3)}$$

$$= \frac{9 - 6}{3} = 1.$$

16. $x^2y - xy = -5$

$y(x^2 - x) = -5$

$$y = \frac{-5}{x^2 - x};$$

for $x = 2$,

$$y = \frac{-5}{(2)^2 - (2)} = \frac{-5}{4 - 2} = \frac{-5}{2};$$

for $x = 4$,

$$y = \frac{-5}{(4)^2 - (4)} = \frac{-5}{16 - 4} = \frac{-5}{12}.$$

18. $x^2y - xy - 5y = -3$

$(x^2 - x - 5)y = -3$

$$y = \frac{-3}{x^2 - x - 5};$$

for $x = -2$,

$$y = \frac{-3}{(-2)^2 - (-2) - 5}$$

$$= \frac{-3}{4 + 2 - 5} = \frac{-3}{1} = -3;$$

for $x = 2$,

$$y = \frac{-3}{(2)^2 - (2) - 5}$$

$$= \frac{-3}{4 - 2 - 5} = \frac{-3}{-3} = 1.$$

20. $3(y^2 + 1) = x$

$3y^2 + 3 = x$

$3y^2 = x - 3$

$y^2 = \frac{x - 3}{3}$

$y = \pm\sqrt{\frac{x - 3}{3}}$ or $\frac{\pm\sqrt{3(x - 3)}}{3}$;

for $x = -2$,

$y = \pm\sqrt{\frac{(-2) - 3}{3}} = \pm\sqrt{\frac{-5}{3}}$

$= \pm\frac{\sqrt{-15}}{3}$

$= \frac{\pm i\sqrt{15}}{3}$;

for $x = 1$,

$y = \pm\sqrt{\frac{(1) - 3}{3}} = \pm\sqrt{\frac{-2}{3}}$

$= \pm\frac{\sqrt{-6}}{3}$

$= \frac{\pm i\sqrt{6}}{3}$.

22. $5x^2 - 4y^2 = 2$

$4y^2 = 5x^2 - 2$

$y^2 = \frac{1}{4}(5x^2 - 2)$

$y = \pm\frac{1}{2}\sqrt{5x^2 - 2}$;

for $x = 2$,

$y = \pm\frac{1}{2}\sqrt{5(2)^2 - 2}$

$= \pm\frac{1}{2}\sqrt{20 - 2} = \frac{1}{2}\sqrt{18}$

$= \frac{\pm 3}{2}\sqrt{2}$;

for $x = 4$,

$y = \pm\frac{1}{2}\sqrt{5(4)^2 - 2}$

$= \pm\frac{1}{2}\sqrt{80 - 2} = \pm\frac{1}{2}\sqrt{78}$

24. $y = |-2| + 5 = 2 + 5 = 7$

The solution is $(-2,7)$.

26. $y = |2(-3) + 1| - |-3|$

$= |-5| - 3 = 5 - 3$

$= 2$

The solution is $(-3,2)$.

28. $y = 2^{-3} = \frac{1}{2^3} = \frac{1}{8}$

The solution is $\left(-3,\frac{1}{8}\right)$.

30. $y = 2^{-(-3)} = 2^3 = 8$

The solution is $(-3,8)$.

32. If $(-6,1)$ is a solution, then

$k(-6) - 2(1) = 6$

$-6k = 6 + 2$

$k = \frac{-8}{6} = \frac{-4}{3}$.

EXERCISE 8.2

2. $x = 0, \quad y = (0) + 3 = 3$
   $y = 0, \ (0) = x + 3, \ x = -3$
   y-intercept: 3;
   x-intercept: -3

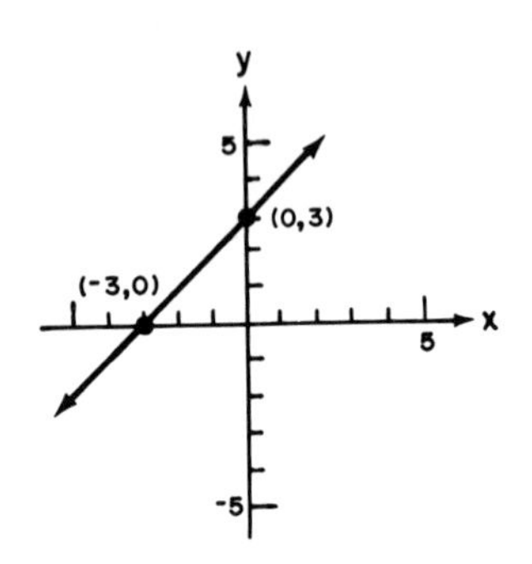

4. $x = 0, \quad y = 4(0) - 8 = -8$
   $y = 0, \ (0) = 4x - 8, \ x = 2$
   y-intercept: -8;
   x-intercept: 2

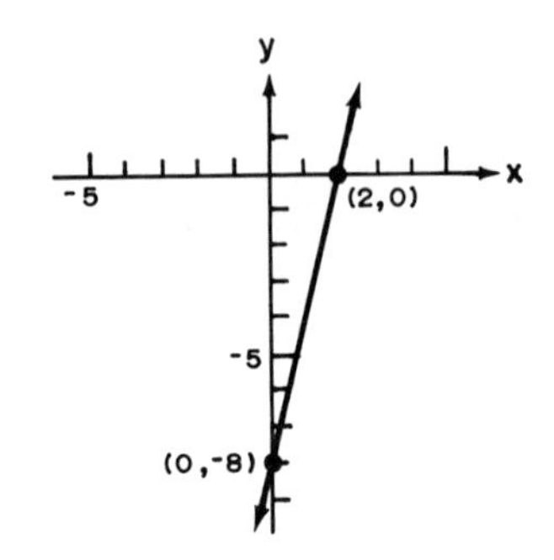

6. $x = 0, \ 2(0) - y = 6,$
   $y = -6$
   $y = 0, \ 2x - 0 = 6,$
   $x = 3$
   y-intercept: -6,
   x-intercept: 3

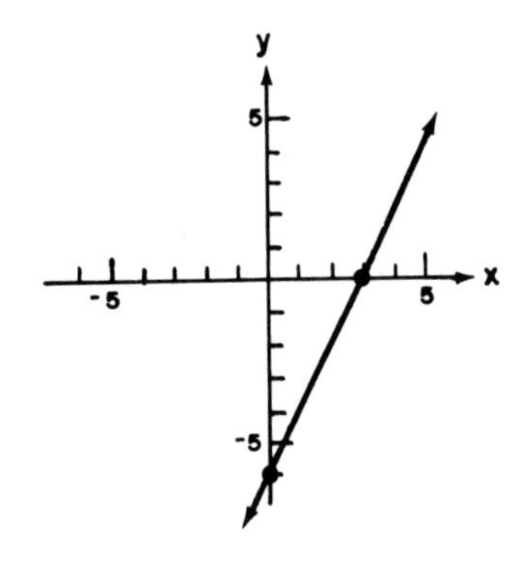

8. $x = 0, \ 2(0) + 6y = 6,$
   $y = 1$
   $y = 0, \ 2x + 6(0) = 6,$
   $x = 3$
   y-intercept: 1;
   x-intercept: 3

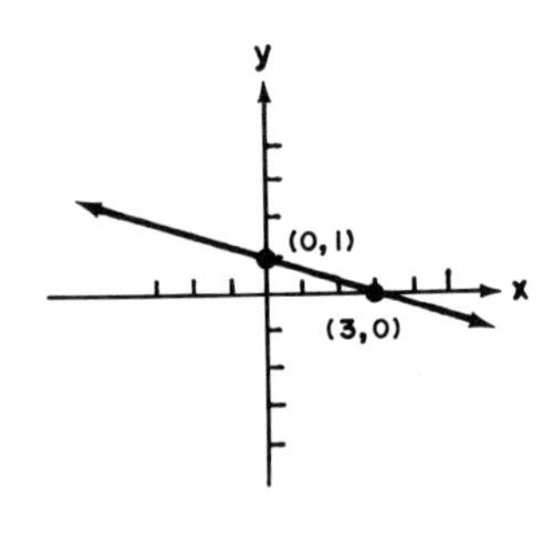

10. $x = 0,\ 4(0) = y - 6,$

$y = 6$

$y = 0,\ 4x = 0 - 6$

$x = \frac{-3}{2}$

y-intercept: 6,

x-intercept: $\frac{-3}{2}$

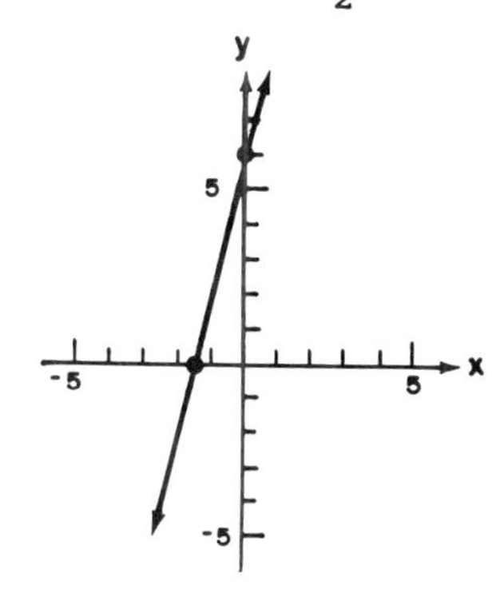

12. 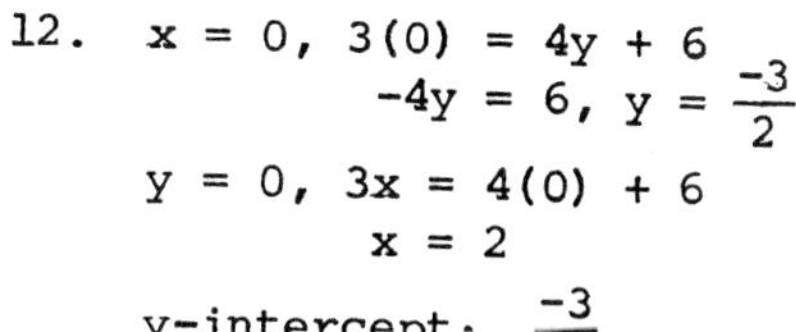

$x = 0,\ 3(0) = 4y + 6$

$-4y = 6,\ y = \frac{-3}{2}$

$y = 0,\ 3x = 4(0) + 6$

$x = 2$

y-intercept: $\frac{-3}{2}$,

x-intercept: 2

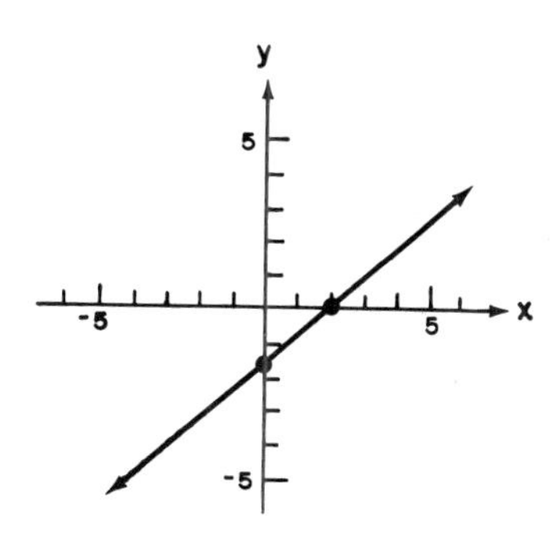

14. $x + 3y = 0$

$x = 0,\ (0) + 3y = 0$

$y = 0$

Both intercepts are at the origin. Arbitrarily, let $y = -1$; then

$x + 3(-1) = 0$

$x = 3.$

Graph (0,0) and (3,-1), or any other solutions of the equation, and draw the line.

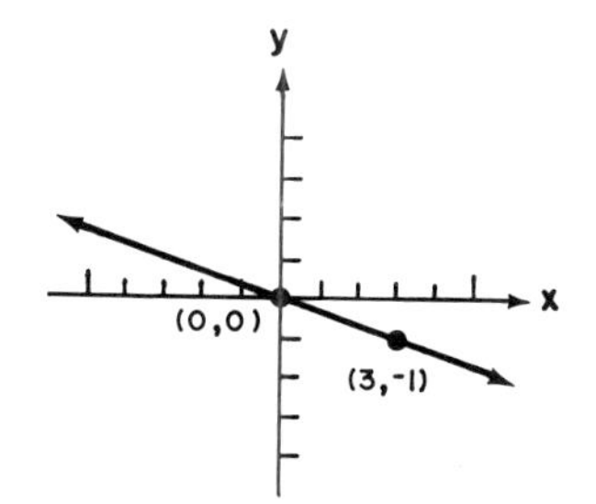

16. 

$y = 0;\ x = 2(0) = 0$

Both intercepts are at the origin. Arbitrarily, let $y = 3$; hence,

$x = 2(3) = 6.$

Graph (0,0) and (6,3), or any other solutions of the equation, and draw the line containing the two points.

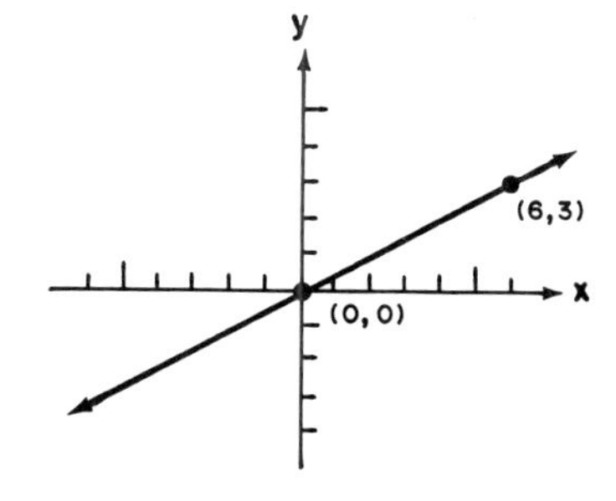

18. $x = 0$; $4(0) + y = 0$, $y = 0$

Both intercepts are at the origin. Arbitrarily, let $x = -1$; hence,

$$4(-1) + y = 0;$$
$$y = 4.$$

Graph (0,0) and (-1,4), or any other solutions of the equation, and draw the line containing the two points.

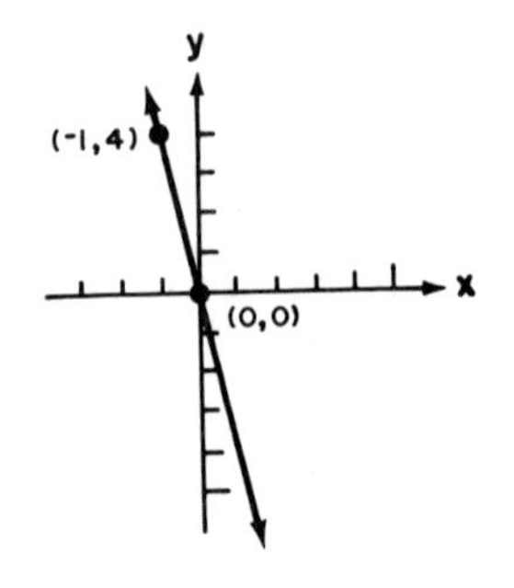

20. $x = 0$; $0 - y = 0$, $y = 0$

Both intercepts are at the origin. Arbitrarily, let $x = 2$; hence,

$$2 - y = 0$$
$$2 = y.$$

Graph (0,0) and (2,2) or any other solutions of the equation and draw the line containing the two points.

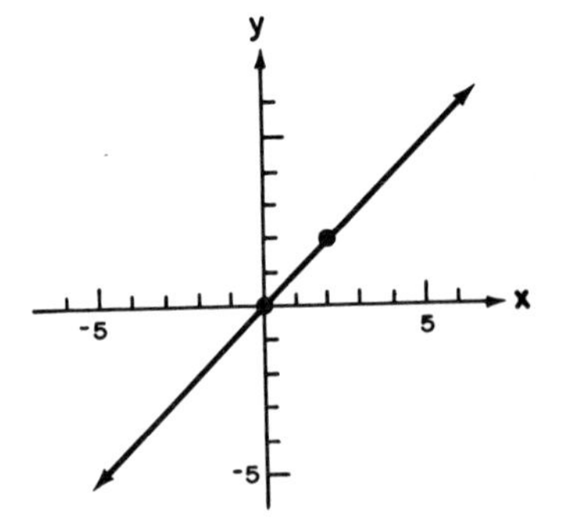

22. $x = -2$ is equivalent to $x + 0y = -2$.

For all values of y, $x = -2$. Graph any two ordered pairs of the form (-2,y) and draw the line containing the two points.

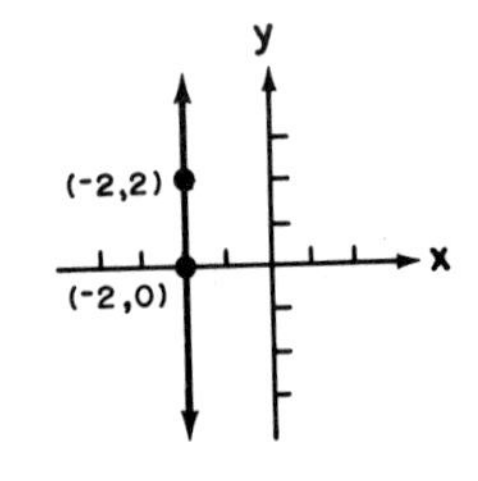

24. Solve for y and obtain $y = 5$, which is equivalent to $0x + y = 5$.

For all values of x, $y = 5$. Graph any two ordered pairs of the form $(x,5)$ and draw the line.

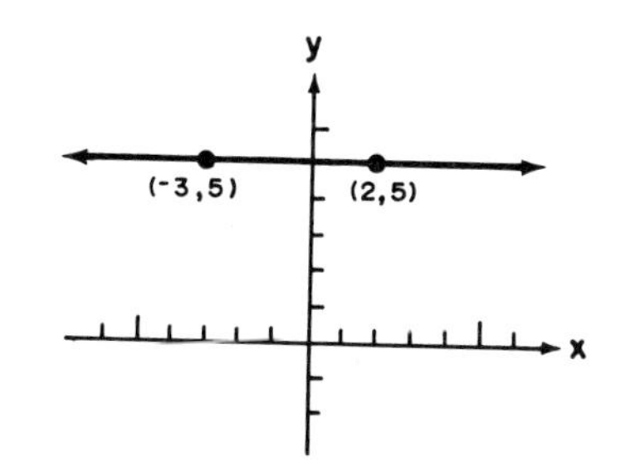

26. $y = 0$ is equivalent to $0x + y = 0$.

For all values of x, $y = 0$. Graph any two ordered pairs of the form $(x,0)$ and draw the line.

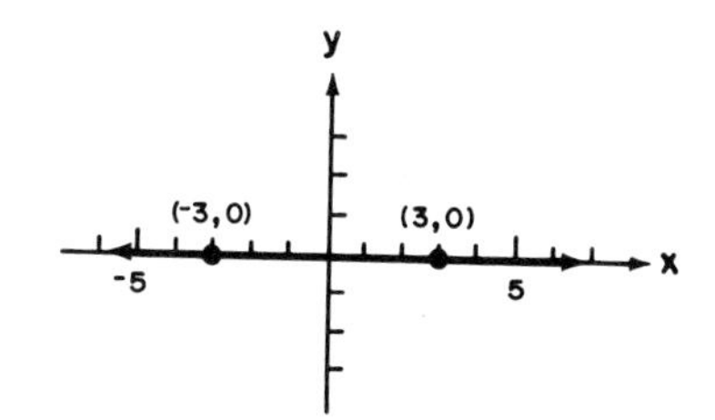

28.

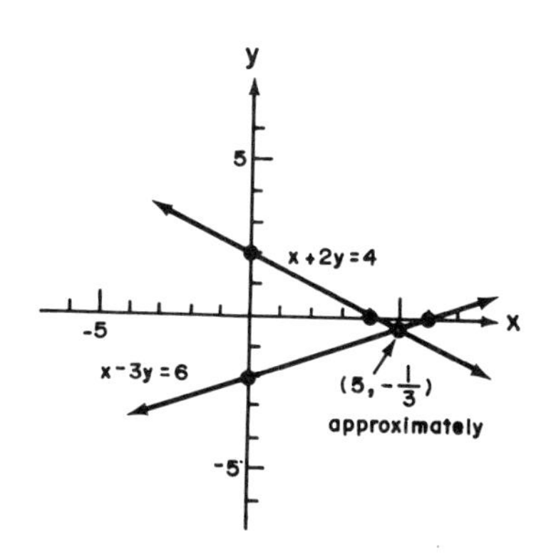

30.

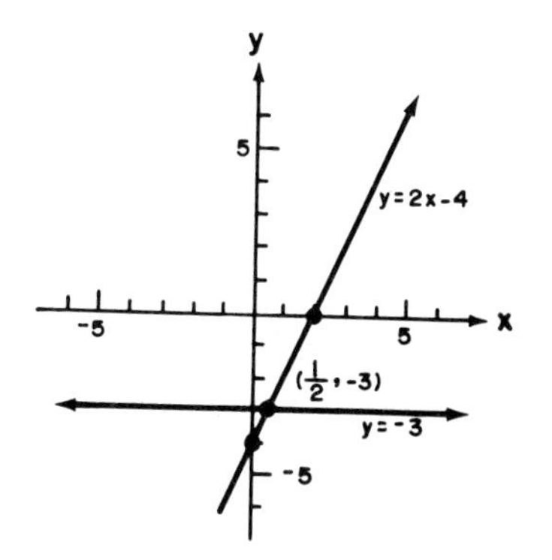

EXERCISE 8.3

2. $p_1 = (-1,1)$ and $p_2 = (5,9)$

$$m = \frac{9 - 1}{5 - (-1)} = \frac{8}{6} = \frac{4}{3}$$

$$d = \sqrt{[5 - (-1)]^2 + (9 - 1)^2}$$

$$= \sqrt{6^2 + 8^2} = \sqrt{100} = 10$$

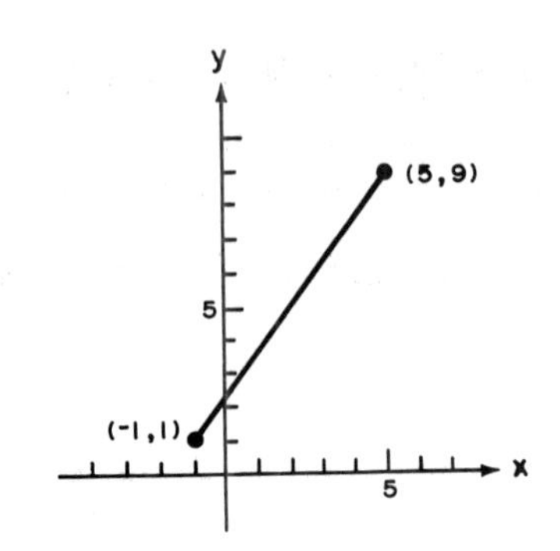

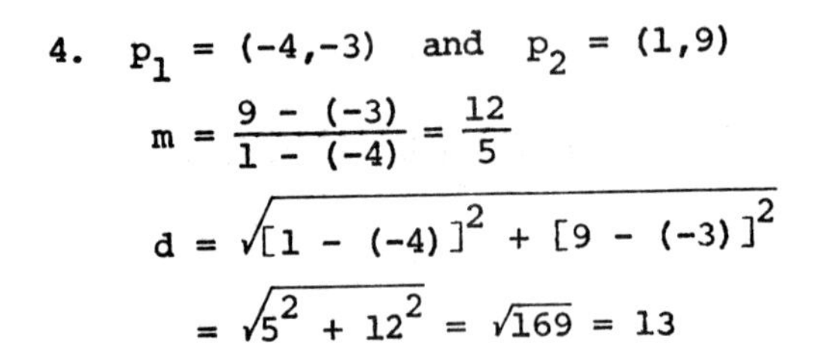

4. $p_1 = (-4,-3)$ and $p_2 = (1,9)$

$$m = \frac{9 - (-3)}{1 - (-4)} = \frac{12}{5}$$

$$d = \sqrt{[1 - (-4)]^2 + [9 - (-3)]^2}$$

$$= \sqrt{5^2 + 12^2} = \sqrt{169} = 13$$

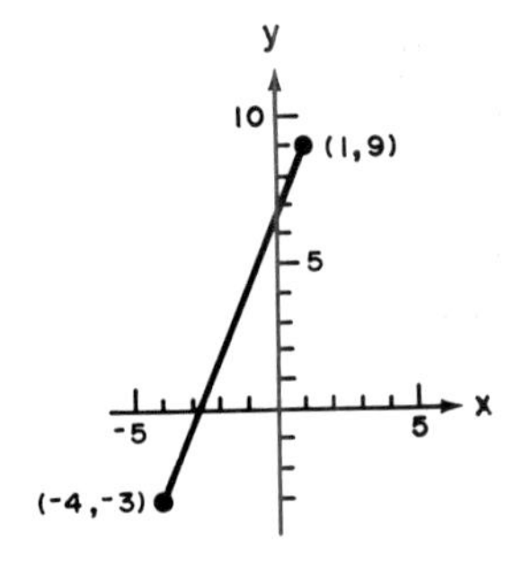

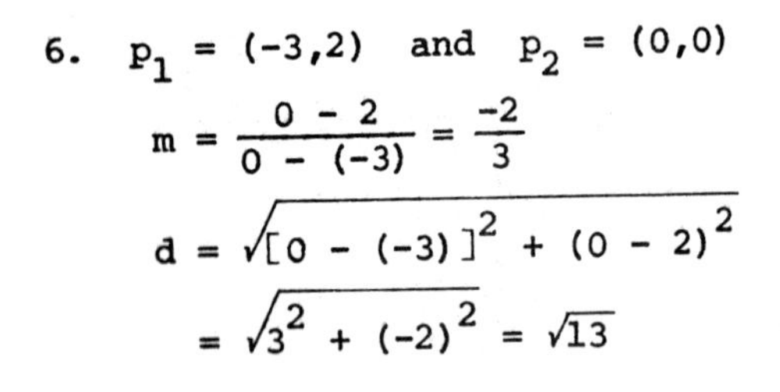

6. $p_1 = (-3,2)$ and $p_2 = (0,0)$

$$m = \frac{0 - 2}{0 - (-3)} = \frac{-2}{3}$$

$$d = \sqrt{[0 - (-3)]^2 + (0 - 2)^2}$$

$$= \sqrt{3^2 + (-2)^2} = \sqrt{13}$$

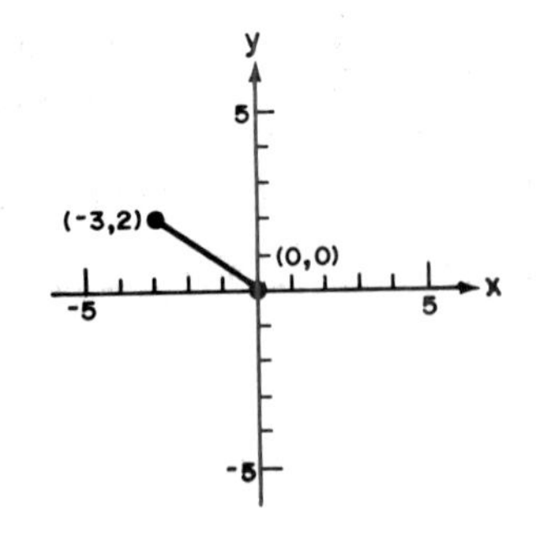

8. $p_1 = (2,-3)$ and $p_2 = (-2,-1)$

$$m = \frac{-1 - (-3)}{-2 - 2} = \frac{2}{-4} = \frac{-1}{2}$$

$$d = \sqrt{(-2 - 2)^2 + [-1 - (-3)]^2}$$

$$= \sqrt{(-4)^2 + 2^2} = \sqrt{20} = 2\sqrt{5}$$

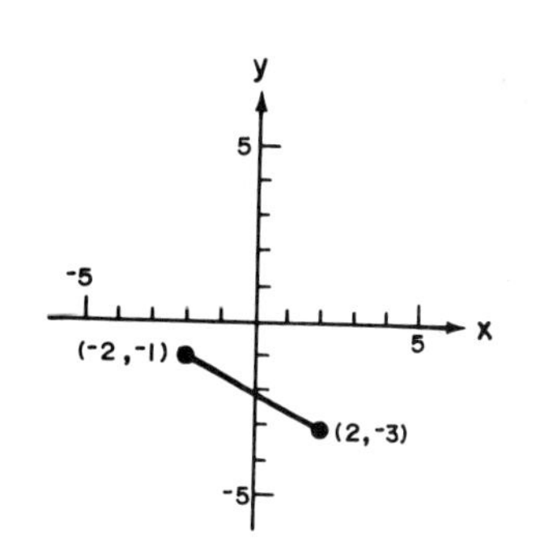

10. $p_1 = (2,0)$ and $p_2 = (-2,0)$

$$m = \frac{0 - 0}{-2 - 2} = 0$$

$$d = \sqrt{(-2 - 2)^2 + (0 - 0)^2}$$

$$= \sqrt{(-4)^2} = 4$$

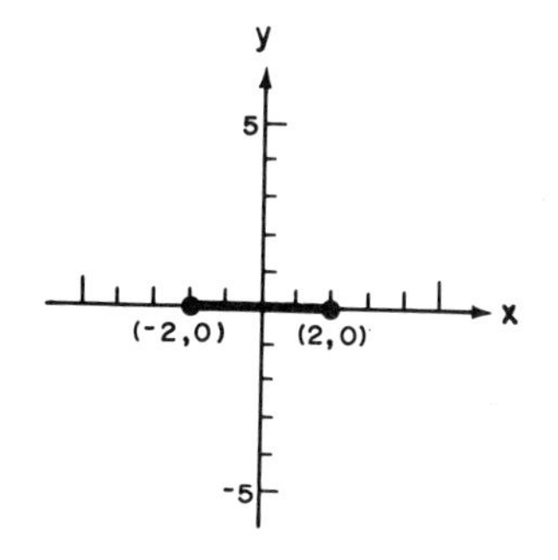

12. $p_1 = (-2,-5)$ and $p_2 = (-2,3)$

$$m = \frac{3 - (-5)}{-2 - (-2)} = \frac{8}{0},$$ so the slope is not defined.

$$d = \sqrt{(-2 - (-2))^2 + (3 - (-5))^2}$$

$$= \sqrt{0^2 + 8^2} = 8$$

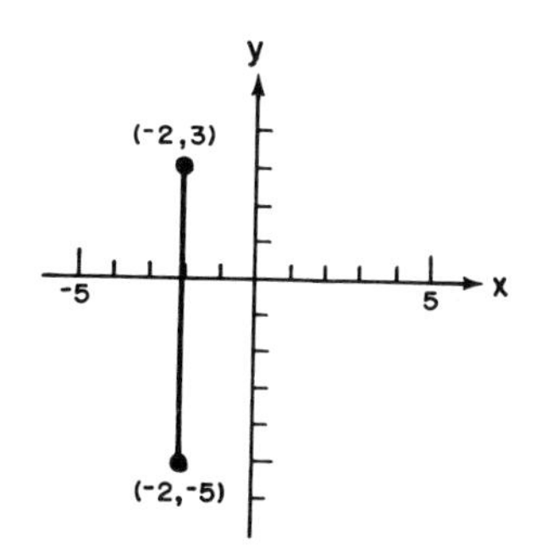

In Problems 14-16, let the points be A, B, and C, respectively, and P the perimeter.

14. $AB = \sqrt{(3 - 10)^2 + (1 - 1)^2}$

$= \sqrt{(-7)^2 + 0^2} = \sqrt{49} = 7$

$AC = \sqrt{(5 - 10)^2 + (9 - 1)^2}$

$= \sqrt{(-5)^2 + 8^2} = \sqrt{89}$

$BC = \sqrt{(5 - 3)^2 + (9 - 1)^2}$

$= \sqrt{2^2 + 8^2} = \sqrt{68} = 2\sqrt{17}$

$P = 7 + \sqrt{89} + 2\sqrt{17}$

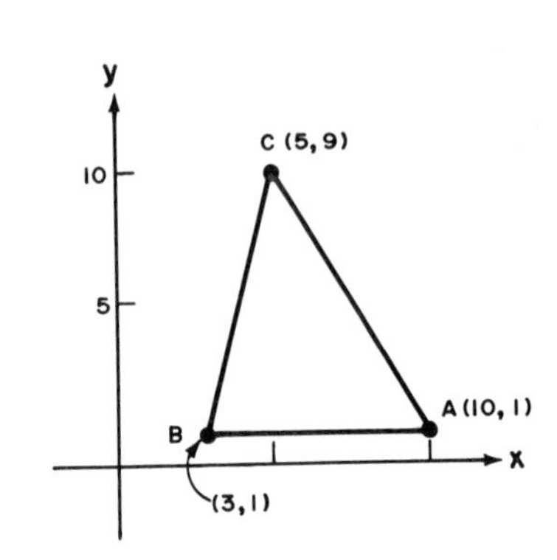

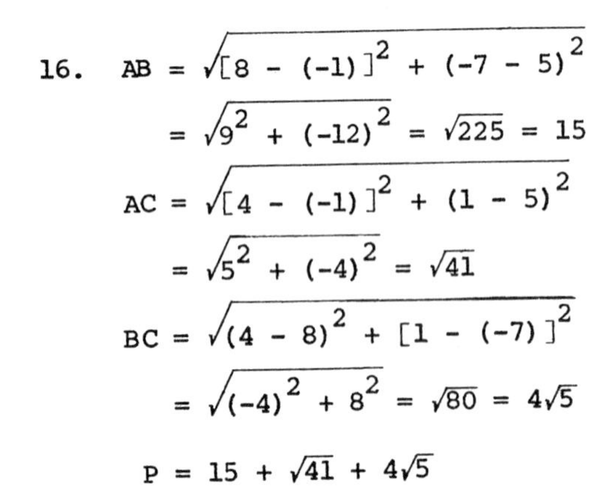

16. $AB = \sqrt{[8 - (-1)]^2 + (-7 - 5)^2}$

$= \sqrt{9^2 + (-12)^2} = \sqrt{225} = 15$

$AC = \sqrt{[4 - (-1)]^2 + (1 - 5)^2}$

$= \sqrt{5^2 + (-4)^2} = \sqrt{41}$

$BC = \sqrt{(4 - 8)^2 + [1 - (-7)]^2}$

$= \sqrt{(-4)^2 + 8^2} = \sqrt{80} = 4\sqrt{5}$

$P = 15 + \sqrt{41} + 4\sqrt{5}$

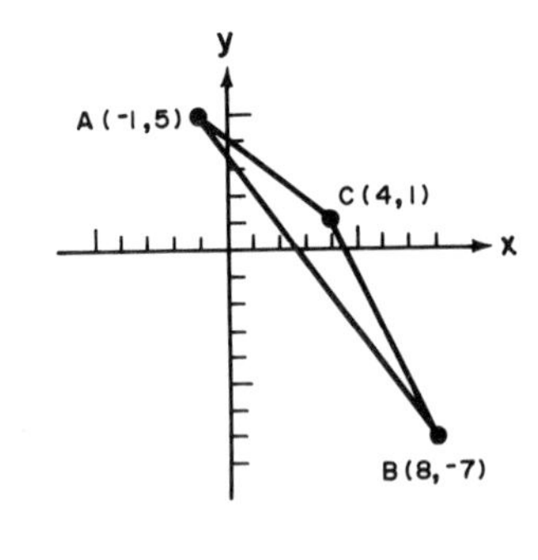

18. Let m_1 be the slope of the first segment, and m_2 the slope of the second segment. Then

$$m_1 = \frac{-2 - 2}{2 - (-4)} = \frac{-4}{6} = \frac{-2}{3}; \quad m_2 = \frac{4 - 0}{-3 - 3} = \frac{4}{-6} = \frac{-2}{3}.$$

Since $m_1 = m_2$, the segments are parallel.

20. Let m_1 be the slope of the segment joining (8,0) and (6,6), so $m_1 = \frac{6 - 0}{6 - 8} = \frac{6}{-2} = -3$; let m_2 be the slope of the segment with endpoints (-3,3) and (6,6), so

$$m_2 = \frac{6 - 3}{6 - (-3)} = \frac{3}{9} = \frac{1}{3}.$$

Since $m_1 m_2 = (-3)\left(\frac{1}{3}\right) = -1$, the line segments are perpendicular.

22. a. $m_1 = \frac{-3 - 0}{0 - (-2)} = -\frac{3}{2}$

b. Since ℓ_2 is vertical, m_2 is undefined.

24. a. $m_1 = \frac{0 - (-5)}{2 - 0} = \frac{5}{2}$

b. $m_2 = \frac{0 - (-2)}{-5 - 0} = -\frac{2}{5}$

26. Let A = (0,0), B = (6,0), and C = (3,3).

$$AB = \sqrt{(6 - 0)^2 + (0 - 0)^2} = \sqrt{36} = 6$$

$$AC = \sqrt{(3 - 0)^2 + (3 - 0)^2} = \sqrt{3^2 + 3^2} = \sqrt{18} = 3\sqrt{2}$$

$$BC = \sqrt{(3 - 6)^2 + (3 - 0)^2} = \sqrt{(-3)^2 + 3^2} = \sqrt{18} = 3\sqrt{2}$$

Since $(3\sqrt{2})^2 + (3\sqrt{2})^2 = 18 + 18 = 36$ and $6^2 = 36$, angle C is a right angle, and therefore, the triangle is a right triangle. AC = BC, so it is an isosceles triangle.

28. Let A, B, C, and D be the given points.

$$\text{slope of AB} = m_1 = \frac{-11 - 4}{7 - (-5)} = \frac{-15}{12} = \frac{-5}{4}$$

$$\text{slope of CD} = m_2 = \frac{40 - 25}{0 - 12} = \frac{15}{-12} = \frac{-5}{4}$$

$$\text{slope of AD} = m_3 = \frac{40 - 4}{0 - (-5)} = \frac{36}{5}$$

$$\text{slope of BC} = m_4 = \frac{25 - (-11)}{12 - 7} = \frac{36}{5}$$

Since $m_1 = m_2$ and $m_3 = m_4$, AB is parallel to CD and AD is parallel to BC so that ABCD is a parallelogram.

30. If the segments are to be perpendicular, then $m_1m_2 = -1$, or

$$(-4)\frac{k - 4}{8} = -1; \quad \frac{k - 4}{-2} = -1; \quad k - 4 = 2; \quad k = 6.$$

EXERCISE 8.4

Use the point-slope form, $y - y_1 = m(x - x_1)$, in Problems 2-12.

2.
$$y - (-5) = -3(x - 2)$$
$$y + 5 = -3x + 6$$
$$3x + y - 1 = 0$$

4.
$$y - (-1) = 4[x - (-6)]$$
$$y + 1 = 4x + 24$$
$$-4x + y - 23 = 0$$
or
$$4x - y + 23 = 0$$

6. $y - 0 = \frac{-1}{3}(x - 2)$

$3y = -x + 2$

$x + 3y - 2 = 0$

8. $y - (-1) = \frac{5}{3}(x - 2)$

$3y + 3 = 5x - 10$

$-5x + 3y + 13 = 0$

or

$5x - 3y - 13 = 0$

10. $y - (-6) = 0(x - 0)$

$y + 6 = 0$

12. $y - 0 = 1(x - 0)$

$y = x$

$-x + y = 0$ or $x - y = 0$

14. $m = \frac{-3 - (-1)}{2 - 5} = \frac{-2}{-3} = \frac{2}{3}$.

Using the point-slope formula with $(x_1,y_1) = (5,-1)$ gives

$y - (-1) = \frac{2}{3}(x - 5)$

$y + 1 = \frac{2}{3}(x - 5)$

$3y + 3 = 2x - 10$

$2x - 3y = 13.$

16. $m = \frac{-4 - 5}{3 - 0} = \frac{-9}{3} = -3.$

Using the point-slope formula with $(x_1,y_1) = (0,5)$ gives

$y - 5 = -3(x - 0)$

$y - 5 = -3x$

$3x + y = 5.$

18. $m = \frac{-3 - 4}{-3 - (-1)} = \frac{-7}{-2} = \frac{7}{2}$.

Using the point-slope formula with $(x_1,y_1) = (-1,4)$ gives

$y - 4 = \frac{7}{2}(x - (-1))$

$2y - 8 = 7x + 7$

$-7x + 2y = 15$

or

$7x - 2y = -15.$

20. $m = \frac{-4 - 0}{0 - (-1)} = -4$

Using the point-slope formula with $(x_1,y_1) = (0,-4)$ gives

$y - (-4) = -4(x - 0)$

$y + 4 = -4x$

$4x + y = -4.$

In Problems 22-30, m = slope and b = y-intercept.

22. a. $y = -2x - 1$

b. $m = -2$ and $b = -1$

24. a. $y = 3x - 7$

b. $m = 3$ and $b = -7$

26. a. $y = \frac{2}{3}x$

b. $m = \frac{2}{3}$ and $b = 0$

28. a. $y = \frac{-1}{2}x + \frac{5}{2}$

b. $m = \frac{-1}{2}$ and $b = \frac{5}{2}$

30. a. $y = 3$

$y = 0x + 3$

b. $m = 0$ and $b = 3$

32. Substituting -4 and 1 for a and b respectively in $\frac{x}{a} + \frac{y}{b} = 1$, we obtain

$$\frac{x}{-4} + \frac{y}{1} = 1.$$

from which by multiplying each member by the LCD, -4,

$$x - 4y = -4.$$

34. Substituting 5 and -3 for a and b respectively in $\frac{x}{a} + \frac{y}{b} = 1$, we obtain

$$\frac{x}{5} + \frac{y}{-3} = 1,$$

from which

$$-3x + 5y = -15$$

or

$$3x - 5y = 15.$$

36. $\frac{x}{-4} + \frac{y}{-2} = 1$

$-2x - 4y = 8$

or

$x + 2y = -4$

38. $\frac{x}{2/3} + \frac{y}{-1/4} = 1$

Simplify each complex fraction:

$\frac{x}{2/3} = \frac{3x}{2}$ and $\frac{y}{-1/4} = -4y$.

Hence, we have

$\frac{3x}{2} + (-4y) = 1,$

from which

$3x - 8y = 2.$

40. Solving $2y - 3x = 5$ explicitly for y gives $y = \frac{3}{2}x + \frac{5}{2}$; so the slope is $\frac{3}{2}$. Applying the point-slope form, $y - y_1 = m(x - x_1)$, using (0,5) and $m = \frac{3}{2}$,

$$y - 5 = \frac{3}{2}(x - 0)$$

$3x - 2y + 10 = 0.$

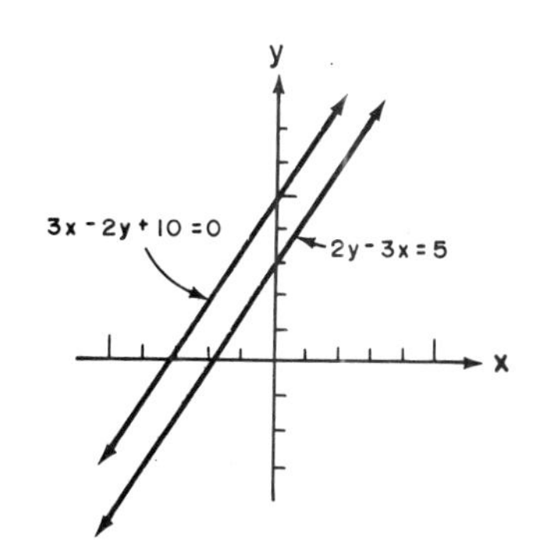

42. Solve $2y - 3x = 5$ for y:
$y = \frac{3}{2}x + \frac{5}{2}$. The slope of this line is $\frac{3}{2}$. The slope of a perpendicular is $\frac{-2}{3}$. Using the point-slope form with $m = \frac{-2}{3}$,

$$y - 5 = \frac{-2}{3}(x - 0)$$
$$2x + 3y - 15 = 0.$$

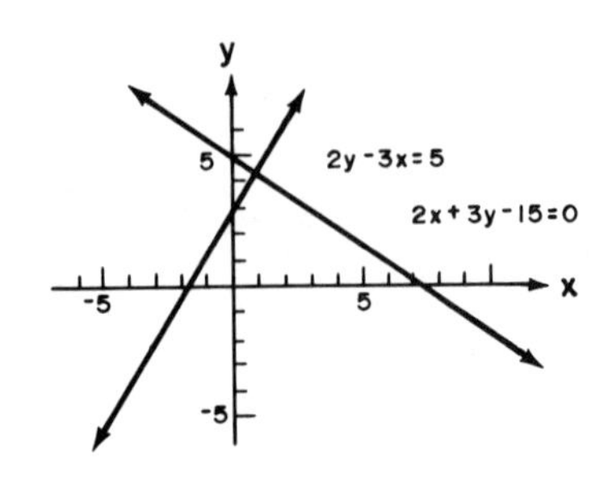

EXERCISE 8.5

Follow the examples in the text. Remember to use a broken line when the < or > symbol is used in the given inequality and a solid line when the ≤ or ≥ symbol is used.

2.

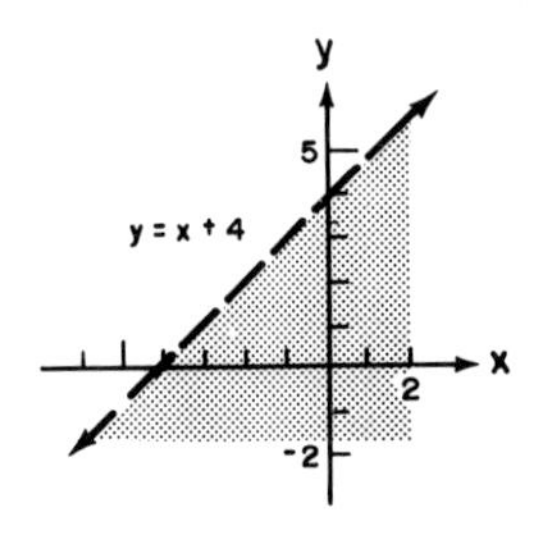

4.

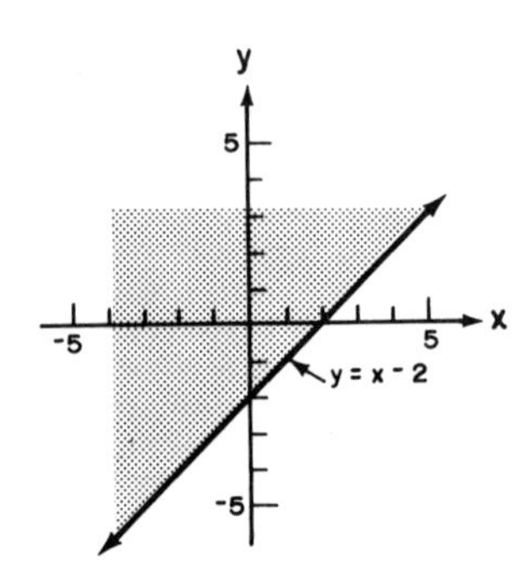

6.

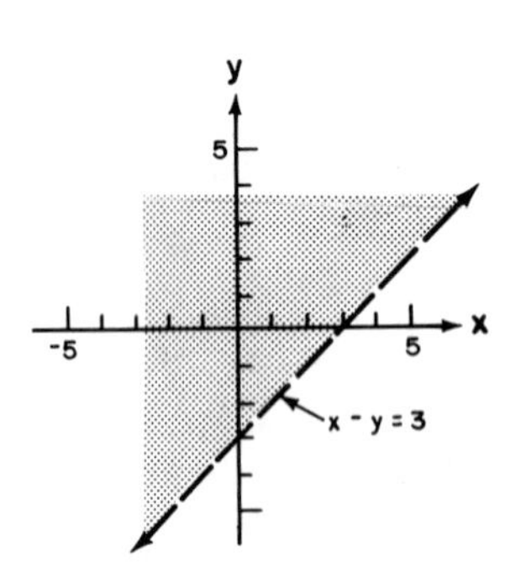

8.

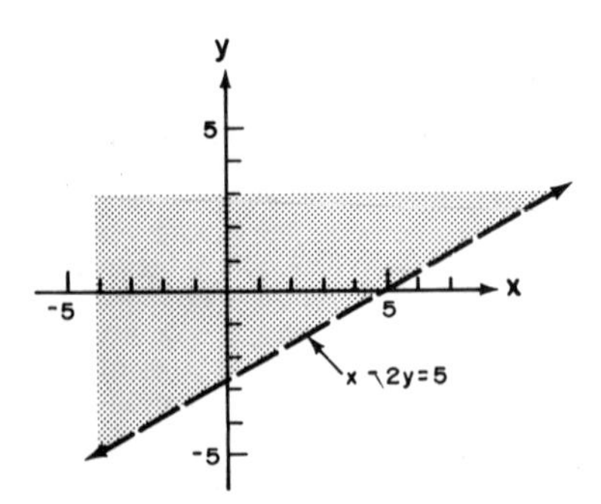

10.

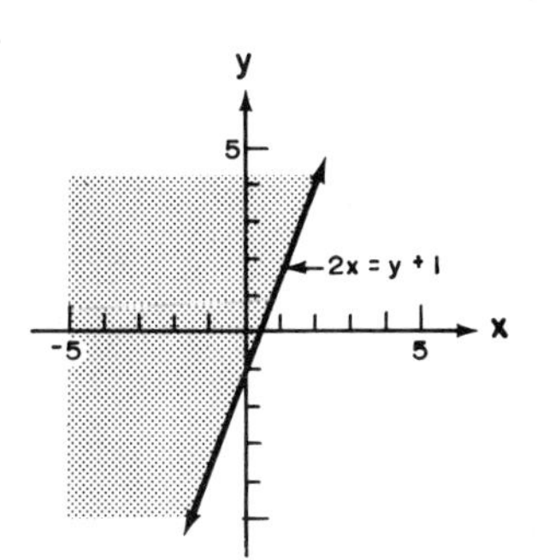

12.

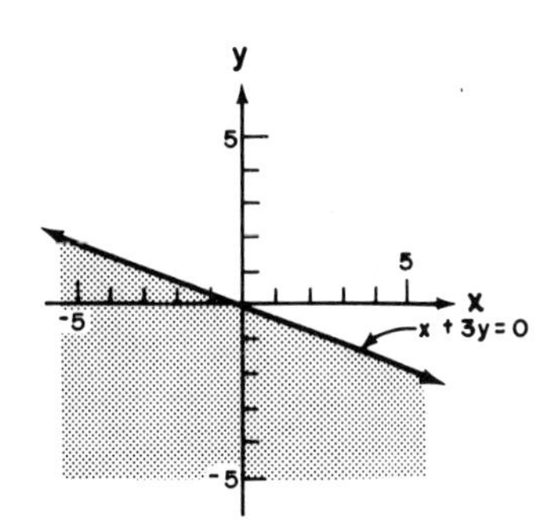

14.

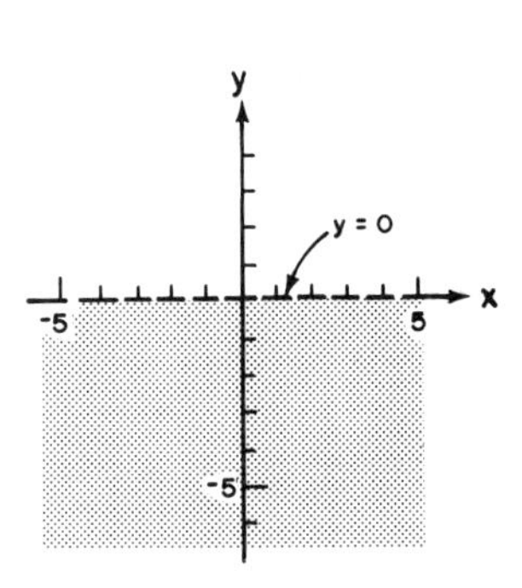

16.

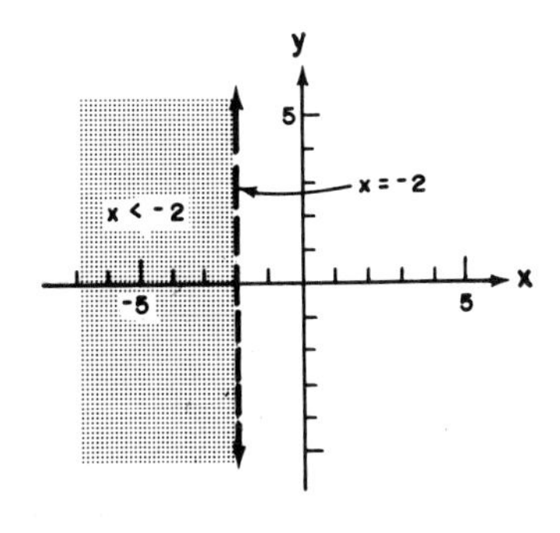

18. $-2 \leq x < 0$ means that x is between -2 and 0 including -2 but not including 0.

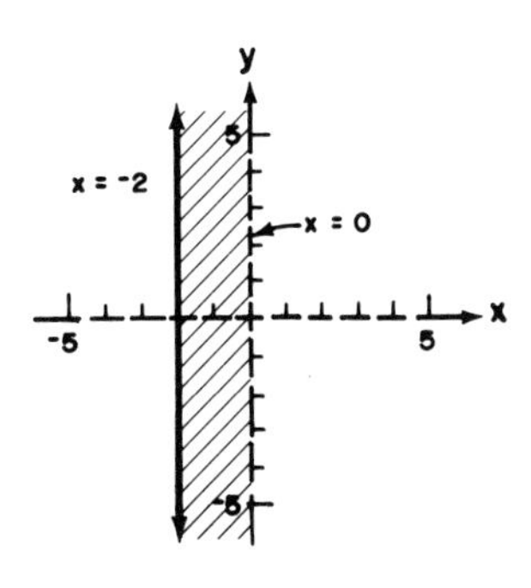

20. $0 \le y \le 1$ means that y is between 0 and 1 including 0 and 1.

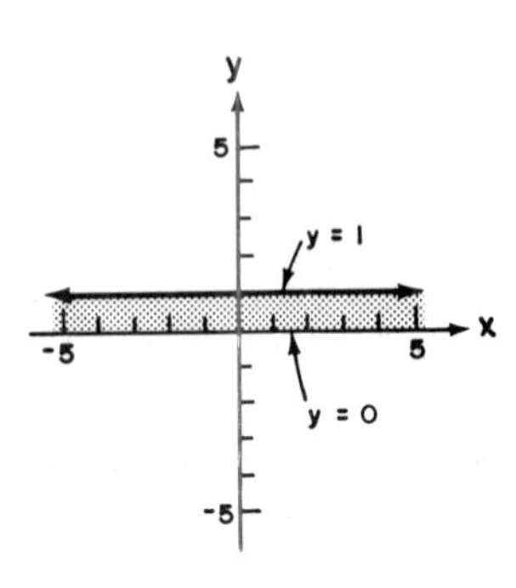

22.

24.

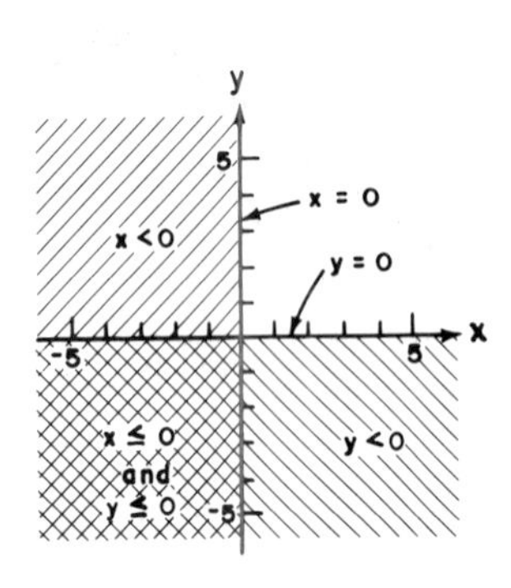

26.

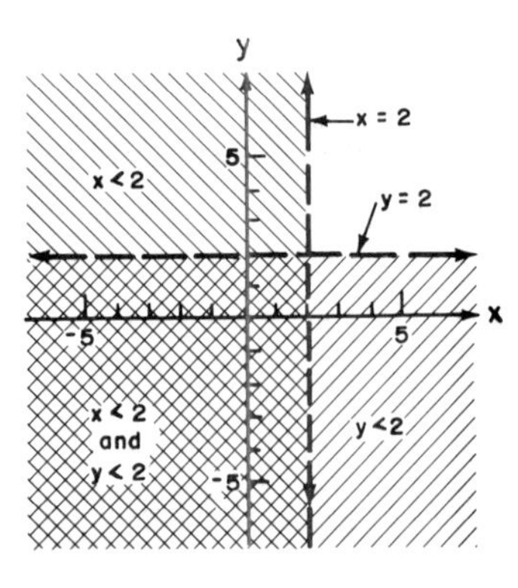

28.

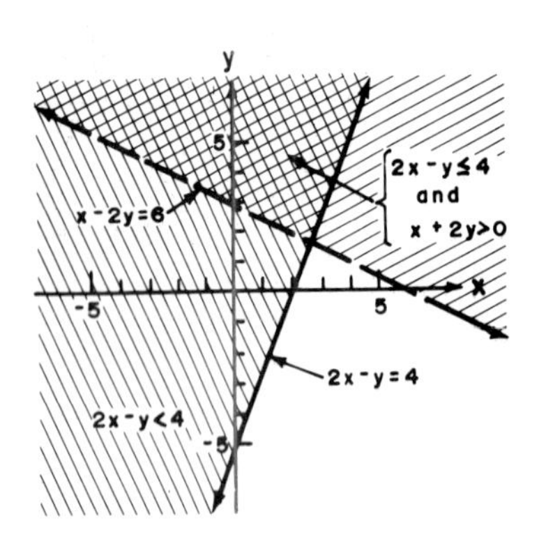

30.

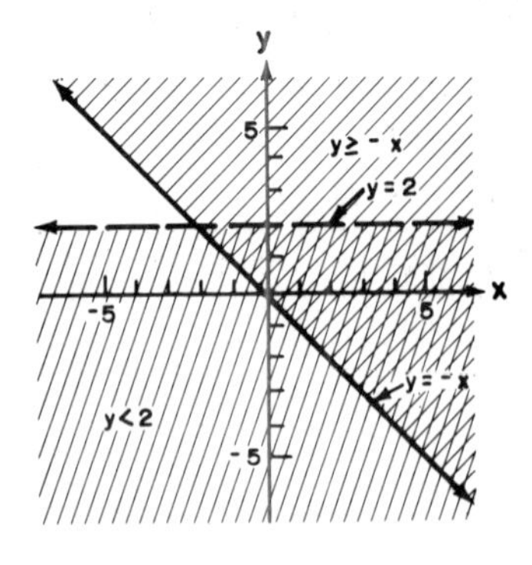

9 Systems of Linear Equations and Inequalities

EXERCISE 9.1

For these problems there are methods of solution other than those shown.

2. Add the two equations:

   $3x = 9$; $x = 3$.

   Substitute 3 for x in the second equation:

   $3 + 3y = 3$; $3y = 0$; $y = 0$.

   The solution set is $\{(3,0)\}$.

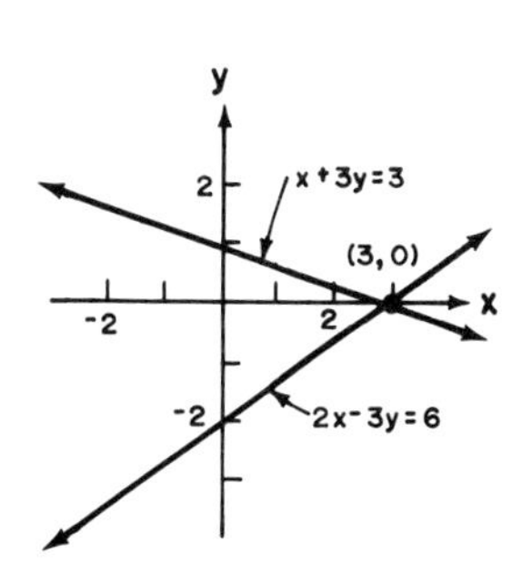

4. Add 2 times the first equation to the second:

$$\begin{array}{l} 4x - 2y = 14 \\ \underline{3x + 2y = 14} \\ 7x \qquad = 28; \; x = 4. \end{array}$$

   Substitute 4 for x in the first equation:

   $2(4) - y = 7$; $1 = y$.

   The solution set is $\{(4,1)\}$.

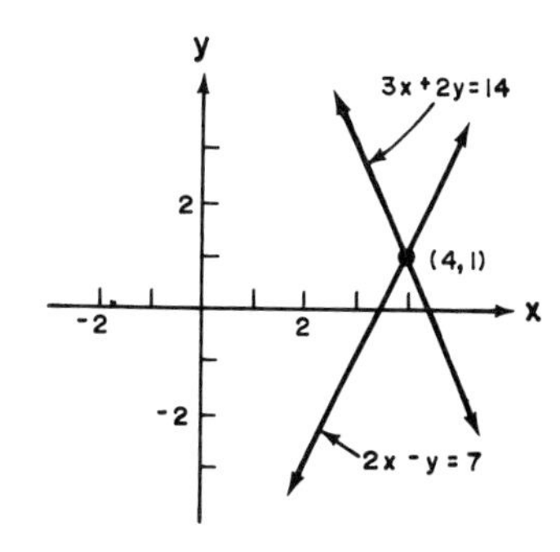

6. Add - 2 times the second equation to the first:

$$\begin{array}{rcr} x + 4y & = & -14 \\ \underline{-6x - 4y} & \underline{=} & \underline{4} \\ -5x \qquad & = & -10; \; x = 2. \end{array}$$

Substitute 2 for x in the first equation:

$2 + 4y = -14;\ 4y = -16$

$y = -4.$

The solution set is $\{(2,-4)\}$.

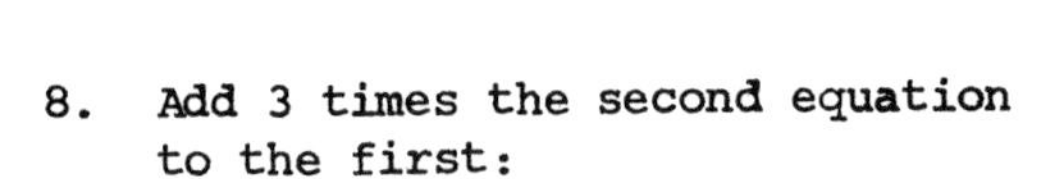

8. Add 3 times the second equation to the first:

$$\begin{array}{rcr} 2x - 3y & = & 8 \\ \underline{3x + 3y} & \underline{=} & \underline{-3} \\ 5x \qquad & = & 5 \; ; \; x = 1. \end{array}$$

Substitute 1 for x in the second equation:

$1 + y = -1;\ y = -2.$

The solution set is $\{(1,-2)\}$.

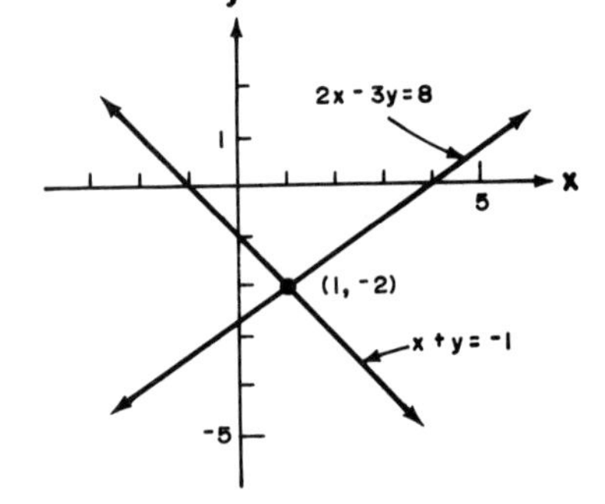

10. Write the equations in standard form.

$$\begin{array}{rcr} 3x - 3y & = & -3 \\ -6x + 2y & = & 14 \end{array}$$

Add 2 times the first equation to the second:

$$\begin{array}{rcr} 6x - 6y & = & -6 \\ \underline{-6x + 2y} & \underline{=} & \underline{14} \\ -4y & = & 8 \; ; \; y = -2. \end{array}$$

Substitute -2 for y in the first equation:

$3x - 3(-2) = -3;\ x = -3.$

The solution set is $\{(-3,-2)\}$.

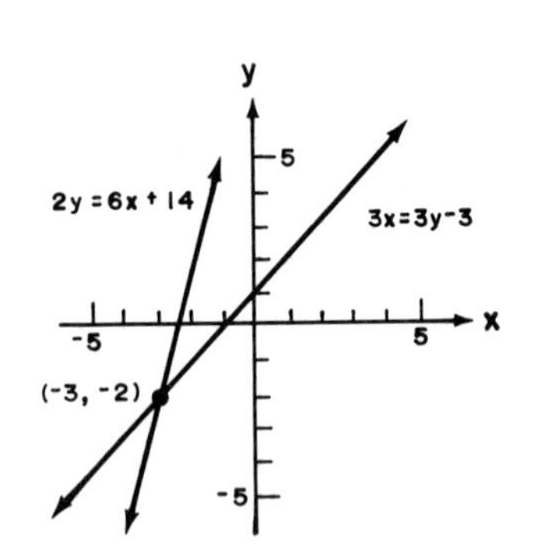

12. Add -2 times the first to 3 times the second:

$$\begin{array}{rl} -6x - 10y &= -2 \\ 6x - 9y &= 21 \\ \hline -19y &= 19;\ y = -1 \end{array}$$

Substitute -1 for y in the first equation:

$3x + 5(-1) = 1;\ x = 2$

The solution set is $\{(2,-1)\}$.

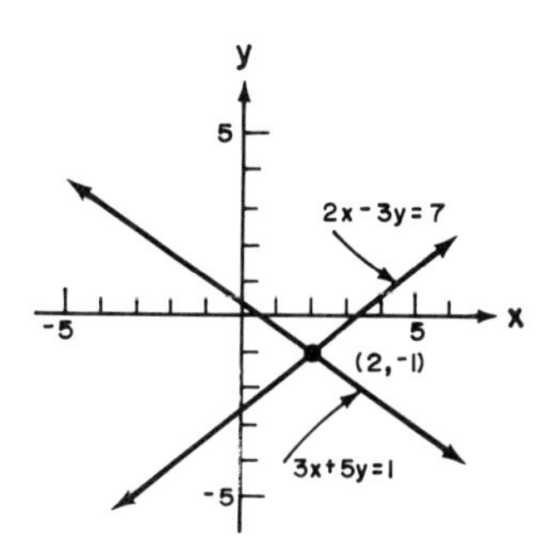

14. Substitute -3 for x in the first equation:

$2(-3) - y = 0;\ -6 = y.$

The solution set is $\{(-3,-6)\}$.

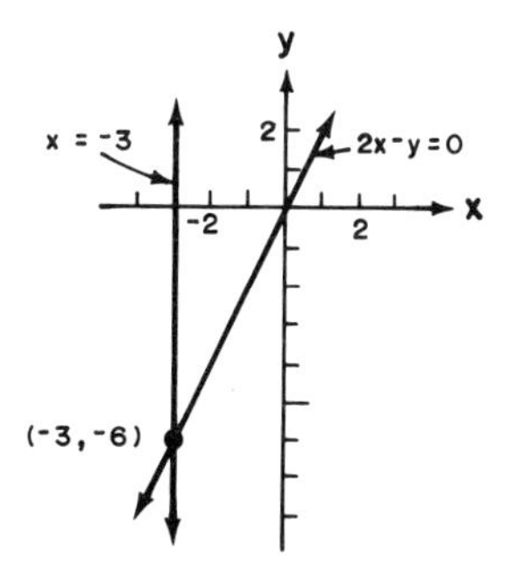

16. Substitute 2 for x in the first equation:

$2 + 2y = 6;\ 2y = 4;\ y = 2.$

The solution set is $\{(2,2)\}$.

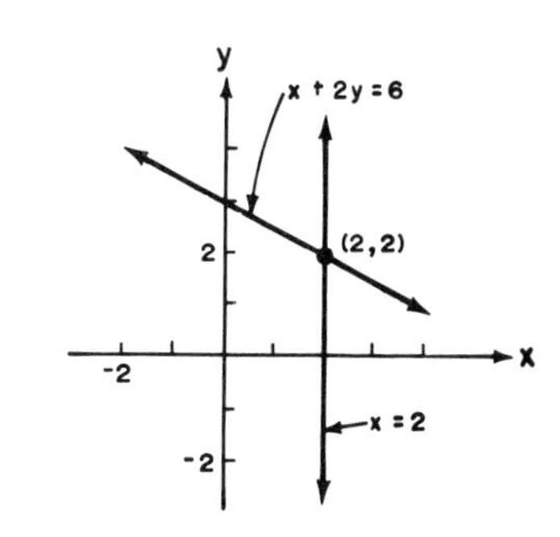

18. $\frac{2}{3}x - y = 4$ (A)

$x - \frac{3}{4}y = 6$ (B).

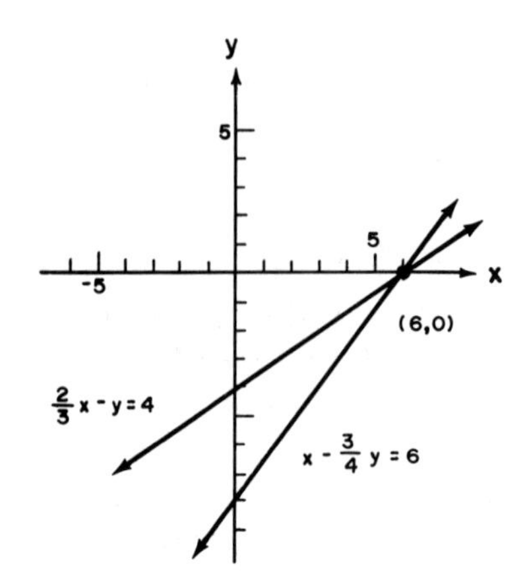

Multiply Equation (A) by 3 and Equation (B) by 4:

$2x - 3y = 12$ (A')
$4x - 3y = 24$ (B')

Add (−1) times Equation (A') to Equation (B'):

$2x = 12;\ x = 6.$

Substitute 6 for x in Equation (A'):

$2(6) - 3y = 12;$
$12 - 12 = 3y;$
$0 = 3y;\ 0 = y.$

The solution set is $\{(6,0)\}$.

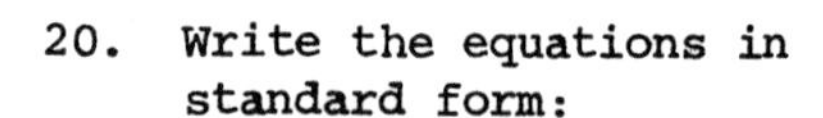

20. Write the equations in standard form:

$\frac{1}{3}x - \frac{2}{3}y = 2$ (A)

$x - 2y = 6$ (B)

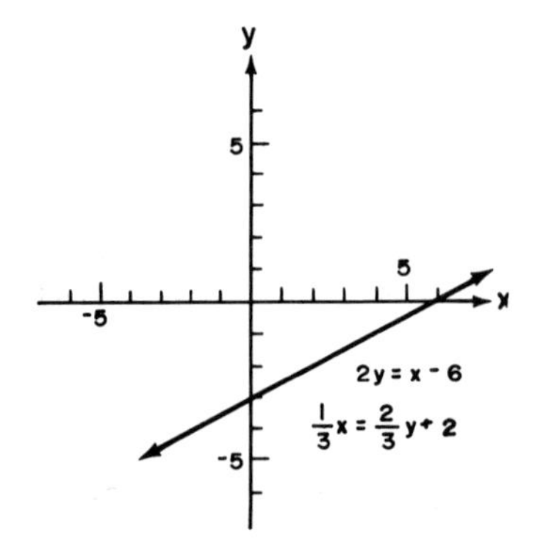

Multiply Equation (A) by 3:

$x - 2y = 6$ (A')
$x - 2y = 6$ (B').

Note that

$$\frac{1}{1} = \frac{-2}{-2} = \frac{6}{6}.$$

Hence, from Property 9, the system has an infinite number of solutions.

22. $3x - 2y = 6$
$6x - 4y = 8$

Note that

$$\frac{3}{6} = \frac{-2}{-4} \neq \frac{6}{8}.$$

Hence, from Property 8, the system has no solutions.

24. $6x + 2y = 1$
$12x + 4y = 3$

Note that

$$\frac{6}{12} = \frac{2}{4} = \frac{1}{2}.$$

Hence, from Property 9, the system has an infinite number of solutions.

26. $\frac{2}{1} \neq \frac{1}{-3}.$

Hence, from Property 7, the system has one solution.

28. In standard form, the system is

$$-3x + y = 4$$
$$6x - 2y = 4.$$
$$\frac{-3}{6} = \frac{1}{-2} \neq \frac{4}{4}.$$

Hence, from Property 8, the system has no solutions.

30. In standard form the system is

$$x - y = -2$$
$$-3x + 3y = 6.$$
$$\frac{1}{-3} = \frac{-1}{3} = \frac{-2}{6}.$$

Hence, from Property 9, the system has an infinite number of solutions.

32. In standard form the system is

$$2x - y = 3$$
$$-x + 2y = -3$$
$$\frac{2}{-1} \neq \frac{-1}{2}.$$

Hence, from Property 7, the system has one solution.

34.

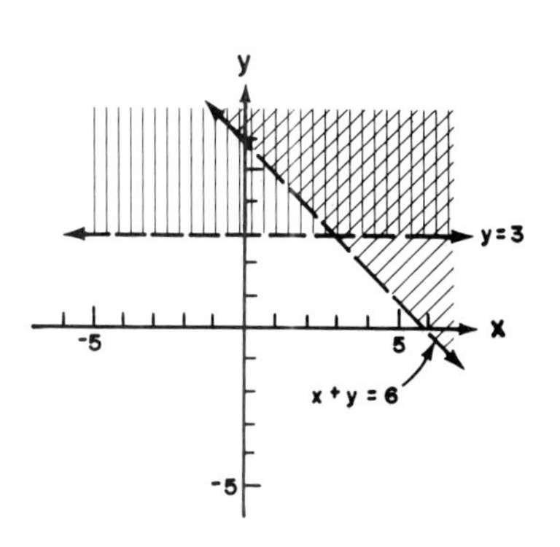

36.

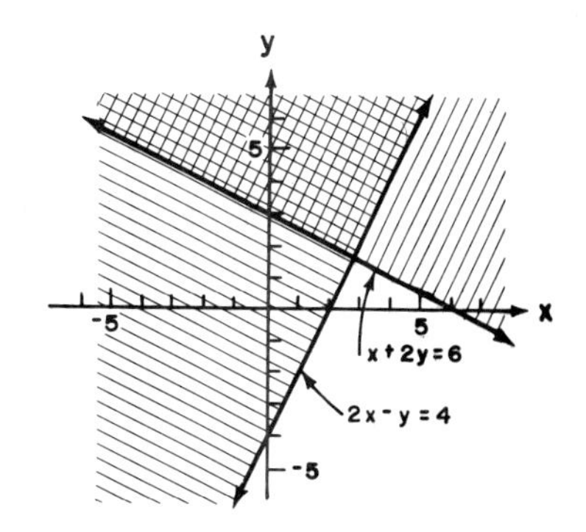

38.

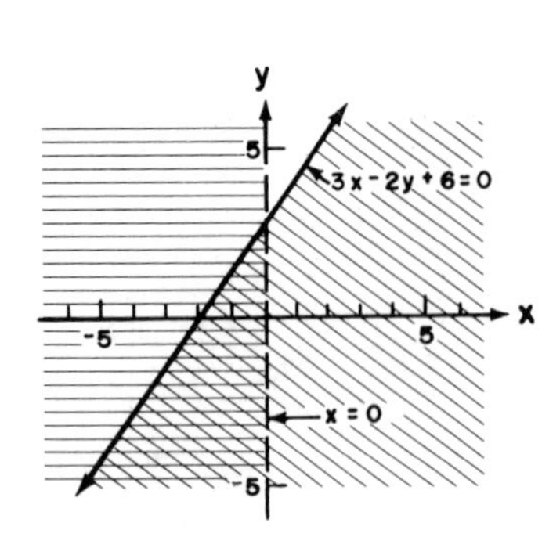

40.

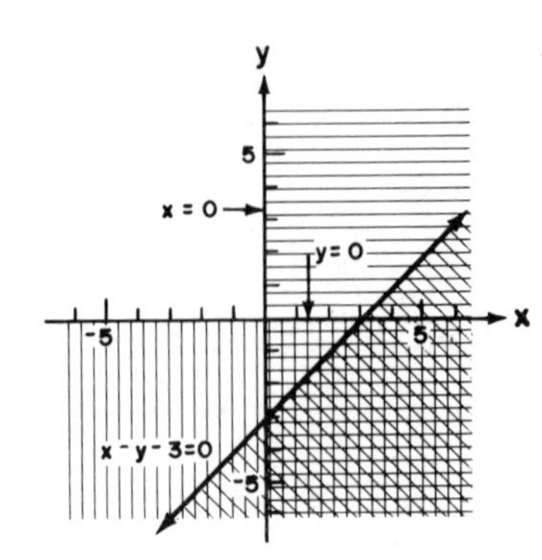

Vertices: (0,-3), (3,0)

42.

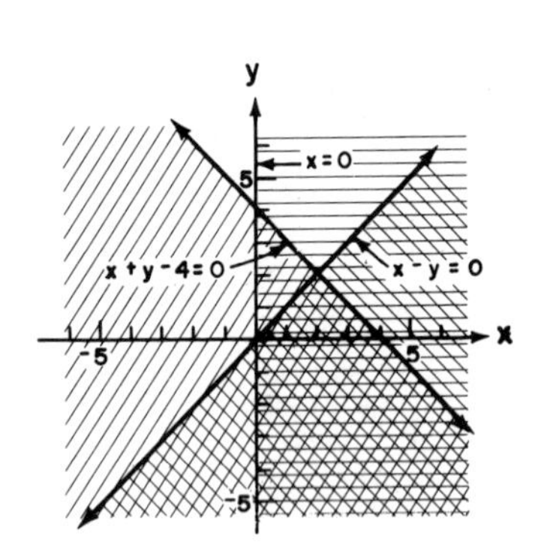

Vertices: (0,0), (2,2)

44.

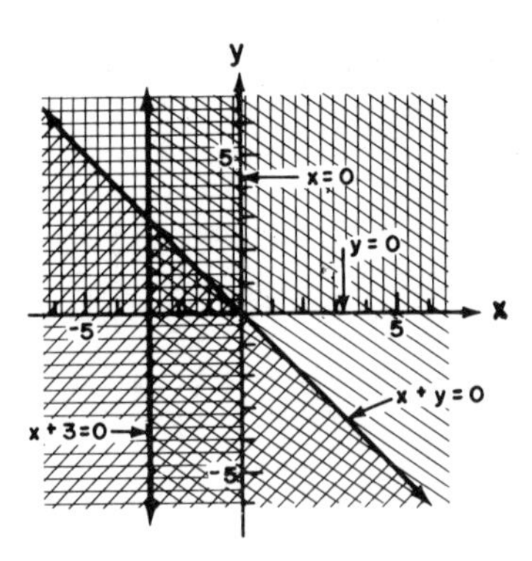

Vertices: (0,0), (-3,0), (-3,3)

46. Substitute u for $\frac{1}{x}$ and v for $\frac{1}{y}$.

$$u + 2v = \frac{-11}{12} \quad \text{or} \quad 12u + 24v = -11 \qquad (A)$$

$$u + v = \frac{-7}{12} \quad \text{or} \quad 12u + 12v = -7 \qquad (B)$$

Multiply Equation (B) by -1 and add to Equation (A):

$12v = -4;\ v = \frac{-1}{3}$.

Substitute $\frac{-1}{3}$ for v in Equation (B):

$12u + 12\left(\frac{-1}{3}\right) = -7;\ u = \frac{-3}{12} = \frac{-1}{4}$.

Thus, $u = \frac{1}{x} = \frac{-1}{4};\ x = -4$, and $v = \frac{1}{y} = \frac{-1}{3};\ y = -3$.

The solution set is $\{(-4,-3)\}$.

48. Substitute u for $\frac{1}{x}$ and v for $\frac{1}{y}$.

$$u + 2v = 11 \quad (A)$$
$$u - 2v = -1 \quad (B)$$

Add Equations (A) and (B): $2u = 10$; $u = 5$.

Substitute 5 for u in Equation (A): $5 + 2v = 11$; $v = 3$.

Thus, $u = \frac{1}{x} = 5$; $x = \frac{1}{5}$, and $v = \frac{1}{y} = 3$; $y = \frac{1}{3}$.

The solution set is $\left\{\left(\frac{1}{5}, \frac{1}{3}\right)\right\}$.

50. Think of each equation as: $\frac{2}{3}\left(\frac{1}{x}\right) + \frac{3}{4}\left(\frac{1}{y}\right) = \frac{7}{12}$; $4\left(\frac{1}{x}\right) - \frac{3}{4}\left(\frac{1}{y}\right) = \frac{7}{4}$

Substitute u for $\frac{1}{x}$ and v for $\frac{1}{y}$.

$$\frac{2}{3}u + \frac{3}{4}v = \frac{7}{12}; \quad 8u + 9v = 7 \quad (A)$$
$$4u - \frac{3}{4}v = \frac{7}{4}; \quad 16u - 3v = 7 \quad (B)$$

Multiply Equation (B) by 3 and add to Equation (A): $56u = 28$; $u = \frac{1}{2}$.

Multiply Equation (A) by -2 and add to Equation (B): $-21v = -7$; $v = \frac{1}{3}$.

Thus, $u = \frac{1}{x} = \frac{1}{2}$; $x = 2$, and $v = \frac{1}{y} = \frac{1}{3}$; $y = 3$.

The solution set is $\{(2,3)\}$.

EXERCISE 9.2

NOTE: For these problems there are methods of solution other than those shown.

2.
$$x + y + z = 1 \quad (1)$$
$$2x - y + 3z = 2 \quad (2)$$
$$2x - y - z = 2 \quad (3)$$

Add Equations (1) and (2): $3x + 4z = 3$ (4)
Add Equations (1) and (3): $3x = 3$; or $x = 1$.

Substitute 1 for x in Equation (4): $3(1) + 4z = 3$; $z = 0$.
Substitute 1 for x and 0 for z in Equation (1): $1 + y + 0 = 1$; $y = 0$.

The solution set is $\{(1,0,0)\}$.

4. $x - 2y + 4z = -3$ (1)
$3x + y - 2z = 12$ (2)
$2x + y - 3z = 11$ (3)

Add 2 times Equation (2) to Equation (1): $7x = 21$; $x = 3$. (4)
Add 2 times Equation (3) to Equation (1): $5x - 2z = 19$. (5)

Substitute 3 for x in Equation (5): $5(3) - 2z = 19$; $-2z = 4$; $z = -2$.
Substitute 3 for x and -2 for z in Equation (2): $3(3) + y - 2(-2) = 12$;
$9 + y + 4 = 12$; $y = -1$.

The solution set is $\{(3,-1,-2)\}$.

6. $x + 5y - z = 2$ (1)
$3x - 9y + 3z = 6$ (2)
$x - 3y - z = -6$ (3)

Add 3 times Equation (1) to Equation (2):

$$6x + 6y = 12 \quad \text{or} \quad x + y = 2 \quad (4)$$

Add Equation (1) to -1 times Equation (3):

$$8y = 8; \; y = 1$$

Substitute 1 for y in Equation (4):

$$x + 1 = 2; \; x = 1$$

Substitute 1 for y and 1 for x in Equation (1):

$$1 + 5(1) - z = 2; \; z = 4.$$

The solution set is $\{(1,1,4)\}$.

8. $x - 2y + 4z = 10$ (1)
$2x + 3y - z = -7$ (2)
$x - y + 2z = 4$ (3)

Add -2 times Equation (1) to Equation (2):

$$7y - 9z = -27 \quad (4)$$

Add -2 times Equation (3) to Equation (2):

$$5y - 5z = -15 \quad (5)$$

Add 5 times Equation (4) to -7 times Equation (5):

$$-10z = -30; \; z = 3$$

Substitute 3 for z in Equation (5):

$$5y - 5(3) = -15; \; y = 0$$

8. cont'd.

Substitute 3 for z and 0 for y in Equation (1):

$$x - 2(0) + 4(3) = 10; \; x = -2$$

The solution set is $\{(-2,0,3)\}$.

10. $$\begin{aligned} 2x + z &= 5 \quad (1) \\ 3y + 2z &= 6 \quad (2) \\ x - 2z &= 10 \quad (3) \end{aligned}$$

Add −2 times Equation (3) to Equation (1):

$$5z = -15; \; z = -3$$

Substitute −3 for z in Equation (3):

$$x - 2(-3) = 10; \; x = 4$$

Substitute −3 for z in Equation (2):

$$3y + 2(-3) = 6; \; y = 4$$

The solution set is $\{(4,4,-3)\}$.

12. $$\begin{aligned} 3y + z &= 3 \quad (1) \\ -2x + 3y &= 7 \quad (2) \\ 3x - 2z &= -6 \quad (3) \end{aligned}$$

Add 2 times Equation (1) to Equation (3):

$$\begin{aligned} 3x + 6y &= 0 \quad (4) \\ -2x + 3y &= 7 \quad (2) \end{aligned}$$

Add −2 times Equation (2) to Equation (4):

$$7x = -14; \; x = -2$$

Substitute −2 for x in Equation (3):

$$3(-2) - 2z = -6; \; z = 0$$

Substitute 0 for z in Equation (1):

$$3y + 0 = 3; \; y = 1$$

The solution set is $\{(-2,1,0)\}$.

14. $x + 2y + \frac{1}{2}z = 0$ (1)

$x + \frac{3}{5}y - \frac{2}{5}z = \frac{1}{5}$ (2)

$4x - 7y - 7z = 6$ (3)

Multiply Equation (1) by 2 and Equation (2) by 5.

$$2x + 4y + z = 0 \quad (4)$$
$$5x + 3y - 2z = 1 \quad (5)$$
$$4x - 7y - 7z = 6 \quad (3)$$

Add 2 times Equation (4) to Equation (5):

$$9x + 11y = 1. \quad (6)$$

Add 7 times Equation (4) to Equation (3):

$$18x + 21y = 6. \quad (7)$$

Add -2 times Equation (6) to Equation (7): $-y = 4$; $y = -4$.

Substitute -4 for y in Equation (6): $9x + 11(-4) = 1$; $9x = 45$; $x = 5$.
Substitute 5 for x and -4 for y in Equation (4): $2(5) + 4(-4) + z = 0$;
$z = 6$.

The solution set is $\{(5,-4,6)\}$.

16. $x + y - 2z = 3$ (1)

$x - \frac{1}{3}y + \frac{1}{3}z = \frac{5}{3}$ (2)

$\frac{1}{2}x - \frac{1}{2}y - z = \frac{3}{2}$ (3)

Multiply Equation (2) by 3 and Equation (3) by 2:

$$x + y - 2z = 3 \quad (1)$$
$$3x - y + z = 5 \quad (4)$$
$$x - y - 2z = 3 \quad (5)$$

Add Equation (1) to Equation (4) and also to Equation (5):

$$4x - z = 8 \quad (6)$$
$$2x - 4z = 6 \quad (7)$$

Add (-2) times Equation (7) to Equation (6):

$$7z = -4; \; z = \frac{-4}{7}$$

Substitute $\frac{-4}{7}$ for z in Equation (6):

$$4x - \left(\frac{-4}{7}\right) = 8; \; 28x + 4 = 56; \quad x = \frac{13}{7}$$

16. cont'd.

Substitute $\frac{13}{7}$ for x and $\frac{-4}{7}$ for z in Equation (1):

$$\frac{13}{7} + y - 2\left(\frac{-4}{7}\right) = 3;\ 13 + 7y + 8 = 21;\ y = 0$$

The solution set is $\left\{\left(\frac{13}{7}, 0, \frac{-4}{7}\right)\right\}$.

18. $x = y + \frac{1}{2}$ (1)

$y = z + \frac{5}{4}$ (2)

$2z = x - \frac{7}{4}$ (3)

Multiply Equations (1), (2), and (3) by 2, 4, and 4, respectively:

$$2x = 2y + 1 \quad (4)$$
$$4y = 4z + 5 \quad (5)$$
$$8z = 4x - 7 \quad (6)$$

Rewrite the system with only the constants on the right side of each equation.

$$2x - 2y = 1 \quad (7)$$
$$4y - 4z = 5 \quad (8)$$
$$-4x + 8z = -7 \quad (9)$$

Add 2 times Equation (8) to Equation (9):

$$-4x + 8y = 3 \quad (10)$$
$$2x - 2y = 1 \quad (7)$$

Add 2 times Equation (7) to Equation (10):

$$4y = 5;\ y = \frac{5}{4}.$$

Substitute $\frac{5}{4}$ for y in Equation (1):

$$x = \frac{5}{4} + \frac{1}{2};\ x = \frac{7}{4}$$

Substitute $\frac{7}{4}$ for x in Equation (3):

$$2z = \frac{7}{4} - \frac{7}{4};\ z = 0.$$

The solution set is $\left\{\left(\frac{7}{4}, \frac{5}{4}, 0\right)\right\}$.

20.
$$\begin{aligned} x + 3y - z &= 4 \quad (1) \\ -2x - 6y + 2z &= 1 \quad (2) \\ x + 2y - z &= 3 \quad (3) \end{aligned}$$

Add 2 times Equation (1) to Equation (2):

$$0 = 9$$

Since the linear combination is a contradiction, there is no unique solution.

22.
$$\begin{aligned} 3x + 6y + 2z &= -2 \quad (1) \\ \frac{1}{2}x - 3y - z &= 1 \quad (2) \\ 4x + y + \frac{1}{3}z &= -\frac{1}{3} \quad (3) \end{aligned}$$

Multiply Equation (2) by 2 and Equation (3) by 3:

$$\begin{aligned} 3x + 6y + 2z &= -2 \quad (1) \\ x - 6y - 2z &= 1 \quad (4) \\ 12x + 3y + z &= -1 \quad (5) \end{aligned}$$

Add Equation (1) to Equation (4):

$$4x = -1 \quad (6)$$

Add 2 times Equation (5) to Equation (4):

$$25x = -1 \quad (7)$$

Add 25 times Equation (6) to -4 times Equation (7):

$$0 = -21$$

Since the linear combination is a contradiction, there is no unique solution.

24.
$$\begin{aligned} x - 2y + z &= 5 \quad (1) \\ -x + y &= -2 \quad (2) \\ y - z &= -3 \quad (3) \end{aligned}$$

Add Equation (1) to Equation (2):

$$\begin{aligned} -y + z &= 3 \quad (4) \\ y - z &= -3 \quad (5) \end{aligned}$$

Add Equations (4) and (5):

$$0 = 0$$

Since the linear combination vanishes, there is no unique solution.

26. $x = y + z$ (1)
$y = 2x - z$ (2)
$z = 3x - y$ (3)

Rewrite the equations:

$$x - y - z = 0 \quad (4)$$
$$-2x + y + z = 0 \quad (5)$$
$$-3x + y + z = 0 \quad (6)$$

Add Equation (4) to Equation (5):

$$-x = 0; \; x = 0 \quad (7)$$

Substitute 0 for x in Equations (4) and (5):

$$-y - z = 0 \quad (8)$$
$$y + z = 0 \quad (9)$$

Add Equations (8) and (9):

$$0 = 0$$

Since the linear combination vanishes, there is no unique solution.

28. $x = \frac{1}{2}y - \frac{1}{2}z + 1$ (1)

$x = 2y + z - 1$ (2)

$x = \frac{1}{2}y - \frac{1}{2}z + \frac{1}{4}$ (3)

Add -1 times Equation (1) to Equation (3):

$$0 = -\frac{3}{4}$$

Since the linear combination is a contradiction, there is no unique solution.

30. $3x + y = 1$ (1)
$2x - y + z = -1$ (2)
$x - 3y - z = \frac{-2}{3}$ (3)

Add Equation (1) and Equation (2):

$$5x + z = 0 \quad (4)$$

Add 3 times Equation (1) to Equation (3):

$$10x - z = \frac{7}{3} \quad (5)$$

Add Equations (4) and (5):

$$15x = \frac{7}{3}; \; x = \frac{7}{45}$$

30. cont'd.

Substitute $\frac{7}{45}$ for x in Equation (1):

$$3\left(\frac{7}{45}\right) + y = 1;\ 21 + 45y = 45;\ y = \frac{8}{15}$$

Substitute $\frac{7}{45}$ for x and $\frac{8}{15}$ for y in Equation (2):

$$2\left(\frac{7}{45}\right) - \frac{8}{15} + z = -1,\ 14 - 24 + 45z = -45;\ z = \frac{-7}{9}$$

The solution set is $\left\{\left(\frac{7}{45}, \frac{8}{15}, \frac{-7}{9}\right)\right\}$.

32. $\frac{4}{x} - \frac{2}{y} + \frac{1}{z} = 4$ (1)

$\frac{3}{x} - \frac{1}{y} + \frac{2}{z} = 0$ (2)

$-\frac{1}{x} + \frac{3}{y} - \frac{2}{z} = 0$ (3)

Let $\frac{1}{x} = u$, $\frac{1}{y} = v$, and $\frac{1}{z} = w$ in Equations (1), (2), and (3).

$$\begin{aligned} 4u - 2v + w &= 4 \quad (4) \\ 3u - v + 2w &= 0 \quad (5) \\ -u + 3v - 2w &= 0 \quad (6) \end{aligned}$$

Add 2 times Equation (4) to Equation (6):

$$7u - v = 8 \quad (7)$$

Add Equation (5) to Equation (6):

$$2u + 2v = 0 \quad (8)$$

Add 2 times Equation (7) to Equation (8):

$$16u = 16;\ u = 1$$

Substitute 1 for u in Equation (8):

$$2(1) + 2v = 0;\ v = -1$$

Substitute 1 for u and -1 for v in Equation (4):

$$4(1) - 2(-1) + w = 4;\ w = -2$$

Substituting 1 for u, -1 for v, and -2 for w in $\frac{1}{x} = u$, $\frac{1}{y} = v$, and $\frac{1}{z} = w$, we have

$$\frac{1}{x} = 1,\ \frac{1}{y} = -1,\ \text{and}\ \frac{1}{z} = -2$$

from which $x = 1$, $y = -1$, and $z = -\frac{1}{2}$.

The solution set is $\{(1,-1,-\frac{1}{2})\}$.

34. $\frac{2}{x} - \frac{1}{y} - \frac{1}{z} = -1$ (1)

$\frac{4}{x} - \frac{2}{y} + \frac{1}{z} = -5$ (2)

$\frac{2}{x} + \frac{1}{y} - \frac{4}{z} = 4$ (3)

Let $\frac{1}{x} = u$, $\frac{1}{y} = v$, and $\frac{1}{z} = w$ in Equations (1), (2), and (3).

$$2u - v - w = -1 \quad (4)$$
$$4u - 2v + w = -5 \quad (5)$$
$$2u + v - 4w = 4 \quad (6)$$

Add Equations (4) and (5):

$$6u - 3v = -6 \quad (7)$$

Add -4 times Equation (4) to Equation (6):

$$-6u + 5v = 8 \quad (8)$$

Add Equations (7) and (8):

$$2v = 2; \ v = 1$$

Substitute 1 for v in Equation (7):

$$6u - 3(1) = -6; \ u = -\frac{1}{2}$$

Substitute 1 for v and $-\frac{1}{2}$ for u in Equation (4):

$$2\left(-\frac{1}{2}\right) - (1) - w = -1; \quad w = -1$$

Substituting $-\frac{1}{2}$ for u, 1 for v, and -1 for w in $\frac{1}{x} = u$, $\frac{1}{y} = v$, and $\frac{1}{z} = w$, we have

$$\frac{1}{x} = -\frac{1}{2}, \frac{1}{y} = 1, \text{ and } \frac{1}{z} = -1$$

from which x = -2, y = 1, and z = -1.

The solution set is $\{(-2,1,-1)\}$.

EXERCISE 9.3

2. $(3)(1) - (4)(-2) = 11$

4. $(1)(2) - (-1)(-2) = 0$

6. $(20)(-2) - (-20)(3) = 20$

8. $(-1)(-6) - (-2)(-5) = -4$

10. $D = \begin{vmatrix} 3 & -4 \\ 1 & -2 \end{vmatrix} = (3)(-2) - (1)(-4) = -2$

$D_x = \begin{vmatrix} -2 & -4 \\ 0 & -2 \end{vmatrix} = (-2)(-2) - (0)(-4) = 4$

$D_y = \begin{vmatrix} 3 & -2 \\ 1 & 0 \end{vmatrix} = (3)(0) - (1)(-2) = 2$

$x = \frac{D_x}{D} - \frac{4}{-2} = -2; \; y = \frac{D_y}{D} = \frac{2}{-2} = -1$

The solution set is $\{(-2,-1)\}$.

12. $D = \begin{vmatrix} 2 & -4 \\ 1 & -2 \end{vmatrix} = (2)(-2) - (1)(-4) = 0$

Since D = 0, the given equations are dependent or inconsistent and there is no unique solution.

14. $D = \begin{vmatrix} \frac{2}{3} & 1 \\ 1 & -\frac{4}{3} \end{vmatrix} = \left(\frac{2}{3}\right)\left(-\frac{4}{3}\right) - (1)(1) = \frac{-17}{9}$

$D_x = \begin{vmatrix} 1 & 1 \\ 0 & -\frac{4}{3} \end{vmatrix} = (1)\left(-\frac{4}{3}\right) - (0)(1) = \frac{-4}{3}$

$D_y = \begin{vmatrix} \frac{2}{3} & 1 \\ 1 & 0 \end{vmatrix} = \left(\frac{2}{3}\right)(0) - (1)(1) = -1$

$x = \frac{D_x}{D} = \frac{\frac{-4}{3}}{\frac{-17}{9}} = \frac{12}{17}; \; y = \frac{D_y}{D} = \frac{-1}{\frac{-17}{9}} = \frac{9}{17}$

The solution set is $\left\{\left(\frac{12}{17}, \frac{9}{17}\right)\right\}$.

16. $D = \begin{vmatrix} \frac{1}{2} & 1 \\ \frac{-1}{4} & -1 \end{vmatrix} = \left(\frac{1}{2}\right)(-1) - \left(\frac{-1}{4}\right)(1) = \frac{-1}{4}$

$D_x = \begin{vmatrix} 3 & 1 \\ -3 & -1 \end{vmatrix} = (3)(-1) - (-3)(1) = 0$

$D_y = \begin{vmatrix} \frac{1}{2} & 3 \\ \frac{-1}{4} & -3 \end{vmatrix} = \left(\frac{1}{2}\right)(-3) - \left(\frac{-1}{4}\right)(3) = \frac{-3}{4}$

$x = \frac{D_x}{D} = \frac{0}{\frac{-1}{4}} = 0; \; y = \frac{D_y}{D} = \frac{\frac{-3}{4}}{\frac{-1}{4}} = 3$

The solution set is $\{(0,3)\}$.

18. $D = \begin{vmatrix} 2 & -3 \\ 1 & 0 \end{vmatrix} = (2)(0) - (1)(-3) = 3$

$D_x = \begin{vmatrix} 12 & -3 \\ 4 & 0 \end{vmatrix} = (12)(0) - (4)(-3) = 12$

$D_y = \begin{vmatrix} 2 & 12 \\ 1 & 4 \end{vmatrix} = (2)(4) - (1)(12) = -4$

$x = \frac{D_x}{D} = \frac{12}{3} = 4; \; y = \frac{D_y}{D} = \frac{-4}{3}$

The solution set is $\left\{\left(4, \frac{-4}{3}\right)\right\}$.

20. $D = \begin{vmatrix} 1 & 1 \\ 1 & -1 \end{vmatrix} = (1)(-1) - (1)(1) = -1 - 1 = -2$

$D_x = \begin{vmatrix} a & 1 \\ b & -1 \end{vmatrix} = (a)(-1) - (b)(1) = -a - b$

$D_y = \begin{vmatrix} 1 & a \\ 1 & b \end{vmatrix} = (1)(b) - (1)(a) = b - a$

$x = \frac{D_x}{D} = \frac{-a - b}{-2} = \frac{a + b}{2}; \; y = \frac{D_y}{D} = \frac{b - a}{-2} = \frac{a - b}{2}$

The solution set is $\left\{\left(\frac{a + b}{2}, \frac{a - b}{2}\right)\right\}$.

22. $-\begin{vmatrix} a_2 & b_2 \\ a_1 & b_1 \end{vmatrix} = -(a_2b_1 - a_1b_2)$

$\begin{vmatrix} a_1 & b_1 \\ a_2 & b_2 \end{vmatrix} = a_1b_2 - a_2b_1 = -(a_2b_1 - a_1b_2) = -\begin{vmatrix} a_2 & b_2 \\ a_1 & b_1 \end{vmatrix}$

24. $\begin{vmatrix} ka_1 & b_1 \\ ka_2 & b_2 \end{vmatrix} = ka_1b_2 - ka_2b_1 = k(a_1b_2 - a_2b_1) = k\begin{vmatrix} a_1 & b_1 \\ a_2 & b_2 \end{vmatrix}$

26. $\begin{vmatrix} a_1 & b_1 \\ ka_2 & kb_2 \end{vmatrix} = ka_1b_2 - ka_2b_1$

$= k(a_1b_2 - a_2b_1)$

$= k\begin{vmatrix} a_1 & b_1 \\ a_2 & b_2 \end{vmatrix}$

28. $D = \begin{vmatrix} a_1 & b_1 \\ a_2 & b_2 \end{vmatrix} = a_1b_2 - a_2b_1$

$D_x = \begin{vmatrix} c_1 & b_1 \\ c_2 & b_2 \end{vmatrix} = c_1b_2 - c_2b_1$; $D_y = \begin{vmatrix} a_1 & c_1 \\ a_2 & c_2 \end{vmatrix} = a_1c_2 - a_2c_1$

If $D = 0$, then $a_1b_2 - a_2b_1 = 0$ or $a_1b_2 = a_2b_1$. Thus,

$$\frac{a_1}{a_2} = \frac{b_1}{b_2} .$$

If $D_x = 0$, then $c_1b_2 - c_2b_1 = 0$ or $c_2b_1 = c_1b_2$. Thus,

$$\frac{b_1}{b_2} = \frac{c_1}{c_2} .$$

Hence, $\frac{a_1}{a_2} = \frac{c_1}{c_2}$ or $a_1c_2 = a_2c_1$. Thus, $a_1c_2 - a_2c_1 = 0$
$= D_y$.

EXERCISE 9.4

2. Expand about the third row:

$$0\begin{vmatrix} 3 & 1 \\ 2 & 1 \end{vmatrix} - 2\begin{vmatrix} 1 & 1 \\ -1 & 1 \end{vmatrix} + 0\begin{vmatrix} 1 & 3 \\ -1 & 2 \end{vmatrix} = 0 - 2(2) + 0 = -4.$$

4. Expand about the third row:

$$4\begin{vmatrix} 4 & -1 \\ 3 & 2 \end{vmatrix} - 0\begin{vmatrix} 2 & -1 \\ -1 & 2 \end{vmatrix} + 2\begin{vmatrix} 2 & 4 \\ -1 & 3 \end{vmatrix} = 4(11) + 0 + 2(10) = 64.$$

6. Expand about the first row:

$$1\begin{vmatrix} 1 & 2 \\ 3 & 4 \end{vmatrix} - 0\begin{vmatrix} 0 & 2 \\ 0 & 4 \end{vmatrix} + 0\begin{vmatrix} 0 & 1 \\ 0 & 3 \end{vmatrix} = 1(-2) + 0 + 0 = -2$$

8. Expand about the second column:

$$-1\begin{vmatrix} 3 & 6 \\ 5 & 10 \end{vmatrix} + 2\begin{vmatrix} 2 & 4 \\ 5 & 10 \end{vmatrix} - (-3)\begin{vmatrix} 2 & 4 \\ 3 & 6 \end{vmatrix} = -1(0) + 2(0) + 3(0) = 0.$$

10. Expand about the second row:

$$-0\begin{vmatrix} 3 & 1 \\ 2 & 1 \end{vmatrix} + 1\begin{vmatrix} 2 & 1 \\ -4 & 1 \end{vmatrix} - 0\begin{vmatrix} 2 & 3 \\ -4 & 2 \end{vmatrix} = 0 + 1(6) + 0 = 6.$$

12. Expand about the first row:

$$a\begin{vmatrix} 2 & 3 \\ 5 & 6 \end{vmatrix} - a\begin{vmatrix} 1 & 3 \\ 4 & 6 \end{vmatrix} + a\begin{vmatrix} 1 & 2 \\ 4 & 5 \end{vmatrix} = a(-3) - a(-6) + a(-3) = -3a + 6a - 3a = 0.$$

14. Expand about the first row:

$$0\begin{vmatrix} x & 0 \\ 0 & 0 \end{vmatrix} - 0\begin{vmatrix} 0 & 0 \\ x & 0 \end{vmatrix} + x\begin{vmatrix} 0 & x \\ x & 0 \end{vmatrix} = 0 + 0 + x(-x^2) = -x^3.$$

16. Expand about the first column:

$$0\begin{vmatrix} 0 & a \\ a & 0 \end{vmatrix} - a\begin{vmatrix} a & b \\ a & 0 \end{vmatrix} + b\begin{vmatrix} a & b \\ 0 & a \end{vmatrix} = 0 - a(-ab) + b(a^2) = 2a^2b.$$

18. Expand about the first row:

$$0\begin{vmatrix} a & b \\ b & 0 \end{vmatrix} - b\begin{vmatrix} b & b \\ 0 & 0 \end{vmatrix} + 0\begin{vmatrix} b & a \\ 0 & b \end{vmatrix} = 0 - b(0) + 0(b^2) = 0.$$

20. $D = \begin{vmatrix} 2 & -6 & 3 \\ 3 & -2 & 5 \\ 4 & 5 & -2 \end{vmatrix} = -129$ $\quad D_x = \begin{vmatrix} -12 & -6 & 3 \\ -4 & -2 & 5 \\ 10 & 5 & -2 \end{vmatrix} = 0$

$D_y = \begin{vmatrix} 2 & -12 & 3 \\ 3 & -4 & 5 \\ 4 & 10 & -2 \end{vmatrix} = -258$ $\quad D_z = \begin{vmatrix} 2 & -6 & -12 \\ 3 & -2 & -4 \\ 4 & 5 & 10 \end{vmatrix} = 0$

$$x = \frac{D_x}{D} = 0; \quad y = \frac{D_y}{D} = 2; \quad z = \frac{D_z}{D} = 0.$$

The solution set is {(0,2,0)}.

22. $D = \begin{vmatrix} 2 & 0 & 5 \\ 4 & 3 & 0 \\ 0 & 3 & -4 \end{vmatrix} = 36$ $\quad D_x = \begin{vmatrix} 9 & 0 & 5 \\ -1 & 3 & 0 \\ -13 & 3 & -4 \end{vmatrix} = 72$

$D_y = \begin{vmatrix} 2 & 9 & 5 \\ 4 & -1 & 0 \\ 0 & -13 & -4 \end{vmatrix} = -108$ $\quad D_z = \begin{vmatrix} 2 & 0 & 9 \\ 4 & 3 & -1 \\ 0 & 3 & -13 \end{vmatrix} = 36$

$$x = \frac{D_x}{D} = 2; \quad y = \frac{D_y}{D} = -3; \quad z = \frac{D_z}{D} = 1$$

The solution set is {(2,-3,1)}.

24. $D = \begin{vmatrix} 4 & 8 & 1 \\ 2 & -3 & 2 \\ 1 & 7 & -3 \end{vmatrix} = 61$ $\quad D_x = \begin{vmatrix} -6 & 8 & 1 \\ 0 & -3 & 2 \\ -8 & 7 & -3 \end{vmatrix} = -122$

$D_y = \begin{vmatrix} 4 & -6 & 1 \\ 2 & 0 & 2 \\ 1 & -8 & -3 \end{vmatrix} = 0$ $\quad D_z = \begin{vmatrix} 4 & 8 & -6 \\ 2 & -3 & 0 \\ 1 & 7 & -8 \end{vmatrix} = 122$

$$x = \frac{D_x}{D} = -2; \quad y = \frac{D_y}{D} = 0; \quad z = \frac{D_z}{D} = 2$$

The solution set is {(-2,0,2)}.

26. $D = \begin{vmatrix} 1 & 1 & -2 \\ 3 & -1 & 1 \\ 3 & 3 & -6 \end{vmatrix} = 0$

Since D = 0, there is no unique solution.

28. $D = \begin{vmatrix} 3 & -2 & 5 \\ 4 & -4 & 3 \\ 5 & -4 & 1 \end{vmatrix} = 22$ $\quad D_x = \begin{vmatrix} 6 & -2 & 5 \\ 0 & -4 & 3 \\ -5 & -4 & 1 \end{vmatrix} = -22$

$D_y = \begin{vmatrix} 3 & 6 & 5 \\ 4 & 0 & 3 \\ 5 & -5 & 1 \end{vmatrix} = 11$ $\quad D_z = \begin{vmatrix} 3 & -2 & 6 \\ 4 & -4 & 0 \\ 5 & -4 & -5 \end{vmatrix} = 44$

$x = \frac{D_x}{D} = -1;\ y = \frac{D_y}{D} = \frac{1}{2};\ z = \frac{D_z}{D} = 2.$

The solution set is $\left\{\left(-1, \frac{1}{2}, 2\right)\right\}$.

30. Multiply the first equation by 3 and the second equation by 12.

$$6x - 2y + 3z = 6$$
$$6x - 4y - 3z = 0$$
$$4x + 5y - 3z = -1$$

$D = \begin{vmatrix} 6 & -2 & 3 \\ 6 & -4 & -3 \\ 4 & 5 & -3 \end{vmatrix} = 288$ $\quad D_x = \begin{vmatrix} 6 & -2 & 3 \\ 0 & -4 & -3 \\ -1 & 5 & -3 \end{vmatrix} = 144$

$D_y = \begin{vmatrix} 6 & 6 & 3 \\ 6 & 0 & -3 \\ 4 & -1 & -3 \end{vmatrix} = 0$ $\quad D_z = \begin{vmatrix} 6 & -2 & 6 \\ 6 & -4 & 0 \\ 4 & 5 & -1 \end{vmatrix} = 288$

$x = \frac{D_x}{D} = \frac{1}{2};\ y = \frac{D_y}{D} = 0;\ z = \frac{D_z}{D} = 1$

The solution set is $\left\{\left(\frac{1}{2}, 0, 1\right)\right\}$.

32. $D = \begin{vmatrix} 2 & 1 & 0 \\ 0 & 1 & 1 \\ 3 & -2 & -5 \end{vmatrix} = -3$ $\quad D_x = \begin{vmatrix} 18 & 1 & 0 \\ -1 & 1 & 1 \\ 38 & -2 & -5 \end{vmatrix} = -21$

$D_y = \begin{vmatrix} 2 & 18 & 0 \\ 0 & -1 & 1 \\ 3 & 38 & -5 \end{vmatrix} = -12$ $\quad D_z = \begin{vmatrix} 2 & 1 & 18 \\ 0 & 1 & -1 \\ 3 & -2 & 38 \end{vmatrix} = 15$

$x = \frac{D_x}{D} = 7;\ y = \frac{D_y}{D} = 4;\ z = \frac{D_z}{D} = -5.$

The solution set is $\{(7,4,-5)\}$.

34. Expand about the first row:

$$0\begin{vmatrix} b & c \\ e & f \end{vmatrix} - 0\begin{vmatrix} a & c \\ d & f \end{vmatrix} + 0\begin{vmatrix} a & b \\ d & e \end{vmatrix} = 0 - 0 + 0 = 0$$

If each entry in a row of a third-order determinant is 0, then the determinant is zero.

36. Expand $\begin{vmatrix} 2 & 0 & 1 \\ 4 & 1 & -2 \\ 6 & 1 & 1 \end{vmatrix}$ about the second column:

$$-0\begin{vmatrix} 4 & -2 \\ 6 & 1 \end{vmatrix} + 1\begin{vmatrix} 2 & 1 \\ 6 & 1 \end{vmatrix} - 1\begin{vmatrix} 2 & 1 \\ 4 & -2 \end{vmatrix} = 0 + 1(-4) - 1(-8) = 4.$$

Expand $\begin{vmatrix} 1 & 0 & 1 \\ 2 & 1 & -2 \\ 3 & 1 & 1 \end{vmatrix}$ about the second column:

$$-0\begin{vmatrix} 2 & -2 \\ 3 & 1 \end{vmatrix} + 1\begin{vmatrix} 1 & 1 \\ 3 & 1 \end{vmatrix} - 1\begin{vmatrix} 1 & 1 \\ 2 & -2 \end{vmatrix} = 0 + 1(-2) - 1(-4) = 2.$$

Thus, $\begin{vmatrix} 2 & 0 & 1 \\ 4 & 1 & -2 \\ 6 & 1 & 1 \end{vmatrix} = 2\begin{vmatrix} 1 & 0 & 1 \\ 2 & 1 & -2 \\ 3 & 1 & 1 \end{vmatrix}$.

If a common factor is factored from each element of a column in a determinant, the resulting determinant multiplied by the common factor equals the original determinant.

38. $$\begin{vmatrix} x & y & 1 \\ 0 & b & 1 \\ 1 & m & 0 \end{vmatrix} = x\begin{vmatrix} b & 1 \\ m & 0 \end{vmatrix} - 0\begin{vmatrix} y & 1 \\ m & 0 \end{vmatrix} + 1\begin{vmatrix} y & 1 \\ b & 1 \end{vmatrix}$$

$$= x(b \cdot 0 - 1 \cdot m) - 0 + 1(y \cdot 1 - b \cdot 1)$$

$$= -mx + y - b.$$

Hence,

$$\begin{vmatrix} x & y & 1 \\ 0 & b & 1 \\ 1 & m & 0 \end{vmatrix} = 0 \text{ is equivalent to}$$

$$-mx + y - b = 0$$

or

$$y = mx + b.$$

EXERCISE 9.5

2.

$$2(\text{row } 1) + \text{row } 2 \quad \begin{bmatrix} -2 & 3 \\ 4 & 1 \end{bmatrix} \sim \begin{bmatrix} -2 & 3 \\ 0 & 7 \end{bmatrix}$$

4.

$$\frac{1}{2}(\text{row } 1) + \text{row } 2 \quad \begin{bmatrix} 6 & 4 \\ -1 & -2 \end{bmatrix} \sim \begin{bmatrix} 6 & 4 \\ 2 & 0 \end{bmatrix}$$

6.

$$\begin{matrix} \\ 2(\text{row } 1) + \text{row } 2 \\ -3(\text{row } 1) + \text{row } 3 \end{matrix} \quad \begin{bmatrix} 2 & -1 & 3 \\ -4 & 0 & 4 \\ 6 & 2 & -1 \end{bmatrix} \sim \begin{bmatrix} 2 & -1 & 3 \\ 0 & -2 & 10 \\ 0 & 5 & -10 \end{bmatrix}$$

8.

$$\begin{matrix} \\ -\frac{1}{4}(\text{row } 1) + \text{row } 3 \\ -\frac{5}{4}(\text{row } 1) + \text{row } 3 \end{matrix} \quad \begin{bmatrix} 3 & -2 & 4 \\ 2 & 2 & 1 \\ -1 & 1 & 5 \end{bmatrix} \sim \begin{bmatrix} 3 & -2 & 4 \\ \frac{5}{4} & \frac{5}{2} & 0 \\ \frac{-19}{4} & \frac{7}{2} & 0 \end{bmatrix}$$

10.

$$\begin{matrix} \\ 4(\text{row } 1) + \text{row } 2 \\ -2(\text{row } 1) + \text{row } 3 \end{matrix} \quad \begin{bmatrix} -1 & 2 & 3 \\ 4 & 0 & 1 \\ -2 & 2 & -3 \end{bmatrix} \sim \begin{bmatrix} -1 & 2 & 3 \\ 0 & 8 & 13 \\ 0 & -2 & -9 \end{bmatrix}.$$

Then,

$$\frac{1}{4}(\text{row } 2) + \text{row } 3 \quad \begin{bmatrix} -1 & 2 & 3 \\ 0 & 8 & 13 \\ 0 & -2 & -9 \end{bmatrix} \sim \begin{bmatrix} -1 & 2 & 3 \\ 0 & 8 & 13 \\ 0 & 0 & \frac{-23}{4} \end{bmatrix}.$$

12. The augmented matrix is

$$\left[\begin{array}{cc|c} 1 & -5 & 11 \\ 2 & 3 & -4 \end{array}\right].$$

An equivalent matrix is

$$\text{row } 2 + [-2 \times \text{row } 1] \longrightarrow \left[\begin{array}{cc|c} 1 & -5 & 11 \\ 0 & 13 & -26 \end{array}\right] \quad \begin{aligned} x - 5y &= 11 \\ 13y &= -26. \end{aligned}$$

Since, from the last equation, $y = -2$, -2 can be substituted for y in the first equation to obtain

$$x - 5(-2) = 11, \; x = 1.$$

The solution set is $\{(1,-2)\}$.

14. The augmented matrix is

$$\left[\begin{array}{cc|c} 1 & 6 & -14 \\ 5 & -3 & -4 \end{array}\right].$$

An equivalent matrix

$$\text{row 2} + [-5 \times \text{row 1}] \longrightarrow \left[\begin{array}{cc|c} 1 & 6 & -14 \\ 0 & -33 & 66 \end{array}\right] \quad \begin{aligned} x + 6y &= -14 \\ -33y &= 66. \end{aligned}$$

From the last equation, $y = -2$; -2 can be substituted for y in the first equation to obtain

$$x + 6(-2) = -14, \; x = -2.$$

The solution set is $\{(-2,-2)\}$.

16. The augmented matrix is

$$\left[\begin{array}{cc|c} 3 & -2 & 16 \\ 4 & 2 & 12 \end{array}\right].$$

An equivalent matrix is

$$\text{row 1} + \text{row 2} \longrightarrow \left[\begin{array}{cc|c} 3 & -2 & 16 \\ 7 & 0 & 28 \end{array}\right] \quad \begin{aligned} 3x - 2y &= 16 \\ 7x \qquad &= 28. \end{aligned}$$

From the last equation, $x = 4$; 4 can be substituted for x in the first equation to obtain

$$3(4) - 2y = 16, \; y = -2.$$

The solution set is $\{(4,-2)\}$.

18. The augmented matrix is

$$\left[\begin{array}{cc|c} 4 & -3 & 16 \\ 2 & 1 & 8 \end{array}\right]$$

An equivalent matrix is

$$-\frac{1}{2}(\text{row 1}) + \text{row 2} \longrightarrow \left[\begin{array}{cc|c} 4 & -3 & 16 \\ 0 & \frac{5}{2} & 0 \end{array}\right] \quad \begin{aligned} 4x - 3y &= 16 \\ \tfrac{5}{2}y &= 0 \end{aligned}$$

18. cont'd.

From the last equation, $y = 0$; 0 can be substituted for y in the first equation to obtain

$$4x - 3(0) = 16, \quad x = 4$$

The solution set is $\{(4,0)\}$.

20. The augmented matrix is

$$\left[\begin{array}{ccc|c} 1 & -2 & 3 & -11 \\ 2 & 3 & -1 & 6 \\ 3 & -1 & -1 & 3 \end{array}\right].$$

An equivalent matrix is

$$\begin{array}{l} \\ \text{row } 2 + [-2 \text{ x row } 1] \longrightarrow \\ \text{row } 3 + [-3 \text{ x row } 1] \longrightarrow \end{array} \left[\begin{array}{ccc|c} 1 & -2 & 3 & -11 \\ 0 & 7 & -7 & 28 \\ 0 & 5 & -10 & 36 \end{array}\right].$$

Equivalent to the above matrix is

$$\begin{array}{l} \\ \\ \text{row } 3 + \left[\frac{-5}{7} \text{ x row } 2\right] \longrightarrow \end{array} \left[\begin{array}{ccc|c} 1 & -2 & 3 & -11 \\ 0 & 7 & -7 & 28 \\ 0 & 0 & -5 & 16 \end{array}\right] \quad \begin{array}{r} x - 2y + 3z = -11 \\ 7y - 7z = 28 \\ -5z = 16. \end{array}$$

From the last equation, $z = \frac{-16}{5}$. In the second equation, substitute $\frac{-16}{5}$ for z and obtain

$$7y - 7\left(\frac{-16}{5}\right) = 28, \quad 7y + \frac{112}{5} = \frac{140}{5}, \quad y = \frac{4}{5}.$$

In the first equation, substitute $\frac{-16}{5}$ for z and $\frac{4}{5}$ for y to obtain

$$x - 2\left(\frac{4}{5}\right) + 3\left(\frac{-16}{5}\right) = -11, \quad x - \frac{8}{5} - \frac{48}{5} = \frac{-55}{5}, \quad x = \frac{1}{5}.$$

The solution set is $\left\{\left(\frac{1}{5}, \frac{4}{5}, \frac{-16}{5}\right)\right\}$.

22. The augmented matrix is

$$\left[\begin{array}{ccc|c} 1 & -2 & -2 & 4 \\ 2 & 1 & -3 & 7 \\ 1 & -1 & -1 & 3 \end{array}\right].$$

An equivalent matrix is

$$\begin{array}{r} \\ \text{row 2} + [-2 \times \text{row 1}] \longrightarrow \\ \text{row 3} + [-1 \times \text{row 1}] \longrightarrow \end{array} \left[\begin{array}{ccc|c} 1 & -2 & -2 & 4 \\ 0 & 5 & 1 & -1 \\ 0 & 1 & 1 & -1 \end{array}\right].$$

Equivalent to the above matrix is

$$\begin{array}{r} \\ \text{row 2} + [-5 \times \text{row 3}] \longrightarrow \\ \\ \end{array} \left[\begin{array}{ccc|c} 1 & -2 & -2 & 4 \\ 0 & 0 & -4 & 4 \\ 0 & 1 & 1 & -1 \end{array}\right] \quad \begin{array}{r} x - 2y - 2z = 4 \\ -4z = 4 \\ y + z = -1. \end{array}$$

Note that in order to avoid introducing fractions, row 2 was chosen as the row to contain two zeros.

From the second equation, $z = -1$. In the third equation, substitute -1 for z and obtain

$$y + (-1) = -1, \; y = 0.$$

In the first equation, substitute -1 for z and 0 for y and obtain

$$x - 2(0) - 2(-1) = 4, \; x = 2.$$

The solution set is $\{(2,0,-1)\}$.

24. The augmented matrix is

$$\left[\begin{array}{ccc|c} 1 & -2 & -5 & 2 \\ 2 & 3 & 1 & 11 \\ 3 & -1 & -1 & 11 \end{array}\right].$$

An equivalent matrix is

$$\begin{array}{r} \\ \text{row 2} + [-2 \times \text{row 1}] \longrightarrow \\ \text{row 3} + [-3 \times \text{row 1}] \longrightarrow \end{array} \left[\begin{array}{ccc|c} 1 & -2 & -5 & 2 \\ 0 & 7 & 11 & 7 \\ 0 & 5 & 14 & 5 \end{array}\right].$$

24. cont'd.

Equivalent to the above matrix is

$$\text{row } 3 + \left[\frac{-5}{7} \text{x row } 2\right] \longrightarrow \left[\begin{array}{ccc|c} 1 & -2 & -5 & 2 \\ 0 & 7 & 11 & 7 \\ 0 & 0 & \frac{43}{7} & 0 \end{array}\right] \quad \begin{aligned} x - 2y - 5z &= 2 \\ 7y + 11z &= 7 \\ \frac{43}{7}z &= 0. \end{aligned}$$

From the last equation, $z = 0$. In the second equation, substitute 0 for z and obtain

$$7y + 11(0) = 7, \ y = 1.$$

In the first equation, substitute 0 for z and 1 for y and obtain

$$x - 2(1) - 5(0) = 2, \ x = 4.$$

The solution is $\{(4,1,0)\}$.

26. The augmented matrix is

$$\left[\begin{array}{ccc|c} 3 & 0 & -1 & 7 \\ 2 & 1 & 0 & 6 \\ 0 & 3 & -1 & 7 \end{array}\right].$$

An equivalent matrix is

$$-\frac{2}{3}(\text{row } 1) + \text{row } 2 \longrightarrow \left[\begin{array}{ccc|c} 3 & 0 & -1 & 7 \\ 0 & 1 & \frac{2}{3} & \frac{4}{3} \\ 0 & 3 & -1 & 7 \end{array}\right].$$

Equivalent to the above matrix is

$$-\frac{1}{3}(\text{row } 3) + \text{row } 2 \longrightarrow \left[\begin{array}{ccc|c} 3 & 0 & -1 & 7 \\ 0 & 0 & 1 & -1 \\ 0 & 3 & -1 & 7 \end{array}\right] \quad \begin{aligned} 3x - z &= 7 \\ z &= -1 \\ 3y - z &= 7 \end{aligned}$$

From the second equation, $z = -1$. In the first and third equations substitute -1 for z and obtain

$$3x - (-1) = 7, \ x = 2.$$
$$3y - (-1) = 7, \ y = 2.$$

The solution set is $\{(2,2,-1)\}$.

EXERCISE 9.6

2. Larger number: x; smaller number: y

$$x - y = 14 \quad (A)$$
$$x = 2y + 1 \quad \text{or} \quad x - 2y = 1 \quad (B)$$

Add (-1) times Equation (B) to Equation (A): $y = 13$. From Equation (B): $x = 2(13) + 1 = 27$.

The numbers are 13 and 27.

4. One integer: x; the next integer: y

$$y = x + 1 \quad \text{or} \quad x - y = -1 \quad (A)$$
$$\frac{1}{2}x + \frac{1}{5}y = 17 \quad \text{or} \quad 5x + 2y = 170 \quad (B)$$

Add 2 times Equation (A) to Equation (B):

$7x = 168$; $x = 24$.

From Equation (A): $y = 25$.

The integers are 24 and 25.

6. Number of votes cast for winner: x; number of votes cast for loser: y

$$x + y = 7179 \quad (A)$$
$$x - 6 = (y + 6) - 1 \quad \text{or} \quad x - y = 11 \quad (B)$$

Add Equation (A) to Equation (B): $2x = 7190$; $x = 3595$.

From Equation (A); $x + 3595 = 7179$; $x = 3584$.

The votes cast for the winner were 3595; for the loser, 3584.

8. Amount invested at 8%: x; amount invested at 12%: y

$$x + y = 1200 \quad (A)$$
$$0.08x = 0.12y + 3 \quad \text{or} \quad 8x - 12y = 300 \quad (B)$$

Multiply Equation (A) by 12 and add the resulting equation to Equation (B):

$$20x = 14700; \ x = 735.$$

From Equation (A): $735 + y = 1200$; $y = 465$.

The amount invested at 8% was \$735; the amount at 12% was \$465.

10. The number of 1-inch screws: x
The number of $\frac{3}{4}$-inch screws: y

$$\begin{aligned} x + \quad y &= 50 \quad \text{(A)} \\ 0.18x + 0.15y &= 8.10 \quad \text{(B)} \end{aligned}$$

Multiply Equation (B) by 100 and Equation (A) by -15.

$$\begin{aligned} -15x - 15y &= -750 \\ 18x + 15y &= 810 \end{aligned}$$

Add the two equations.

$$3x = 60; \; x = 20$$

Substitute 20 for x in Equation (A) and solve for y.

$$20 + y = 50; \; y = 30$$

20 1-inch and 30 $\frac{3}{4}$-inch screws were purchased.

12. The rate of the fast car: x
The rate of the slow car: y

One equation uses the fact that

$$\text{distance} = (\text{rate}) \times (\text{time})$$

and that

$$\begin{pmatrix}\text{distance fast} \\ \text{car travels}\end{pmatrix} - \begin{pmatrix}\text{distance slow} \\ \text{car travels}\end{pmatrix} = 96$$

$$3x - 3y = 96 \quad \text{and also,}$$

$$x = 2y$$

The system can be written as

$$\begin{aligned} x - \quad y &= 32 \\ -x + 2y &= 0 \end{aligned}$$

from which, by adding the equations, $y = 32$. Substituting 32 for y in $x = 2y$, $x = 2(32) = 64$.

The cars are traveling at 32 miles per hour and 64 miles per hour.

14. First number: x; second number: y; third number: z

$$\begin{aligned} x + y + z &= 2 && (1) \\ x = y + z \quad \text{or} \quad x - y - z &= 0 && (2) \\ z = y - x \quad \text{or} \quad x - y + z &= 0 && (3) \end{aligned}$$

Add Equation (2) to -1 times Equation (3): $-2z = 0$; $z = 0$.

Substitute 0 for z in Equations (1) and (2):

$$\begin{aligned} x + y &= 2; && (4) \\ x - y &= 0. && (5) \end{aligned}$$

Add Equations (4) and (5): $2x = 2$; $x = 1$.

Substitute 1 for x and 0 for z in Equation (1): $1 + y + 0 = 2$; $y = 1$.

The numbers are 1, 1, and 0.

16. The number of ten-dollar bills: T
The number of five-dollar bills: F
The number of one-dollar bills: S

$$\begin{aligned} 10T + 5F + S &= 446 \\ T + F + S &= 94 \\ -T + F &= 10. \end{aligned}$$

Subtracting the second equation from the first, $9T + 4F = 352$. Adding this equation to -4 times the third equation, we have $13T = 312$ or $T = 24$. Then, $F = 34$ and $S = 36$.

18. The smallest ("a second") angle: x
"One angle" of the triangle: y
The third angle: z

$$\begin{aligned} y - x &= 10 \\ z &= 6x + 10 \\ x + y + z &= 180 \end{aligned}$$

Rewrite the system as

$$\begin{aligned} x - y &= -10 \\ -6x + z &= 10 \\ x + y + z &= 180 \end{aligned}$$

Add the first and third equations and form the system

$$\begin{aligned} 2x + z &= 170 \\ -6x + z &= 10 \end{aligned}$$

18. cont'd.

Solving this system, we find $x = 20$, $z = 130$.

Substitute 20 for x in $y - x = 10$ and obtain $y = 30$.

The measures of the angles are 20°, 30°, and 130°.

20. The age of the son in 1984: x
The age of the daughter in 1984: y

In 1984: $x = 3y$
In 1989: $x + 5 = 2(y + 5)$

Write the system as:

$$\begin{aligned} x - 3y &= 0 \\ x - 2y &= 5 \end{aligned}$$

Solve the system and obtain $x = 15$, $y = 5$

When the father died, the son was 15, the daughter was 5.

22. The production cost: C
The revenue: R
The number of records produced: x

$$\begin{aligned} C &= 0.40x + 20 \\ R &= 0.60x \\ C &= R \end{aligned}$$

Write the system as:

$$\begin{aligned} C \qquad - 0.40x &= 20 \\ R - 0.60x &= 0 \\ C - R \qquad &= 0 \end{aligned}$$

Add the second and third equation and form the system:

$$\begin{aligned} C - 0.60x &= 0 \\ -C + 0.40x &= -20 \end{aligned}$$

Solve the above system and obtain

$$-0.20x = -20 \text{ from which } x = 100$$

To break even, 100 records must be produced.

24. In each equation substitute 1 for x and 2 for y.

$$a + 2b = 4$$
$$b - 2a = -3$$

Solve the system and obtain $a = 2$ and $b = 1$.

26. In $ax + by + cz = 1$ substitute the coordinates of each solution for x, y, and z, respectively and obtain the system:

$$\begin{aligned} a(0) + b(4) + c(2) &= 1 \\ a(-1) + b(3) + c(0) &= 1 \\ a(-1) + b(0) + c(2) &= 1 \end{aligned}$$

This can be written as:

$$\begin{aligned} 4b + 2c &= 1 \\ a - 3b &= -1 \\ a - 2c &= -1 \end{aligned}$$

Add the first and third equations and write the system

$$a + 4b = 0$$
$$a - 3b = -1$$

Solve this system and obtain $b = \frac{1}{7}$ and $a = \frac{-4}{7}$.

Substitute $\frac{-4}{7}$ for a in $a - 2c = -1$ and obtain

$$\frac{-4}{7} - 2c = -1 \text{ from which } c = \frac{3}{14} .$$

Hence $a = \frac{-4}{7}$, $b = \frac{1}{7}$, and $c = \frac{3}{14}$.

10 Quadratic Equations and Inequalities in Two Variables

EXERCISE 10.1

2. The graph is a circle with the center at the origin and radius 4.

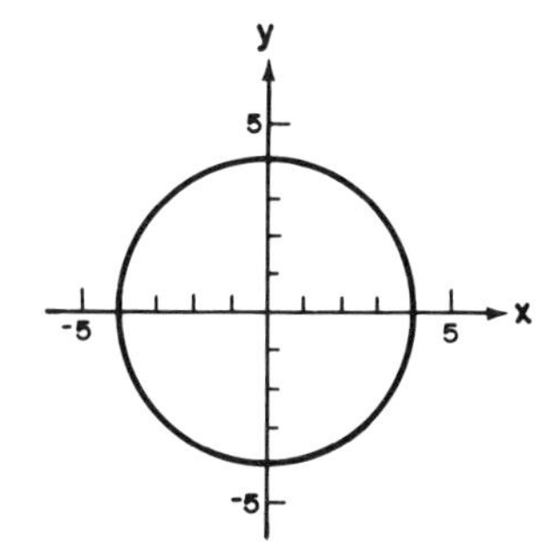

4. $2x^2 + 2y^2 = 18$ is equivalent to $x^2 + y^2 = 9$. The graph is a circle with center at the origin and radius 3.

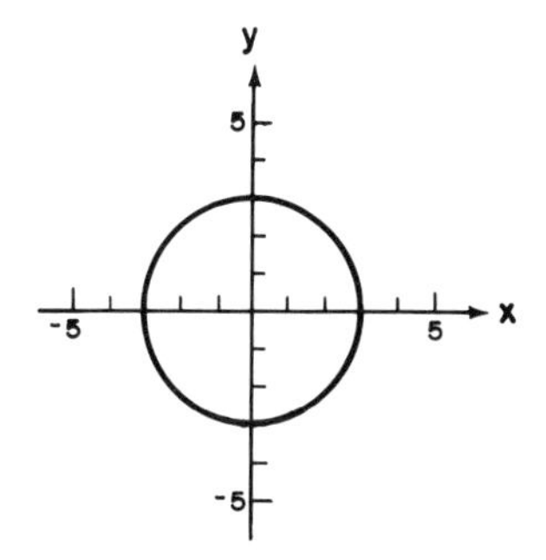

6. The graph is a circle with center at (1,3) and radius 4.

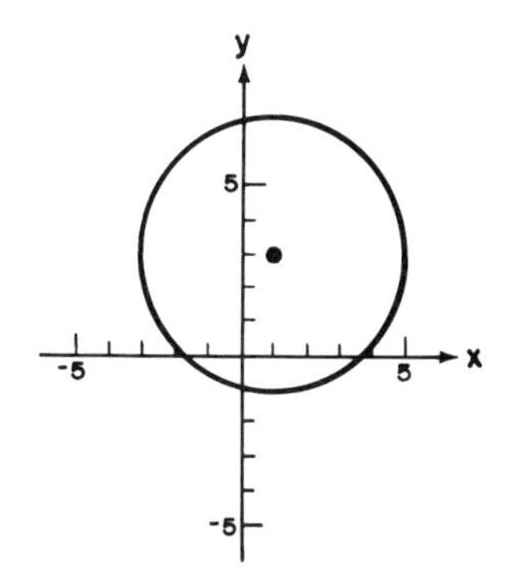

8. The graph is a circle with center at (0,-4) and radius $\sqrt{12} = 2\sqrt{3} \approx 3.5$

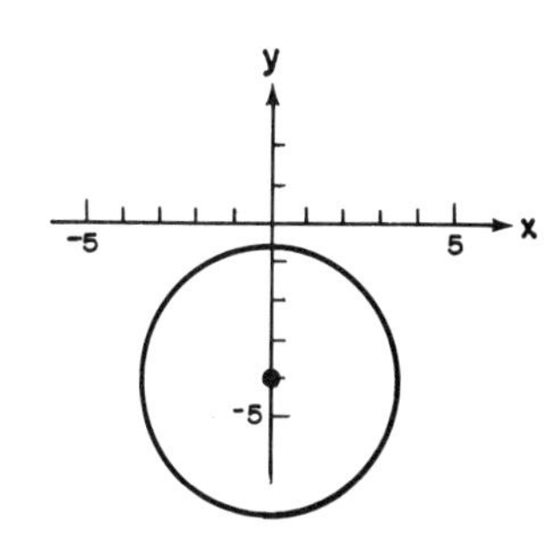

10. The graph is a circle with center at (-2,0) and radius $\sqrt{18} = 3\sqrt{2} \approx 4.2$.

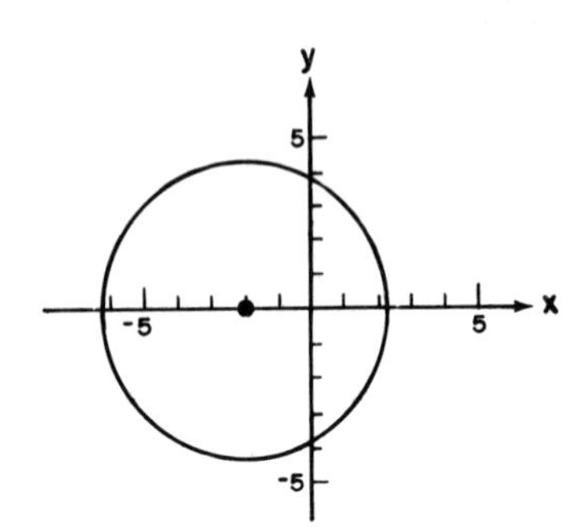

12. The graph is a circle with center at (-6,3) and radius $\sqrt{8} = 2\sqrt{2} \approx 2.8$.

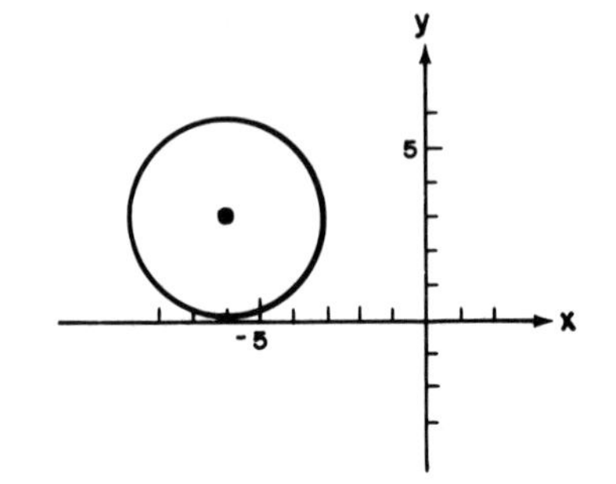

14. $x^2 - 6x + ? + y^2 + 2y + ? = 4 + ? + ?$

$x^2 - 6x + 9 + y^2 + 2y + 1 = 4 + 9 + 1$

$(x - 3)^2 + (y + 1)^2 = 14$

The graph is a circle with center at (3,-1) and radius $\sqrt{14} \approx 3.7$.

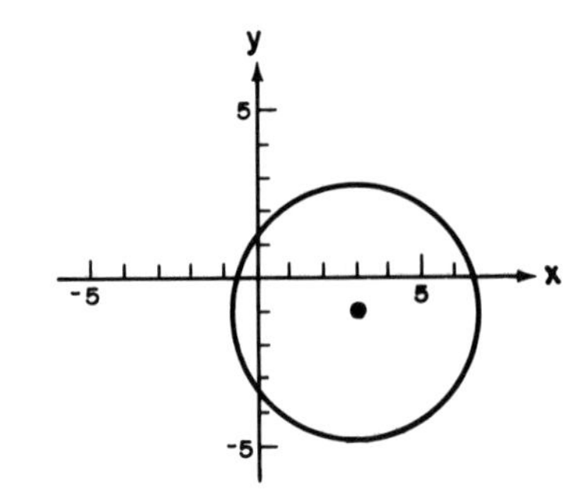

16. $x^2 + y^2 - 10y + ? = 2 + ?$

$x^2 + y^2 - 10y + 25 = 2 + 25$

$(x - 0)^2 + (y - 5)^2 = 27$

The graph is a circle with center at (0,5) and radius $\sqrt{27} = 3\sqrt{3} \approx 5.2$.

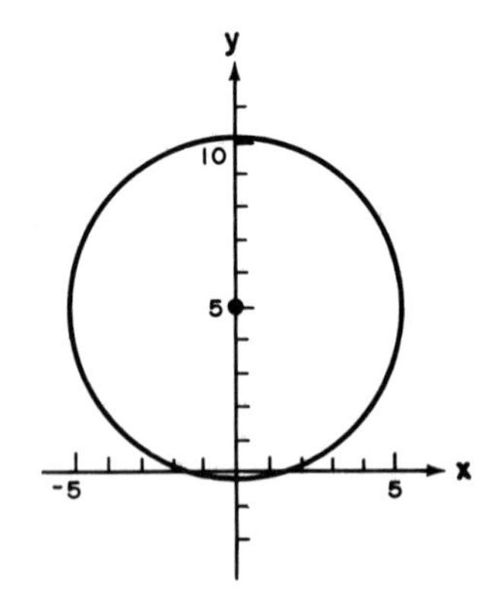

18. $x^2 - 6x + ? + y^2 = 0 + ?$

$x^2 - 6x + 9 + y^2 = 0 + 9$

$(x - 3)^2 + (y - 0)^2 = 9$

The graph is a circle with center at (3,0) and radius 3.

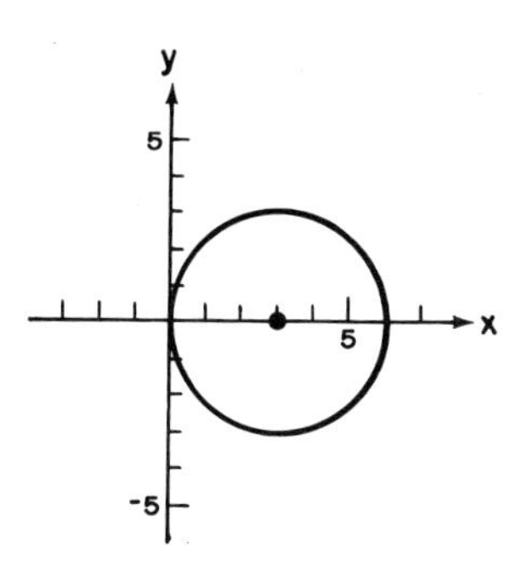

20. The graph is an ellipse with center at the origin. The x-intercepts are 3 and -3. The y-intercepts are 4 and -4.

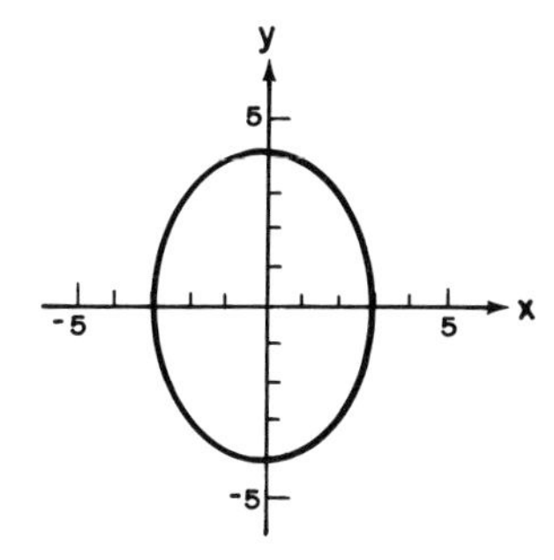

22. The graph is an ellipse with center at the origin. The x-intercepts are 4 and -4. The y-intercepts are $\sqrt{12}$ and $-\sqrt{12}$. Note: $\sqrt{12} - 2\sqrt{3} \approx 3.5$.

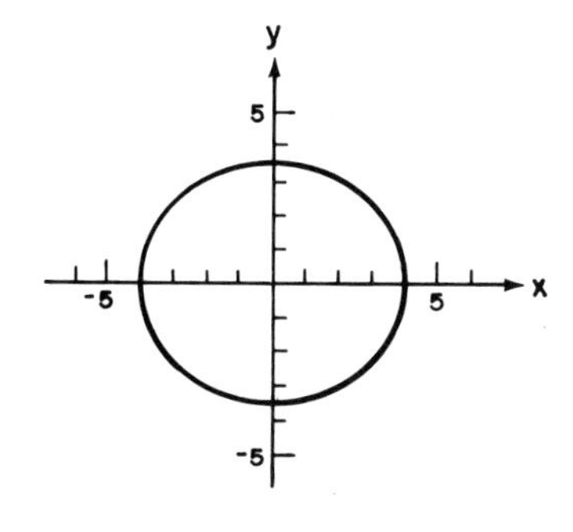

24. $\frac{x^2}{8} + \frac{y^2}{1} = 1$

The graph is an ellipse with center at the origin. The x-intercepts are $\sqrt{8}$ and $-\sqrt{8}$. The y-intercepts are 1 and -1. Note: $\sqrt{8} = 2\sqrt{2} \approx 2.8$.

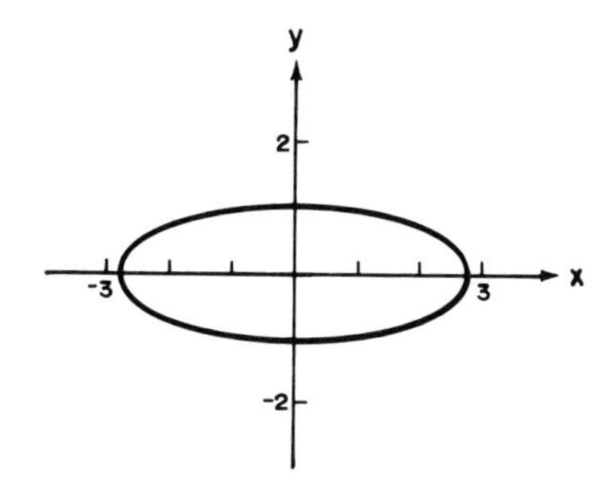

26. The graph is an ellipse with center at (2,5). The x'-intercepts are 2 and -2. The y'-intercepts are 5 and -5.

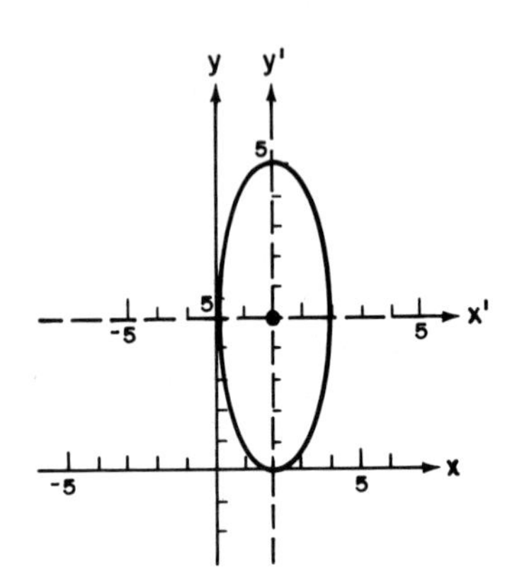

28. The graph is an ellipse with center at (5,-3). The x'-intercepts are $\sqrt{15}$ and $-\sqrt{15}$. The y'-intercepts are $\sqrt{8}$ and $-\sqrt{8}$. Note: $\sqrt{15} \approx 3.9$.
$\sqrt{8} = 2\sqrt{2} \approx 2.8$.

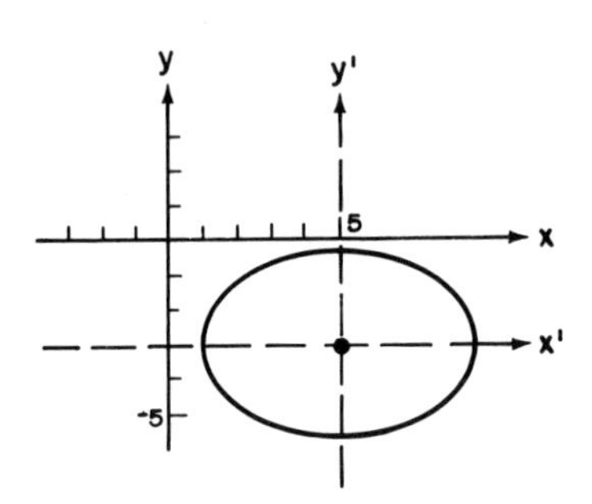

30. The graph is an ellipse with center at (5,0). The x'-intercepts are $\sqrt{15}$ and $-\sqrt{15}$. The y'-intercepts are 5 and -5. Note: $\sqrt{15} \approx 3.9$.

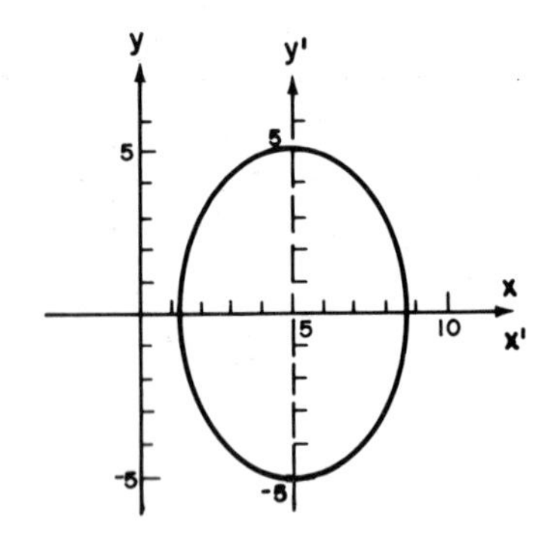

32. $x^2 + 6x + ? + 16y^2 = 7$

$x^2 + 6x + 9 + 16(y - 0)^2 = 7 + 9$

$(x + 3)^2 + 16(y - 0)^2 = 16$

$$\frac{(x + 3)^2}{16} + \frac{(y - 0)^2}{1} = 1$$

The graph is an ellipse with center at (-3,0). The x'-intercepts are 4 and -4. The y'intercepts are 1 and -1.

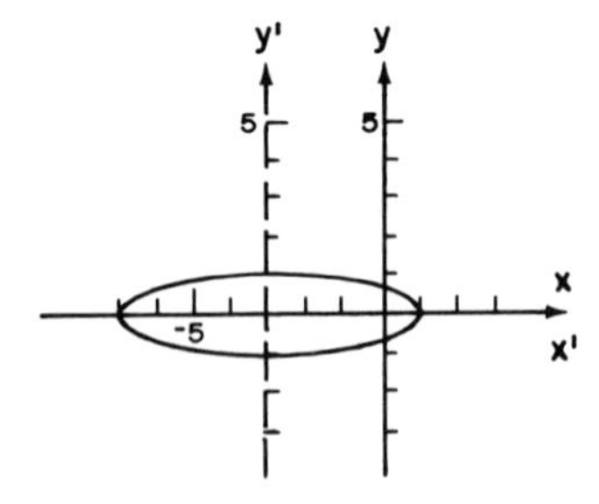

34. $$16x^2 + 64x + 9y^2 - 18y = 71$$
$$16(x^2 + 4x + ?) + 9(y^2 - 2y + ?) = 71 + ? + ?$$
$$16(x^2 + 4x + 4) + 9(y^2 - 2y + 1) = 7 + 64 + 9$$
$$16(x + 2)^2 + 9(y - 1)^2 = 80$$
$$\frac{(x + 2)^2}{5} + \frac{9(y - 1)^2}{80} = 1$$
$$\frac{(x + 2)^2}{5} + \frac{(y - 1)^2}{80/9} = 1$$

The graph is an ellipse with center at $(-2,1)$. The x'-intercepts are $\sqrt{5}$ and $-\sqrt{5}$. The y'-intercepts are $\frac{\sqrt{80}}{3}$ and $\frac{-\sqrt{80}}{3}$.

Note: $\sqrt{5} \approx 2.2$

$$\frac{\sqrt{80}}{3} = \frac{4\sqrt{5}}{3} \approx 3.0$$

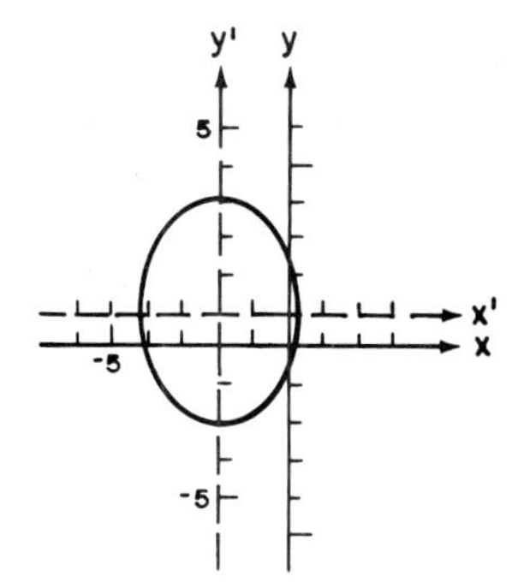

36. $$2x^2 - 16x + y^2 + 6y = -11$$
$$2(x^2 - 8x + ?) + (y^2 + 6y + ?) = -11 + ? + ?$$
$$2(x^2 - 8x + 16) + (y^2 + 6y + 9) = -11 + 32 + 9$$
$$2(x - 4)^2 + (y + 3)^2 = 30$$
$$\frac{(x - 4)^2}{15} + \frac{(y + 3)^2}{30} = 1$$

The graph is an ellipse with center at $(4,-3)$. The x'-intercepts are $\sqrt{15}$ and $-\sqrt{15}$. The y'-intercepts are $\sqrt{30}$ and $-\sqrt{30}$.
Note: $\sqrt{15} \approx 3.9$ and $\sqrt{30} \approx 5.5$

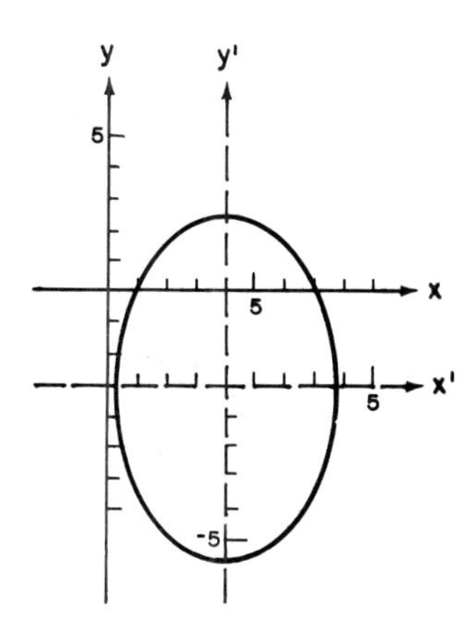

38.

$$5x^2 - 20x + 8y^2 + 16y = 12$$
$$5(x^2 - 4x + ?) + 8(y^2 + 2y + ?) = 12 + ? + ?$$
$$5(x^2 - 4x + 4) + 8(y^2 + 2y + 1) = 12 + 20 + 8$$
$$5(x - 2)^2 + 8(y + 1)^2 = 40$$
$$\frac{(x - 2)^2}{8} + \frac{(y + 1)^2}{5} = 1$$

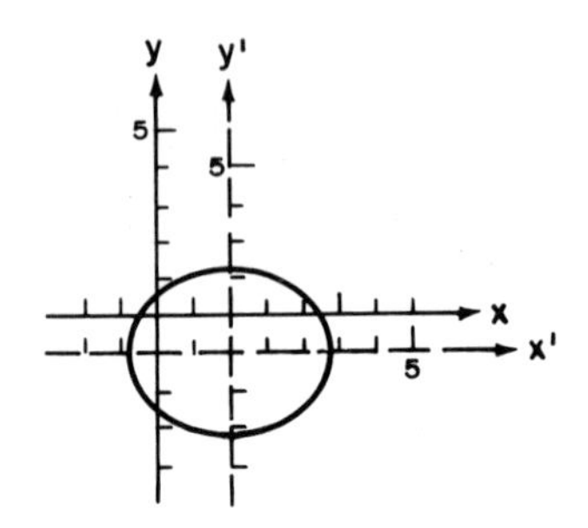

The graph is an ellipse with center at (2,-1). The x'-intercepts are $\sqrt{8}$ and $-\sqrt{8}$. The y'-intercepts are $\sqrt{5}$ and $-\sqrt{5}$.

Note: $\sqrt{8} = 2\sqrt{2} \approx 2.8$ and $\sqrt{5} \approx 2.2$

40.

$$x^2 + 4x + 10y^2 + 20y = -4$$
$$x^2 + 4x + ? + 10(y^2 + 2y + ?) = -4 + ? + ?$$
$$x^2 + 4x + 4 + 10(y^2 + 2y + 1) = -4 + 4 + 10$$
$$(x + 2)^2 + 10(y + 1)^2 = 10$$
$$\frac{(x + 2)^2}{10} + \frac{(y + 1)^2}{1} = 1$$

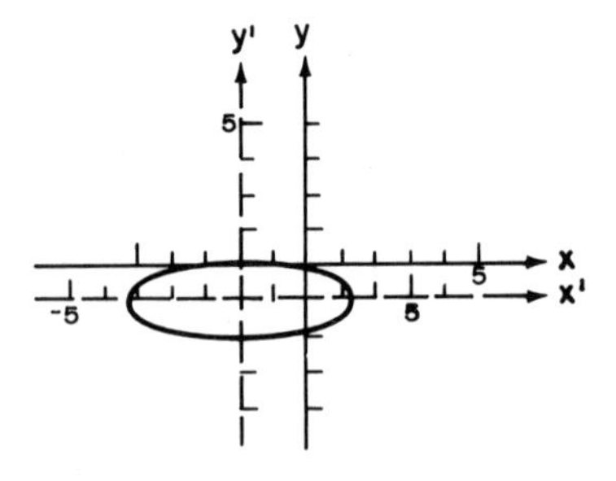

The graph is an ellipse with center at (-2,-1). The x'-intercepts are $\sqrt{10}$ and $-\sqrt{10}$. The y'-intercepts are 1 and -1.

Note: $\sqrt{10} \approx 3.2$.

EXERCISE 10.2

2\. a. Since $y = x^2 + 5x + 6$ is of the form $y = ax^2 + bx + c$ with $a = 1 > 0$, the graph opens upward.

b. To find the y-intercept let $x = 0$ and solve for y.

$$y = 0^2 + 5(0) + 6 = 6$$

The y-intercept is 3.
To find the x-intercepts let $y = 0$ and solve for

$$0 = x^2 + 5x + 6$$
$$0 = (x + 3)(x + 2)$$
$$x = -3 \quad \text{or} \quad x = -2$$

The x-intercepts are -3 and -2.

4\. a. $a = -1 < 0$. Hence the graph opens downward.

b. When $x = 0$, $y = 12$ which is the y-intercept.
When $y = 0$, we have

$$0 = -x^2 - x + 12$$

which is equivalent to

$$x^2 + x - 12 = 0$$
$$(x + 4)(x - 3) = 0$$
$$x = -4 \quad \text{or} \quad x = 3$$

which are the x-intercepts.

6\. a. $a = 3 > 0$. Hence the graph opens upward.

b. When $x = 0$, $y = -2$ which is the y-intercept.
When $y = 0$, we have

$$3x^2 - x - 2 = 0$$
$$(3x + 2)(x - 1) = 0$$
$$x = \frac{-2}{3} \quad \text{or} \quad x = 1$$

which are the x-intercepts.

8\. a. $a = -2 < 0$. Hence the graph opens downward.

b. When $x = 0$, $y = 6$ which is the y-intercept.
When $y = 0$, we have

$$-2x^2 + x + 6 = 0$$
$$2x^2 - x - 6 = 0$$
$$(2x + 3)(x - 2) = 0$$
$$x = \frac{-3}{2} \quad \text{or } x = 2$$

which are the x-intercepts.

10\. a. $a = 3 > 0$. Hence the graph opens upward.

b. When $x = 0$, $y = 0$ which is the y-intercept.
When $y = 0$, we have

$$3x^2 - 12x = 0$$
$$3x(x - 4) = 0$$
$$x = 0 \quad \text{or} \quad x = 4$$

which are the x-intercepts.

12\. a. $a = -2 < 0$. Hence the graph opens downward.

b. When $x = 0$, $y = -6$ which is the y-intercept.
When $y = 0$, we have

$$-2x^2 - 6 = 0$$
$$-2x^2 = 6$$
$$x^2 = -3$$
$$x = \pm i\sqrt{3}$$

Since the solutions are imaginary, there are no x-intercepts.

14. $y = 3(x - 2)^2 - 5;\ (2,-5)$

16. $y = 2x^2 - 12$ can be written as

$y = 2(x - 0)^2 - 12;\ (0,-12)$

18. $y = -3(x - 2)^2$ can be thought of as

$y = -3(x - 2)^2 + 0;\ (2,0)$

20. Completing the square in x we have:

$y = x^2 - 8x + ? + 2 - ?$

$y = x^2 - 8x + 16 + 2 - 16$

$y = (x - 4)^2 - 14;\ (4,-14)$

22. Completing the square in x we have:

$y = 3x^2 - 6x \quad + 4$

$y = 3(x^2 - 2x + ?) + 4 - (3 \cdot ?)$

$y = 3(x^2 - 2x + 1) + 4 - (3 \cdot 1)$

$y = 3(x - 1)^2 + 1;\ (1,1)$

24. Completing the square in x we have:

$y = -2x^2 + 5x \quad -1$

$y = -2(x^2 - \frac{5}{2}x + ?) - 1 + (-2) \cdot ?$

$y = -2(x^2 - \frac{5}{2}x + \frac{25}{16}) - 1 - [(-2)(\frac{25}{16})]$

$y = -2(x - \frac{5}{4})^2 + \frac{17}{8};\ \left(\frac{5}{4}, \frac{17}{8}\right)$

26. 1) Since $a = 1 > 0$, the curve opens upward.

2) Complete the square in x.

$y = x^2 + x - 6$

$= x^2 + x + \frac{1}{4} - 6 - \frac{1}{4}$

$= (x + \frac{1}{2})^2 - \frac{25}{4}$

The vertex is at $\left(-\frac{1}{2}, \frac{-25}{4}\right)$.

3) When $x = 0$, $y = -6$. The y-intercept is -6.
When $y = 0$, then $x^2 + x - 6 = 0$

$(x + 3)(x - 2) = 0$.

The x-intercepts are -3 and 2.

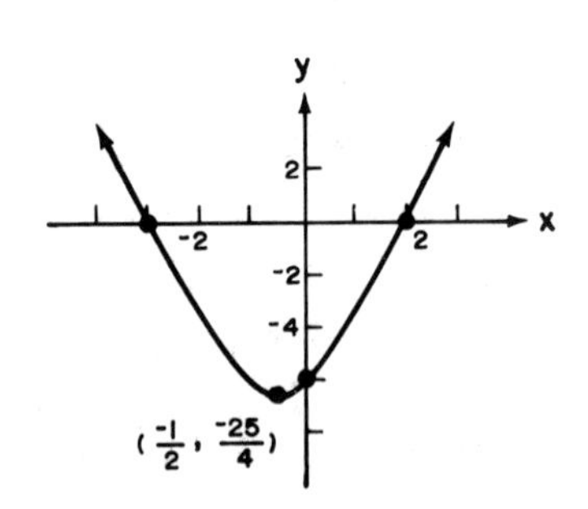

28. 1) Since $a = 1 > 0$, the curve opens upward.

2) Complete the square in x.

$$y = x^2 + 6x + ?$$
$$= x^2 + 6x + 9 - 9$$
$$= (x + 3)^2 - 9$$

The vertex is at $(-3,-9)$.

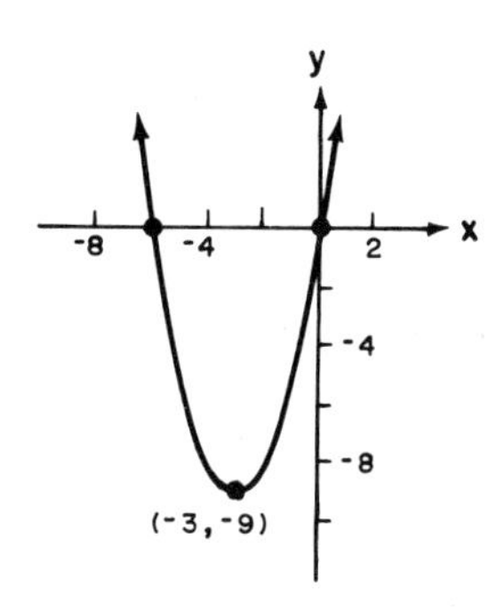

3) When $x = 0$, $y = 0$. The y-intercept is 0.
When $y = 0$, then

$$x^2 + 6x = 0$$
$$x(x + 6) = 0.$$

The x-intercepts are -6 and 0.

30. 1) Since $a = 1 > 0$, the graph opens upward.

2) Complete the square in x.

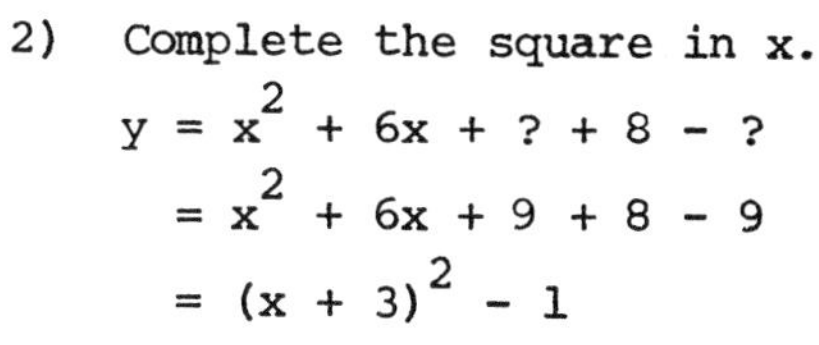

$$y = x^2 + 6x + ? + 8 - ?$$
$$= x^2 + 6x + 9 + 8 - 9$$
$$= (x + 3)^2 - 1$$

The vertex is at $(-3,-1)$.

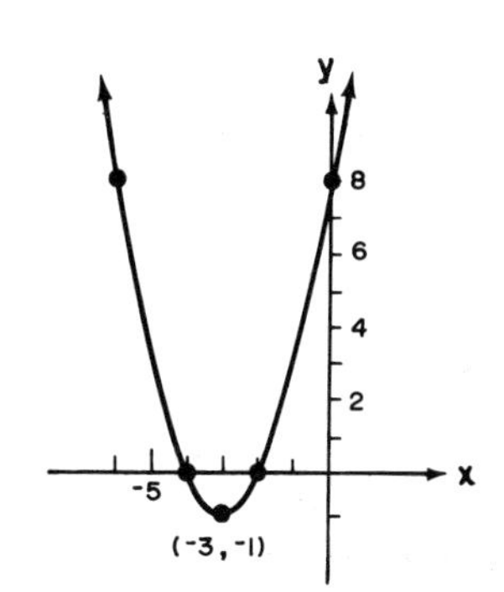

3) When $x = 0$, $y = 8$. The y-intercept is 8.
When $y = 0$, then

$$x^2 + 6x + 8 = 0$$
$$(x + 4)(x + 2) = 0.$$

The x-intercepts are -4 and -2.

32. The graph opens upward. Since the equation can be written as $y = (x - 0)^2 + 4$, the vertex is at (0,4) and the y-axis is the axis of symmetry.
When $y = 0$, we have

$$x^2 + 4 = 0$$

which has no real solutions. Hence there are no x-intercepts. Obtain a point on the graph, say, when $x = 3$. Thus obtain the point (3,13). Graph this point and use symmetry.

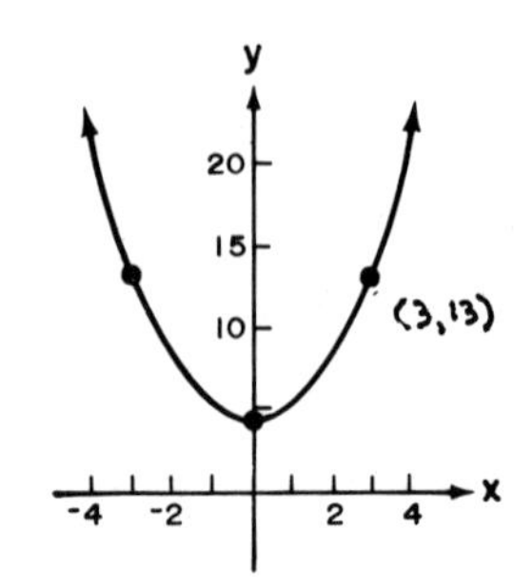

34. The graph opens upward. The vertex is at (0,-5).
When $y = 0$, we have $x^2 - 5 = 0$ from which we have x-intercepts of $-\sqrt{5}$ and $\sqrt{5}$.
Note: $\sqrt{5} \approx 2.2$.

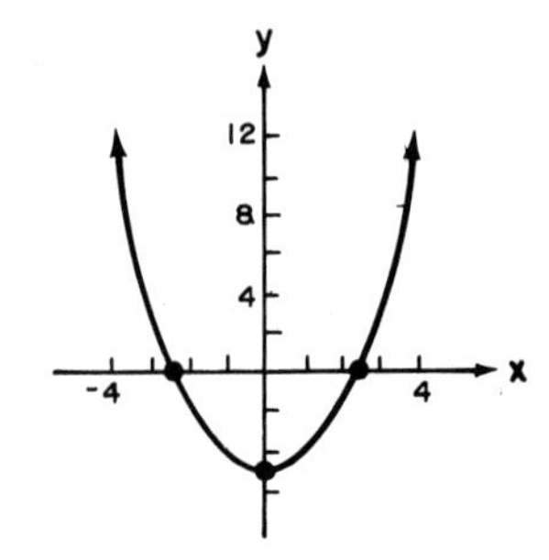

36. The graph opens downward. The vertex is at (0,5).
When $y = 0$, we have $-x^2 + 5 = 0$ from which we have x-intercepts of $-\sqrt{5}$ and $\sqrt{5}$. Note: $\sqrt{5} \approx 2.2$.

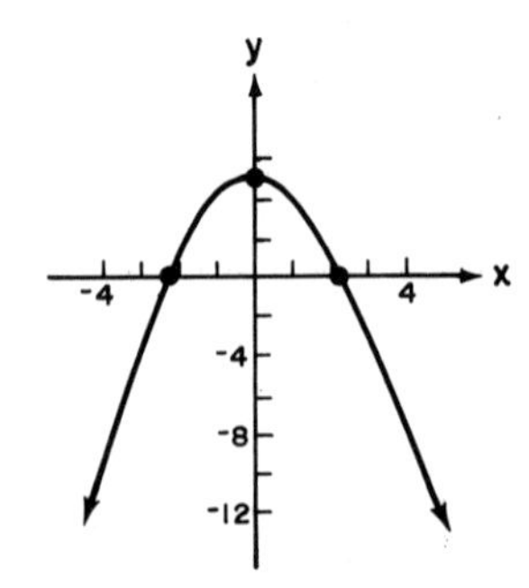

38. The graph opens upward.
The y-intercept is 0.
Complete the square in x:

$$y = 5(x^2 - \frac{1}{5}x + ?) - 5 \cdot ?$$
$$= 5(x^2 - \frac{1}{5}x + \frac{1}{100}) - 5(\frac{1}{100})$$
$$= 5(x - \frac{1}{10})^2 - \frac{1}{20}$$

The vertex is at $\left(\frac{1}{10}, -\frac{1}{20}\right)$.
The x-intercepts will be too close to the vertex to be useful.
Another pair of points are at (-3,48),(3,42)

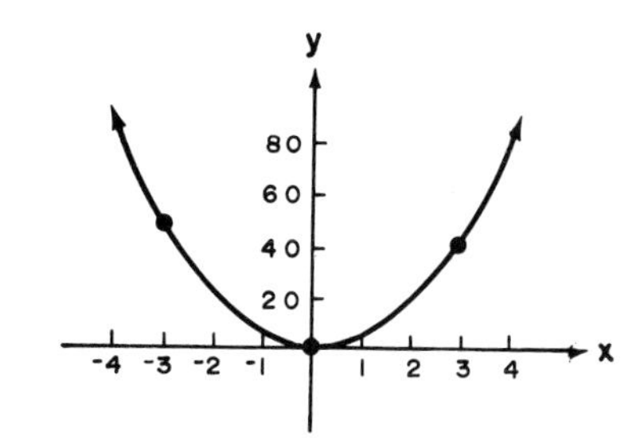

40. The graph opens upward.
The y-intercept is 1.
The equation can be written in the form $y = (x - 1)^2 + 0$ from which it is seen that the vertex is at (1,0).
From the above it is seen that there are no additional x-intercepts.
Find an additional point, say (4,9), and use symmetry.

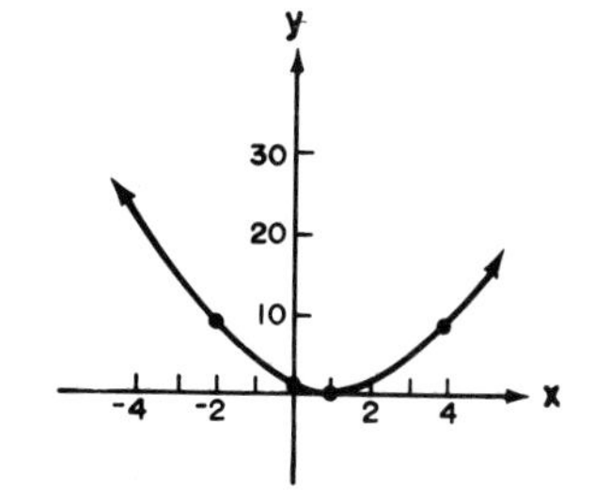

42. The graph opens downward.
The y-intercept is -1.
The equation can be written in the form

$$y = -(x^2 - 2x + 1)$$
$$= -(x - 1)^2 + 0$$

from which is seen that the vertex is at (1,0).
From the above it is seen that there are no additional x-intercepts.
Find an additional point, say (4,-9) and use symmetry.

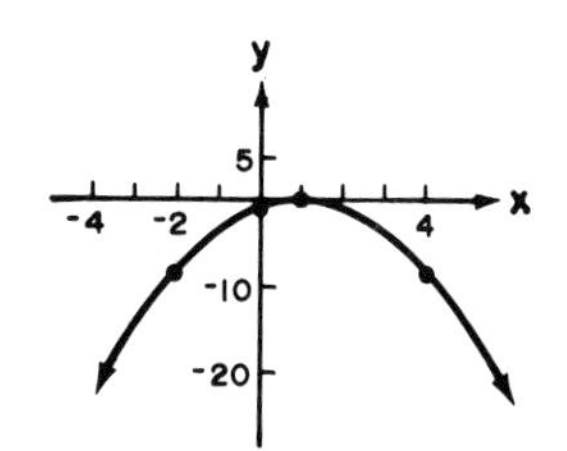

44. The graph opens downward.
The y-intercept is -2.
Complete the square in x:

$$y = -3(x^2 - \frac{1}{3}x + ?) + (-3) \cdot ?$$
$$= -3(x^2 - \frac{1}{3}x + \frac{1}{36}) - (-3)\left(\frac{1}{36}\right)$$
$$= -3(x - \frac{1}{6})^2 + \frac{1}{12}$$

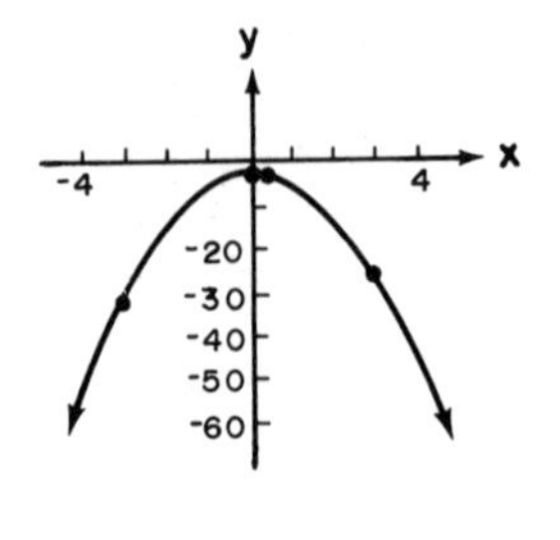

The vertex is at $\left(\frac{1}{6}, \frac{1}{12}\right)$.

The x-intercepts are too close to the vertex to be useful.
Another pair of points are at (-3,-32) and (3,-26).

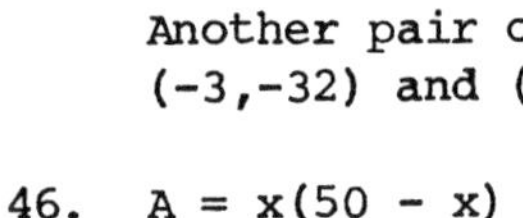

46. $A = x(50 - x)$
$= -x^2 + 50x$

The graph of the equation is a parabola opening downward.
Completing the square we have

$$A = -1(x^2 - 50x + ?) - ?$$
$$= -1(x^2 - 50x + 625) - (-1)(625)$$
$$= -(x - 25)^2 + 625$$

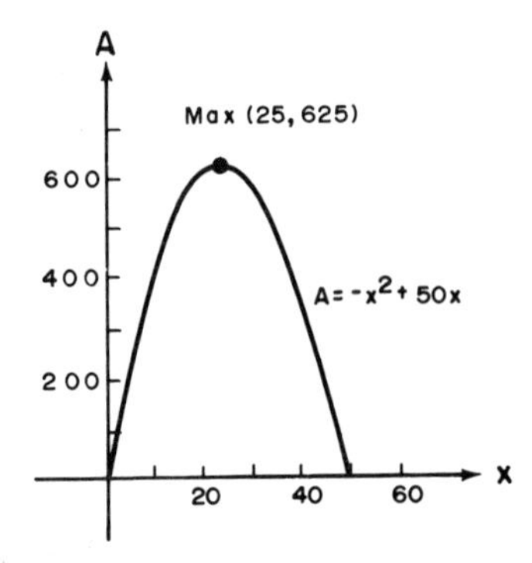

Hence, the coordinates of the vertex (the point where A has its greatest value) are (25,625).
The maximum area is 625 square inches.

48. The object will strike the ground when d = 0. Hence, substituting 0 for d,

$$32t - 8t^2 = 0$$
$$8t(4 - t) = 0.$$

Therefore, t = 0 or t = 4. This means that the object was at ground level (d = 0) at 0 seconds and again at 4 seconds. Hence, the object was in the air for 4 seconds.

50. for $k = -1$, $y = -x^2$;

$k = -2$, $y = -2x^2$;

$k = -3$, $y = -3x^2$;

$k = -4$, $y = -4x^2$.

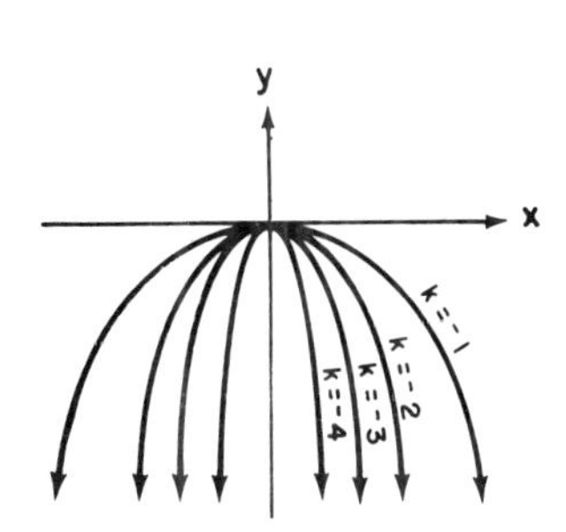

52. For $k = -2$, $y = x^2 - 2x$;

$k = 0$, $y = x^2$;

$k = 2$, $y = x^2 + 2x$;

$k = 4$, $y = x^2 + 4x$.

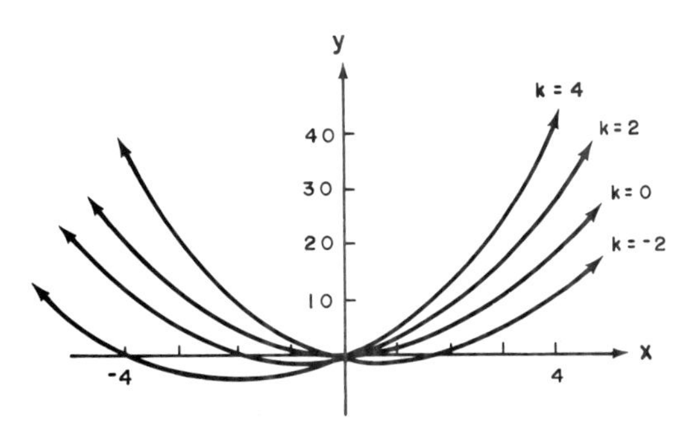

54. If the graph contains the points, then their coordinates must satisfy the equation.

Substitute 1 for x and 0 for y:

$$0 = a(1)^2 + b(1) + c;\ a + b + c = 0. \quad (1)$$

Substitute 3 for x and -2 for y:

$$-2 = a(3)^2 + b(3) + c;\ 9a + 3b + c = -2. \quad (2)$$

Substitute 5 for x and 4 for y:

$$4 = a(5)^2 + b(5) + c;\ 25a + 5b + c = 4. \quad (3)$$

Add -1 times Equation (1) to Equation (2):

$$8a + 2b = -2 \quad \text{or} \quad 4a + b = -1. \quad (4)$$

Add -1 times Equation (1) to Equation (3):

$$24a + 4b = 4 \quad \text{or} \quad 6a + b = 1. \quad (5)$$

Add -1 times Equation (4) to Equation (5):

$$2a = 2 \quad \text{or} \quad a = 1.$$

54. cont'd.

If $a = 1$, then from Equation (4),

$$4(1) + b = -1 \quad \text{or} \quad b = -5.$$

Substitute 1 for a and -5 for b in Equation (1):

$$(1) + (-5) + c = 0 \quad \text{or} \quad c = 4.$$

The required values are: $a = 1$, $b = -5$, and $c = 4$.

EXERCISE 10.3

2. The graph is a hyperbola with center at the origin. Since $a^2 = 4$ and $b^2 = 16$, then $a = 2$ and $b = 4$. The y-intercepts are -2 and 2. The central rectangle has dimensions $2a = 4$ and $2b = 8$.

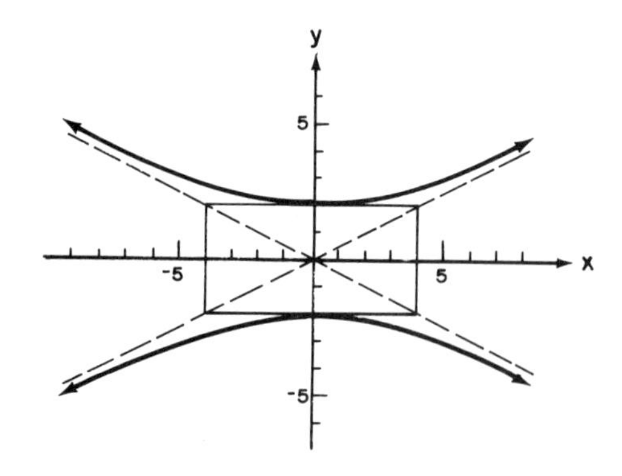

4. The graph is a hyperbola with center at the origin. Since $a^2 = 15$ and $b^2 = 10$, then $a = \sqrt{15}$ and $b = \sqrt{10}$. The x-intercepts are $-\sqrt{15}$ and $\sqrt{15}$. The central rectangle has dimensions $2a = 2\sqrt{15}$ and $2b = 2\sqrt{10}$. Note: $\sqrt{15} \approx 3.9$ and $\sqrt{10} \approx 3.2$.

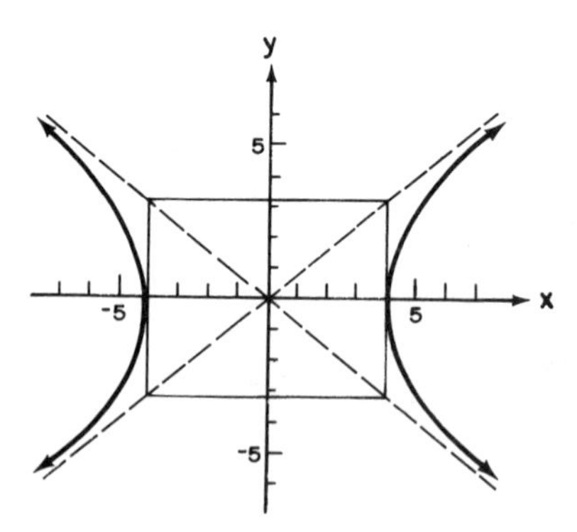

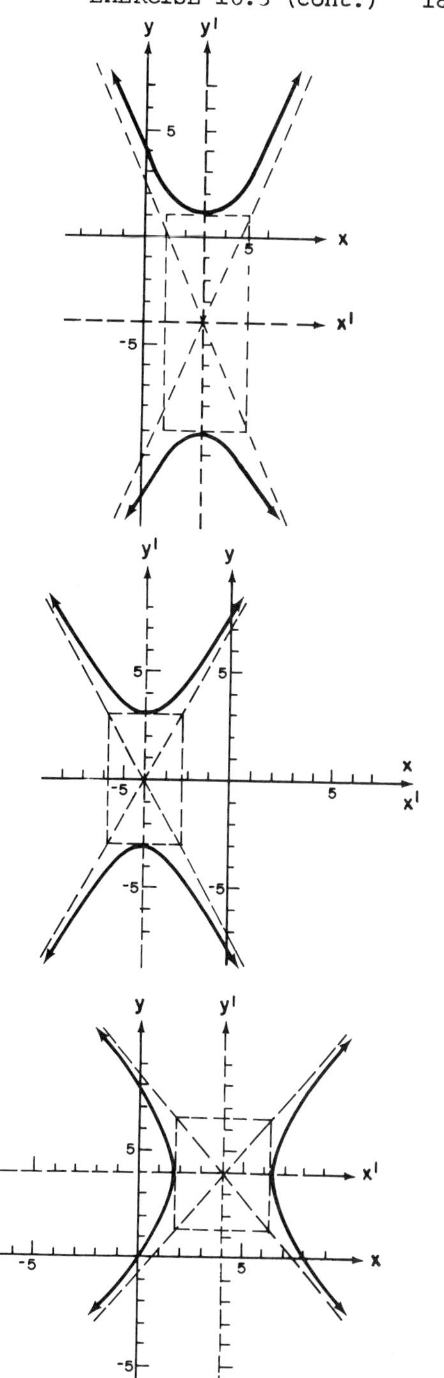

6. The graph is a hyperbola with center at $(3,-4)$. Since $a^2 = 25$ and $b^2 = 4$ then $a = 5$ and $b = 2$. The y'-intercepts are -5 and 5. The central rectangle has dimensions $2a = 10$ and $2b = 4$.

8. The graph is a hyperbola with center at $(-4,0)$. Since $a^2 = 9$ and $b^2 = 12$, then $a = 3$ and $b = \sqrt{12} = 2\sqrt{3}$. The y'-intercepts are -3 and 3. The central rectangle has dimensions $2a = 6$ and $2b = 4\sqrt{3}$. Note: $\sqrt{3} \approx 1.7$.

10. The graph is a hyperbola with center at $(4,4)$. Since $a^2 = 5$ and $b^2 = 8$, then $a = \sqrt{5}$ and $b = \sqrt{8} = 2\sqrt{2}$. The x'-intercepts are $-\sqrt{5}$ and $\sqrt{5}$. The central rectangle dimensions are $2a = 2\sqrt{5}$ and $2b = 4\sqrt{2}$. Note: $\sqrt{5} \approx 2.2$ and $\sqrt{2} \approx 1.4$.

12. The graph is a hyperbola with center at $(0,-2)$. Since $a^2 = 12$ and $b^2 = 7$, then $a = \sqrt{12} = 2\sqrt{3}$ and $b = \sqrt{7}$. The x'-intercepts are $-2\sqrt{3}$ and $2\sqrt{3}$. The central rectangle has dimensions $2a = 4\sqrt{3}$ and $2b = 2\sqrt{7}$. Note: $\sqrt{3} \approx 1.7$ and $\sqrt{7} \approx 2.6$.

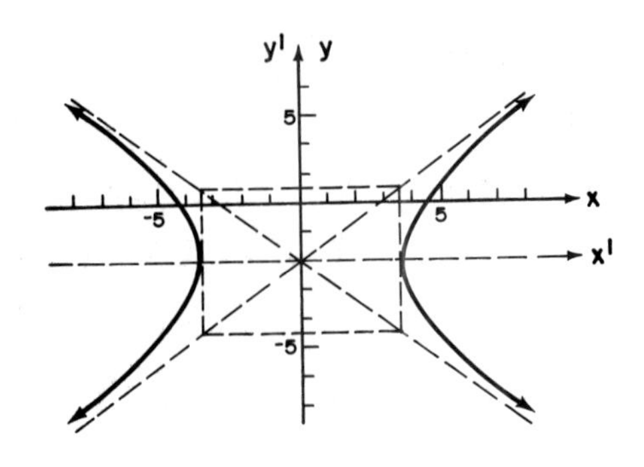

14.
$$9y^2 - 72y \qquad - 4x^2 - 24x \qquad = -72$$
$$9(y^2 - 8y + \ ?) - 4(x^2 + 6x + ?) = -72 + 9(?) - 4(?)$$
$$9(y^2 - 8y + 16) - 4(x^2 + 6x + 9) = -72 + 9(16) - 4(9)$$
$$9(y - 4)^2 - 4(x + 3)^2 = 36$$
$$\frac{(y - 4)^2}{4} - \frac{(x + 3)^2}{9} = 1$$

The graph is a hyperbola with center at $(-3,4)$. Since $a^2 = 4$ and $b^2 = 9$, then $a = 2$ and $b = 3$. The y'-intercepts are -2 and 2. The central rectangle has dimensions $2a = 4$ and $2b = 6$.

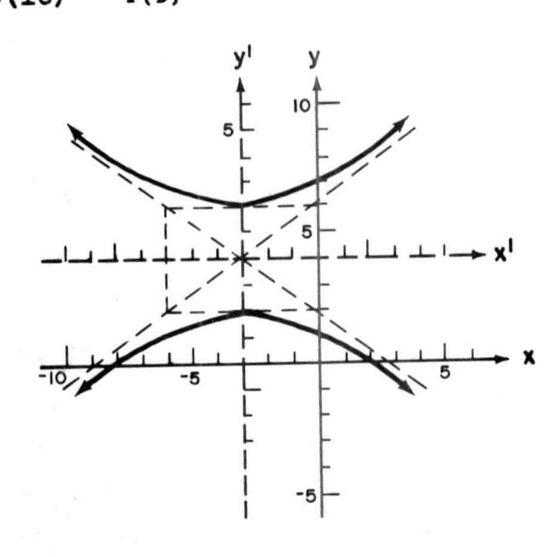

16.
$$16x^2 - 9y^2 + 54y \qquad = 225$$
$$16x^2 - 9(y^2 - 6y + ?) = 225 - 9(?)$$
$$16x^2 - 9(y^2 - 6y + 9) = 225 - 9(9)$$
$$16(x - 0)^2 - 9(y - 3)^2 = 144$$
$$\frac{(x - 0)^2}{9} - \frac{(y - 3)^2}{16} = 1$$

Center: $(0,3)$; $a = 3$, $b = 4$;
x'-intercepts: -3 and 3;
Central rectangle: $2a = 6$ and $2b = 8$

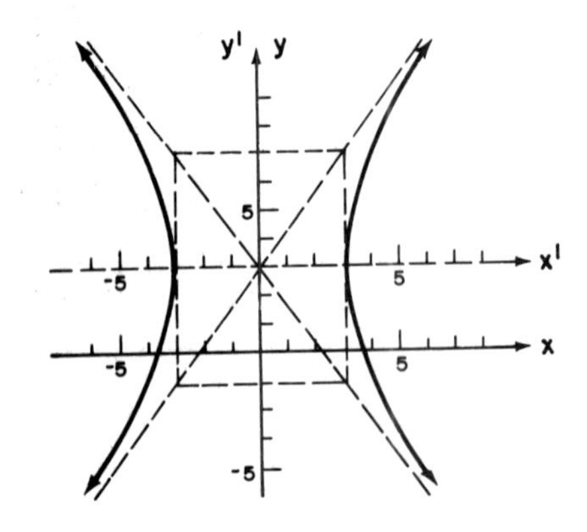

18. $$9y^2 + 72y - 8x^2 + 16x = -64$$
$$9(y^2 + 8y + \ ?) - 8(x^2 - 2x + ?) = -64 + 9(?) - 8(?)$$
$$9(y^2 + 8y + 16) - 8(x^2 - 2x + 1) = -64 + 9(16) - 8(1)$$
$$9(y + 4)^2 - 8(x - 1)^2 = 72$$
$$\frac{(y + 4)^2}{8} - \frac{(x - 1)^2}{9} = 1$$

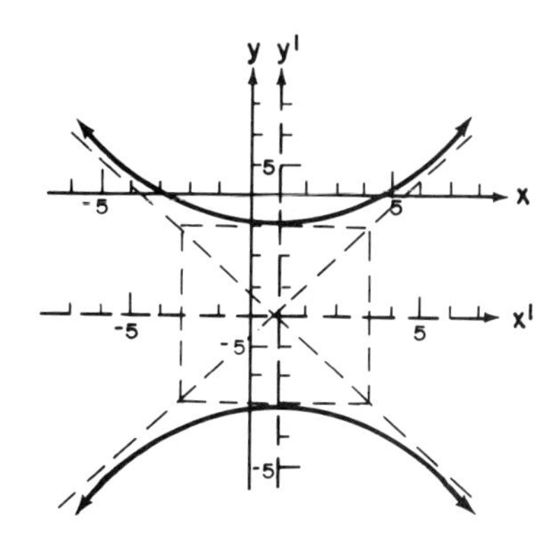

Center: $(1,-4)$; $a = \sqrt{8} = 2\sqrt{2}$ and $b = 3$

y'-intercepts: $-2\sqrt{2}$ and $2\sqrt{2}$

Central rectangle: $2a = 4\sqrt{2}$ and $2b = 6$

Note: $\sqrt{2} \approx 1.4$

20. $$10y^2 - 5x + 30x = 95$$
$$10y^2 - 5(x^2 - 6x + ?) = 95 - 5(?)$$
$$10y^2 - 5(x^2 - 6x + 9) = 95 - 5(9)$$
$$10(y - 0)^2 - 5(x - 3)^2 = 50$$
$$\frac{(y - 0)^2}{5} - \frac{(x - 3)^2}{10} = 1$$

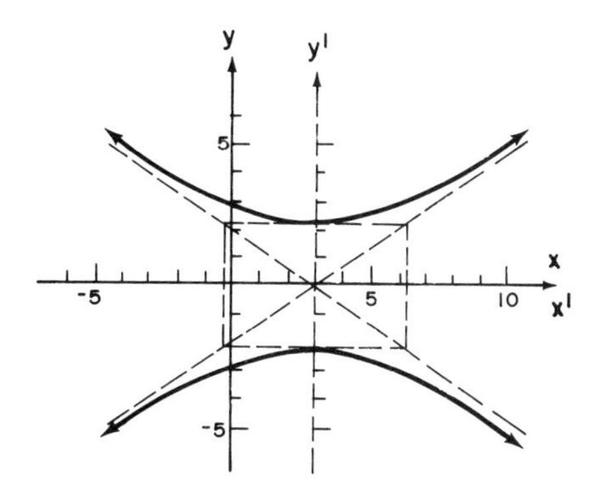

Center: $(3,0)$; $a = \sqrt{5}$, $b = \sqrt{10}$

y'-intercepts: $-\sqrt{5}$ and $\sqrt{5}$

Central rectangle: $2a = 2\sqrt{5}$ and $2b = 2\sqrt{10}$

Note: $\sqrt{5} \approx 2.2$ and $\sqrt{10} \approx 3.2$

In the solutions for Exercise 22-36, the given equations are written in standard form from which the graph is identified.

22. $4x^2 + y^2 = 6$

$$\frac{4x^2}{6} + \frac{y^2}{6} = 1$$

$$\frac{x^2}{3/2} + \frac{y^2}{6} = 1$$

The graph is an ellipse.

24. $2y = -x^2 + 4$

$$y = -\frac{1}{2}x^2 + 2$$

or

$$y = -\frac{1}{2}(x - 0)^2 + 2$$

The graph is a parabola.

26. $6x^2 + 6y^2 = 8$

$$x^2 + y^2 = \frac{4}{3}$$

The graph is a circle.

28. $2x^2 - 4y^2 = 5$

$$\frac{2x^2}{5} - \frac{4y^2}{5} = 1$$

$$\frac{x^2}{5/2} - \frac{y^2}{5/4} = 1$$

The graph is a hyperbola.

30. $\frac{2}{3}x^2 + y^2 = 6$

$$\frac{2}{18}x^2 + \frac{y^2}{6} = 1$$

$$\frac{x^2}{9} + \frac{y^2}{6} = 1$$

The graph is an ellipse.

32. $\frac{1}{4}x^2 - 6y^2 = 4$

$$\frac{x^2}{16} - \frac{6y^2}{4} = 1$$

$$\frac{x^2}{16} - \frac{y^2}{2/3} = 1$$

The graph is a hyperbola.

34. $\frac{(y - 2)^2}{4} - \frac{(x - (-3))^2}{8} = 1$

The graph is a hyperbola.

36. $\frac{(x - (-3))^2}{4} + \frac{y^2}{12} = 1$

The graph is an ellipse.

38. Since the coefficients of x^2 and y^2 are of opposite sign, the graph is a hyperbola.

40. Since the equation is first degree in y and second degree in x, the graph is a parabola.

42. $4x^2 - y^2 = 0$ is equivalent to $(2x - y)(2x + y) = 0$. Hence, we graph $y = 2x$ and $y = -2x$.

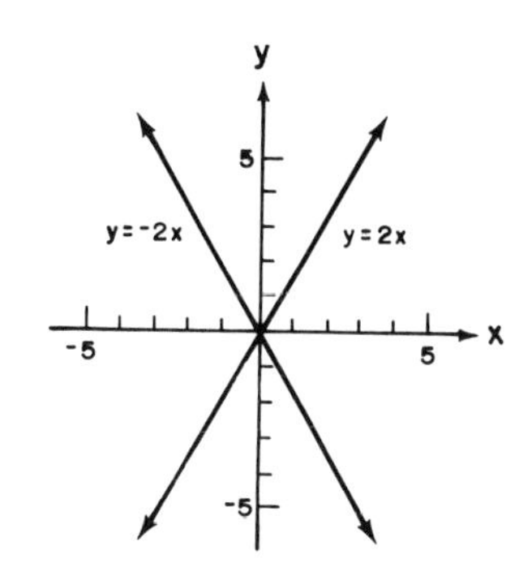

44.

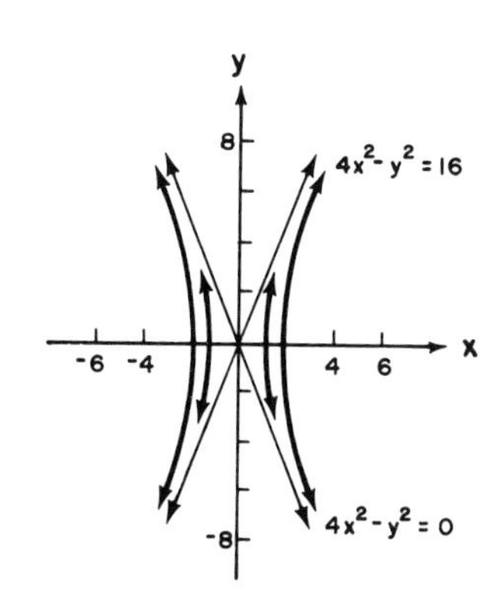

46. The graph of $x^2 + y^2 + ax + by + c = 0$ is a circle. Its graph will contain the points (2,3), (3,2), and (-4,-5) if and only if these pairs are solutions of $x^2 + y^2 + ax + by + c = 0$. Hence, we substitute these pairs into the equation and obtain

$$2^2 + 3^2 + 2a + 3b + c = 0$$
$$3^2 + 2^2 + 3a + 2b + c = 0$$
$$(-4)^2 + (-5)^2 - 4a - 5b + c = 0$$

This system can be written as:

$$2a + 3b + c = -13$$
$$3a + 2b + c = -13$$
$$4a + 5b - c = 41$$

Solve the system.

46. cont'd.

Add the first and third equations: $6a + 8b = 28$
or
$3a + 4b = 14$

Add the second and third equations: $7a + 7b = 28$
or
$a + b = 4$

Solve the system

$$3a + 4b = 14$$
$$a + b = 4$$

Add -3 times the second equation to the first equation and obtain $b = 2$. Substitute 2 for b in $a + b = 4$ and obtain $a = 2$. Substitute 2 for a and b in $2a + 3b + c = -13$ and obtain $c = -23$. Hence the required equation is

$$x^2 + y^2 + 2x + 2y - 23 = 0.$$

EXERCISE 10.4

2. $y = x^2 - 2x + 1$ (1)
   $y + x = 3$ (2)

   Solve Equation (2) explicitly for y: $y = 3 - x$.

   Substitute $3 - x$ for y in Equation (1):

   $3 - x = x^2 - 2x + 1$; $0 = x^2 - x - 2$; $0 = (x - 2)(x + 1)$; $x = 2$; $x = -1$.

   Substitute these values for x in Equation (2):

   $y + 2 = 3$; $y = 1$.
   $y + (-1) = 3$; $y = 4$.

   The solution set is $\{(2,1), (-1,4)\}$.

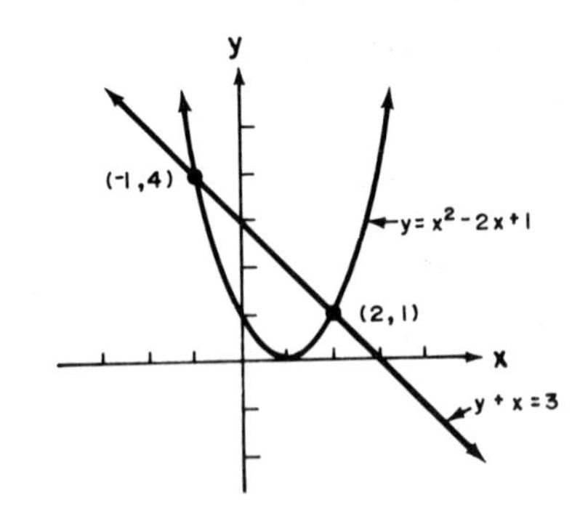

4. $x^2 + 2y^2 = 12$ (1)
$2x - y = 2$ (2)

Solve Equation (2) explicitly for y: $2x - 2 = y$.

Substitute $2x - 2$ for y in Equation (1):

$x^2 + 2(2x - 2)^2 = 12$; $x^2 + 2(4x^2 - 8x + 4) = 12$; $9x^2 - 16x - 4 = 0$;

$(9x + 2)(x - 2) = 0$; $x = \frac{-2}{9}$; $x = 2$.

Substitute these values for x in Equation (2):

$2\left(\frac{-2}{9}\right) - y = 2$; $y = \frac{-22}{9}$;

$2(2) - y = 2$; $y = 2$.

The solution set is

$\left\{\left(\frac{-2}{9}, \frac{-22}{9}\right), (2,2)\right\}$.

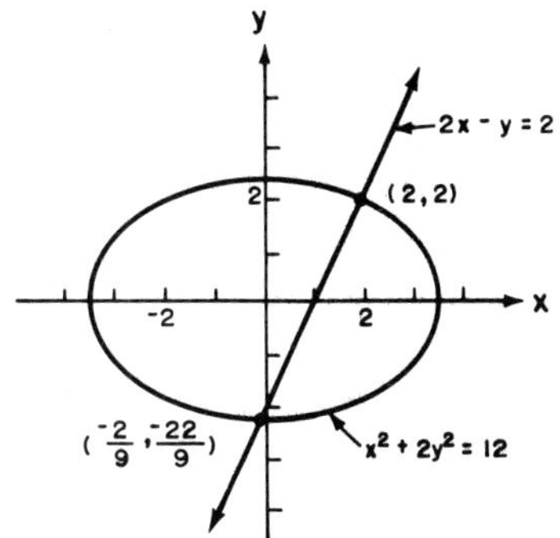

6. $2x - y = 9$ (1)
$xy = -4$ (2)

Solve Equation (1) for y: $2x - 9 = y$.

Substitute $2x - 9$ for y in Equation (2):

$x(2x - 9) = 4$; $2x^2 - 9x + 4 = 0$; $(2x - 1)(x - 4) = 0$; $x = \frac{1}{2}$; $x = 4$.

Substitute these values for x in Equation (1):

$2\left(\frac{1}{2}\right) - y = 9$; $y = -8$;

$2(4) - y = 9$; $y = -1$.

The solution set is

$\left\{\left(\frac{1}{2}, -8\right), (4,-1)\right\}$.

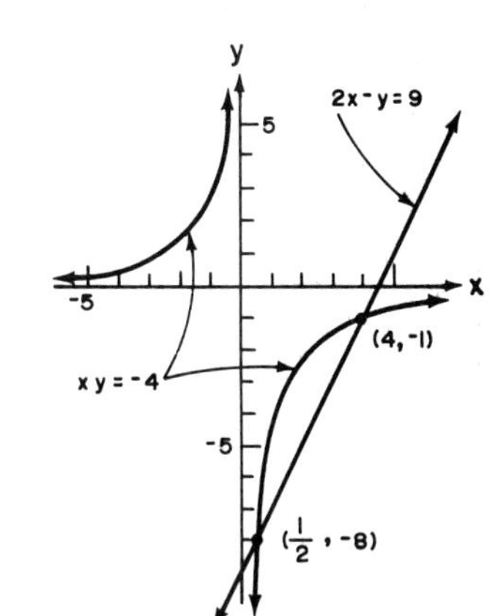

8. $x^2 - y^2 = 35$ (1)
$xy = 6$ (2)

Solve Equation (2) for y: $y = \frac{6}{x}$.

Substitute $\frac{6}{x}$ for y in Equation (1):

$x^2 - \left(\frac{6}{x}\right)^2 = 35$; $x^2 - \frac{36}{x^2} = 35$; $x^4 - 35x^2 - 36 = 0$; $(x^2 - 36)(x^2 + 1) = 0$;

$x = \pm 6$; $x = \pm i$.

Substitute these values for x in Equation (2):

$6y = 6$ or $y = 1$; $-6y = 6$ or $y = -1$; $iy = 6$; $y = \frac{6}{i}$ or $y = -6i$;

$-iy = 6$; $y = \frac{6}{-i}$ or $y = 6i$.

The solution set is $\{(6,1), (-6,-1), (i,-6i), (-i,6i)\}$.

Imaginary solutions do not show on the graph.

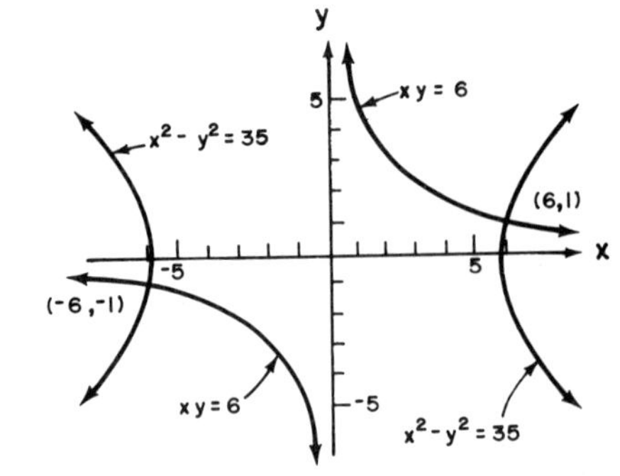

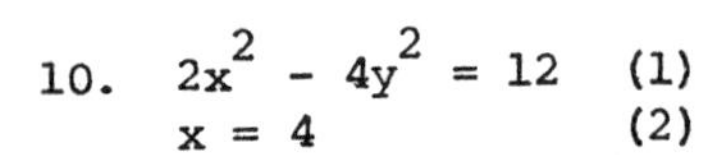

10. $2x^2 - 4y^2 = 12$ (1)
$x = 4$ (2)

Substitute 4 for x in Equation (1):

$2(4^2) - 4y^2 = 12$;

$32 - 4y^2 = 12$;

$5 = y^2$; $y = \pm\sqrt{5}$.

The solution set is $\{(4,\sqrt{5}), (4,-\sqrt{5})\}$.

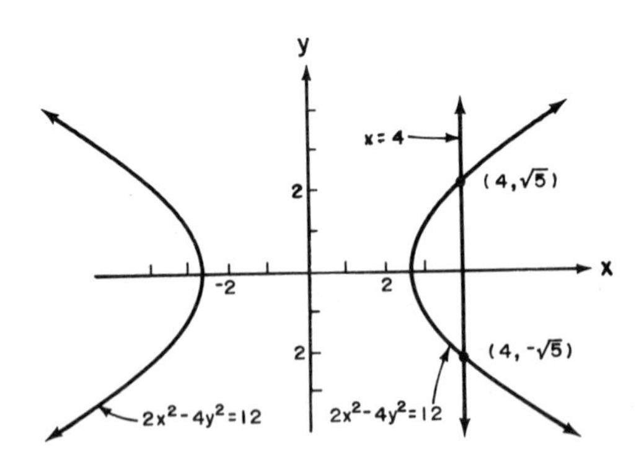

12. $x^2 + 9y^2 = 36$ (1)
$x - 2y = -8$ (2)

Solve Equation (2) for x: $x = 2y - 8$. (3)

Substitute $2y - 8$ for x in Equation (1);

$$(2y - 8)^2 + 9y^2 = 36;\ 13y^2 - 32y + 28 = 0.$$

Use the quadratic formula:

$$y = \frac{32 \pm \sqrt{-432}}{26} = \frac{16 \pm 6i\sqrt{3}}{13}.$$

Substitute these values for y in Equation (3):

$$x = 2\left(\frac{16 \pm 6i\sqrt{3}}{13}\right) - 8 = \frac{-72 \pm 12i\sqrt{3}}{13}.$$

The solution set is

$$\left\{\left(\frac{-72 + 12i\sqrt{3}}{13}, \frac{16 + 6i\sqrt{3}}{13}\right),\right.$$

$$\left.\left(\frac{-72 - 12i\sqrt{3}}{13}, \frac{16 - 6i\sqrt{3}}{13}\right)\right\}.$$

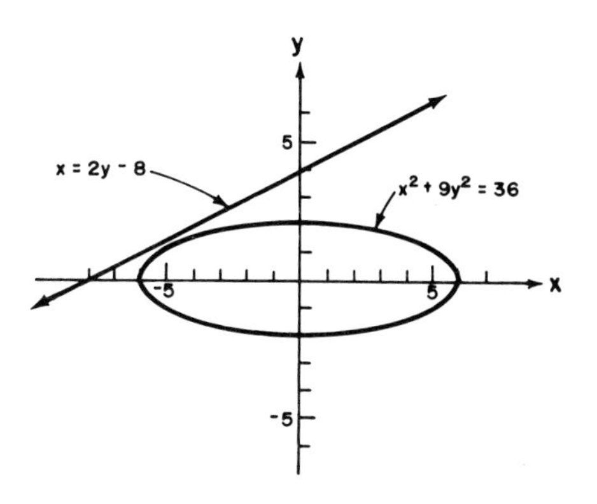

14. $x^2 - 2x + y^2 = 3$ (1)
$2x + y = 4$ (2)

Solve Equation (2) for y: $y = 4 - 2x$.

Substitute $4 - 2x$ for y in Equation (1):

$x^2 - 2x + (4 - 2x)^2 = 3;\ 5x^2 - 18x + 13 = 0;\ (5x - 13)(x - 1) = 0;$
$x = \frac{13}{5};\ x = 1.$

Substitute these values for x in Equation (2);

$$2(1) + y = 4;\ y = 2;\ 2\left(\frac{13}{5}\right) + y = 4;\ y = \frac{-6}{5}.$$

The solution set is $\left\{\left(\frac{13}{5}, \frac{-6}{5}\right), (1,2)\right\}$.

16. $2x^2 + xy + y^2 = 9$ (1)
$-x + 3y = 9$ (2)

Solve Equation (2) for x: $3y - 9 = x$.

Substitute $3y - 9$ for x in Equation (1):

$2(3y - 9)^2 + (3y - 9)y + y^2 = 9$; $22y^2 - 117y + 153 = 0$;
$(22y - 51)(y - 3) = 0$; $y = \frac{51}{22}$; $y = 3$.

Substitute these values for y in Equation (2):

$$-x + 3\left(\frac{51}{22}\right) = 9;\ x = \frac{-45}{22};\quad -x + 3(3) = 9;\ x = 0.$$

The solution set is $\left\{\left(\frac{-45}{22}, \frac{51}{22}\right), (0,3)\right\}$.

18. One of the numbers: x
The other number: y

$$x + y = 6 \quad (1)$$
$$xy = \frac{35}{4} \quad (2)$$

Solve Equation (1) for y: $y = 6 - x$.

Substitute $6 - x$ for y in Equation (2):

$x(6 - x) = \frac{35}{4}$; $24x - 4x^2 = 35$; $0 = (2x - 7)(2x - 5)$; $x = \frac{7}{2}$; $x = \frac{5}{2}$.

Substitute these values in Equation (1):

$$\frac{7}{2} + y = 6;\ y = \frac{5}{2};\ \frac{5}{2} + y = 6;\ y = \frac{7}{2}.$$

The numbers are $\frac{5}{2}$ and $\frac{7}{2}$. The order does not matter.

20. The length: x and the width: y

$$xy = 216 \quad (1)$$
$$2x + 2y = 60 \quad (2)$$

Solve Equation (2) for y: $2y = 60 - 2x$; $y = 30 - x$.

Substitute $30 - x$ for y in Equation (1):

$x(30 - x) = 216$; $0 = x^2 - 30x + 216$; $0 = (x - 18)(x - 12)$.
$x = 18$; $x = 12$.

Substitute these values for x in Equation (2);

$$2(18) + 2y = 60;\ y = 12;\ 2(12) + 2y = 60;\ y = 18.$$

The rectangle is 18 feet by 12 feet.

22. The original pressure: x
The original volume: y

$$xy = 30 \quad (1)$$

$(x + 4)(y - 2) = 30$ which is equivalent to $xy + 4y - 2x = 38 \quad (2)$

Solve Equation (1) for y and obtain $y = \frac{30}{x}$.

In Equation (2) substitute $\frac{30}{x}$ for y and solve for x.

$$x\left(\frac{30}{x}\right) + 4\left(\frac{30}{x}\right) - 2x = 38$$

$$30 + \frac{120}{x} - 2x = 38$$

$$\frac{120}{x} - 2x = 8$$

$$120 - 2x^2 = 8x$$

$$2x^2 + 8x - 120 = 0$$

$$2(x^2 + 4x - 60) = 0$$

$$2(x + 10)(x - 6) = 0; \quad \{-10, 6\}$$

Since the pressure cannot be negative, -10 is rejected.

To determine y substitute 6 for x in $xy = 30$ from which $y = 5$.

Hence, the original pressure was 6 pounds per square inch and the original volume was 5 cubic inches.

EXERCISE 10.5

2. $x^2 + 4y^2 = 52 \quad (1)$
$x^2 + y^2 = 25 \quad (2)$

Adding (-1) times Equation (2) to Equation (1):

$$3y^2 = 27; \; y^2 = 9; \; y = \pm 3.$$

Substituting these values for y in Equation (2):

$$x^2 + (3)^2 = 25; \; x^2 = 16; \; x = \pm 4;$$

$$x^2 + (-3)^2 = 25; \; x^2 = 16; \; x = \pm 4.$$

The solution set is $\{(4,3), (-4,3), (4,-3), (-4,-3)\}$.

4. $9x^2 + 16y^2 = 100$ (1)
   $x^2 + y^2 = 8$ (2)

   Adding -9 times Equation (2) to Equation (1):

$$7y^2 = 28;\ y^2 = 4;\ y = \pm 2.$$

   Substituting these values for y in Equation (2):

$$x^2 + (2)^2 = 8;\ x^2 = 4;\ x = \pm 2;$$
$$x^2 + (-2)^2 = 8;\ x^2 = 4;\ x = \pm 2.$$

   The solution set is $\{(2,2), (-2,2), (2,-2), (-2,-2)\}$.

6. $x^2 + 4y^2 = 25$ (1)
   $4x^2 + y^2 = 25$ (2)

   Adding -4 times Equation (2) to Equation (1):

$$-15x^2 = -75;\ x^2 = 5;\ x = \pm\sqrt{5}.$$

   Substituting these values for x in Equation (2):

$$4(\sqrt{5})^2 + y^2 = 25;\ y^2 = 5;\ y = \pm\sqrt{5};$$
$$4(-\sqrt{5})^2 + y^2 = 25;\ y^2 = 5;\ y = \pm\sqrt{5}.$$

   The solution set is $\{(\sqrt{5},\sqrt{5}), (\sqrt{5},-\sqrt{5}), (-\sqrt{5},\sqrt{5}), (-\sqrt{5},-\sqrt{5})\}$.

8. $4x^2 + 3y^2 = 12$ (1)
   $x^2 + 3y^2 = 12$ (2)

   Adding -1 times Equation (2) to Equation (1):

$$3x^2 = 0;\ x = 0.$$

   Substituting 0 for x in Equation (2):

$$0 + 3y^2 = 12;\ y^2 = 4;\ y = \pm 2.$$

   The solution set is $\{(0,2), (0,-2)\}$.

10. $16y^2 + 5x^2 - 26 = 0$ (1)
$25y^2 - 4x^2 - 17 = 0$ (2)

Adding 4 times Equation (1) to 5 times Equation (2):

$$189y^2 - 189 = 0;\ y^2 = 1;\ y = \pm 1.$$

Substituting these values for y in Equation (1):

$$16(1)^2 + 5x^2 - 26 = 0;\ x^2 = 2;\ x = \pm\sqrt{2};$$
$$16(-1)^2 + 5x^2 - 26 = 0;\ x^2 = 2;\ x = \pm\sqrt{2}.$$

The solution set is $\{(\sqrt{2},1), (-\sqrt{2},1), (\sqrt{2},-1), (-\sqrt{2},-1)\}$.

12. $x^2 + 2xy - y^2 = 14$ (1)
$x^2 - y^2 = 8$ (2)

Adding -1 times Equation (2) to Equation (1):

$$2xy = 6;\ xy = 3. \quad (3)$$

Solving this equation for y: $y = \frac{3}{x}$.

Substituting $\frac{3}{x}$ for y in Equation (2):

$x^2 - \left(\frac{3}{x}\right)^2 = 8;\ x^4 - 8x^2 - 9 = 0;\ (x^2 - 9)(x^2 + 1) = 0;\ x^2 - 9 = 0;$
$x^2 + 1 = 0;\ x = \pm 3;\ x = \pm i.$

Substituting these values for x in Equation (3):

$3y = 3,\ y = 1,\ -3y = 3,\ y = -1.$

$(i)y = 3,\ y = \frac{3}{i} = -3i.\ (-i)y = 3,\ y = \frac{3}{-i} = 3i.$

The solution set is $\{(3,1), (-3,-1), (i,-3i), (-i,3i)\}$.

14. $2x^2 + xy - 2y^2 = 16$ (1)
$x^2 + 2xy - y^2 = 17$ (2)

Adding -2 times Equation (2) to Equation (1):

$$-3xy = -18;$$
$$xy = 6. \quad (3)$$

Solving this equation for y: $y = \frac{6}{x}$.

14. cont'd.

Substituting $\frac{6}{x}$ for y in Equation (2):

$$x^2 + 2x\left(\frac{6}{x}\right) - \left(\frac{6}{x}\right)^2 = 17;\ x^4 - 5x^2 - 36 = 0;$$

$$(x^2 - 9)(x^2 + 4) = 0;$$

$$x^2 - 9 = 0;\ x^2 + 4 = 0;\ x = \pm 3;\ x = \pm 2i.$$

Substituting these values for x in Equation (3):

$3y = 6$, $y = 2$. $-3y = 6$, $y = -2$.

$2iy = 6$, $y = \frac{6}{2i} = -3i$. $-2iy = 6$, $y = \frac{6}{-2i} = 3i$.

The solution set is $\{(3,2), (-3,-2), (2i,-3i), (-2i,3i)\}$.

16. $3x^2 - 2xy + 3y^2 = 34$ (1)

$x^2 + y^2 = 17$ (2)

Adding -3 times Equation (2) to Equation (1):

$$-2xy = -17 \quad (3)$$

Solve Equation (3) for y: $y = \frac{17}{2x}$ (4)

Substitute $\frac{17}{2x}$ for y in Equation (2):

$$x^2 + \left(\frac{17}{2x}\right)^2 = 17$$

$$x^2 + \frac{289}{4x^2} = 17$$

$$4x^4 + 289 = 68x^2$$

$$4x^4 - 68x^2 + 289 = 0$$

$$(2x^2 - 17)(2x^2 - 17) = 0;\quad x = \pm\sqrt{\frac{17}{2}} = \pm\frac{\sqrt{34}}{2}.$$

Substituting these values for x in Equation (4):

$$y = \frac{17}{\not{2}\left(\frac{\sqrt{34}}{\not{2}}\right)} = \frac{17}{\sqrt{34}} = \frac{\sqrt{34}}{2}.$$

$$y = \frac{17}{\not{2}\left(\frac{-\sqrt{34}}{\not{2}}\right)} = \frac{-17}{\sqrt{34}} = \frac{-\sqrt{34}}{2}.$$

16. cont'd.

The solution set is $\left\{\left(\frac{\sqrt{34}}{2},\frac{\sqrt{34}}{2}\right), \left(\frac{-\sqrt{34}}{2},\frac{-\sqrt{34}}{2}\right)\right\}$.

18. $x^2 - xy + y^2 = 21$ (1)

$x^2 + 2xy - 8y^2 = 0$ (2)

Factor the left side of Equation (2)

$$(x + 4y)(x - 2y) = 0$$

Hence $x + 4y = 0$ or $x - 2y = 0$

Solve the two systems

$$\begin{aligned} x^2 - xy + y^2 &= 21 \\ x + 4y &= 0 \end{aligned} \quad \text{and} \quad \begin{aligned} x^2 - xy + y^2 &= 21 \\ x - 2y &= 0. \end{aligned}$$

Solving the first of these systems:

Substitute $-4y$ for x from the second equation into $x^2 - xy + y^2 = 21$ and obtain

$$(-4y)^2 - (-4y)y + y^2 = 21$$
$$21y^2 = 21; \quad y^2 = 1; \; y = \pm 1.$$

Using $x + 4y = 0$, when $y = 1$, $x + 4(1) = 0$, $x = -4$.
when $y = -1$, $x + 4(-1) = 0$, $x = 4$.

Solving the second system:

Substitute $2y$ for x from the second equation into $x^2 - xy + y^2 = 21$ and obtain

$$(2y)^2 - (2y)y + y^2 = 21$$
$$3y^2 = 21, \; y^2 = 7, \; y = \pm\sqrt{7}.$$

Using $x - 2y = 0$, when $y = \sqrt{7}$, $x - 2(\sqrt{7}) = 0$, $x = 2\sqrt{7}$.
when $y = -\sqrt{7}$, $x - 2(-\sqrt{7}) = 0$, $x = -2\sqrt{7}$.

The solution set is $\{(-4,1), (4,-1), (2\sqrt{7},\sqrt{7}), (-2\sqrt{7},-\sqrt{7})\}$.

20. Add -1 times the first equation to the second equation and obtain

$$4x^2 = 1 \text{ from which } x = \pm\frac{1}{2}.$$

Substitute $\frac{1}{2}$ for x in the first equation:

$$2\left(\frac{1}{2}\right)^2 + \frac{1}{2}y - y^2 = 0$$

$$-y^2 + \frac{1}{2}y + \frac{1}{2} = 0 \quad (1)$$

$$2y^2 - y - 1 = 0$$

$$(2y + 1)(y - 1) = 0 \quad \text{from which } y = \frac{-1}{2} \text{ or } y = 1.$$

Thus we have $\left(\frac{1}{2}, \frac{-1}{2}\right)$ and $\left(\frac{1}{2}, 1\right)$.

Substitute $-\frac{1}{2}$ for x in the first equation:

$$2\left(-\frac{1}{2}\right)^2 + \left(-\frac{1}{2}\right)y - y^2 = 0$$

$$\frac{1}{2} - \frac{1}{2}y - y^2 = 0$$

$$2y^2 + y - 1 = 0$$

$$(2y - 1)(y + 1) = 0 \quad \text{from which } y = \frac{1}{2} \text{ or } y = -1.$$

Thus we have $\left(-\frac{1}{2}, \frac{1}{2}\right)$ and $\left(-\frac{1}{2}, -1\right)$.

The solution set is $\left\{\left(-\frac{1}{2}, \frac{1}{2}\right), \left(\frac{1}{2}, 1\right), \left(\frac{1}{2}, \frac{-1}{2}\right), \left(-\frac{1}{2}, -1\right)\right\}$.

22. The graphs have only two points in common, (2,2) and (-2,-2). Hence, these ordered pairs are the only ordered pairs of real numbers in the solution set of the given system of equations. Any solution of $xy = 4$ must have both x- and y-components with the same sign. This is not the case with $x^2 + y^2 = 8$.

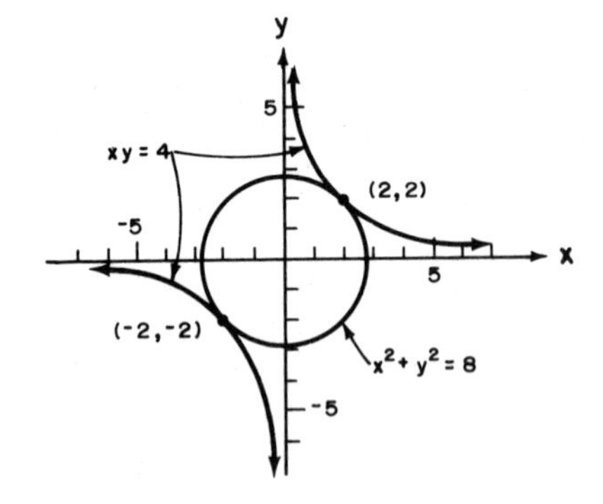

EXERCISE 10.6

2. First graph $y = x^2$ which is in standard form. The graph is a parabola opening upward with vertex at the origin. Plot another point, say (2,4), and use symmetry to complete the graph. Since the parabola is not part of the graph of $y < x^2$, use a dashed line. Substitute the coordinates of a point not on the parabola, say (1,0), into $y < x^2$ and obtain $0 < 1^2$ which is true. So that part of the plane including (1,0) is shaded.

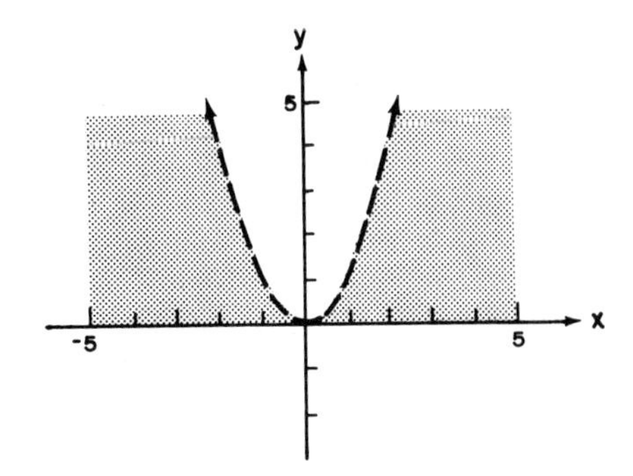

4. First graph $y = x^2 - 4$ which is in standard form. The graph is a parabola opening upward at (0,-4) with x-intercepts at -2 and 2. Since the parabola is part of the graph of $y \geq x^2 - 4$, use a solid line. Substitute the coordinates of a point not on the parabola, say (0,0), into $y \geq x^2 - 4$ and obtain $0 \geq 0^2 - 4$ which is true So that part of the plane including (0,0) is shaded.

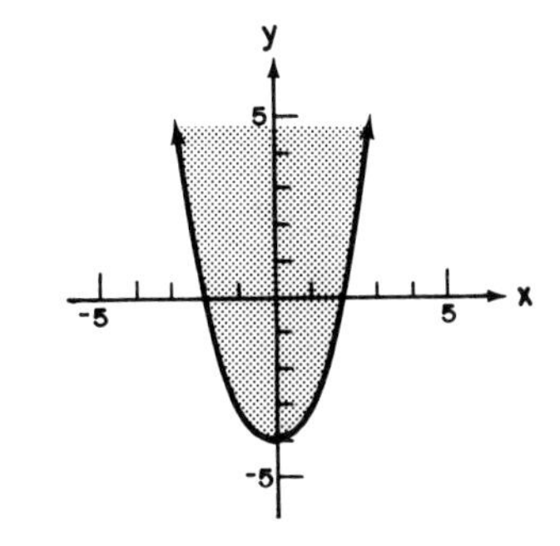

6.

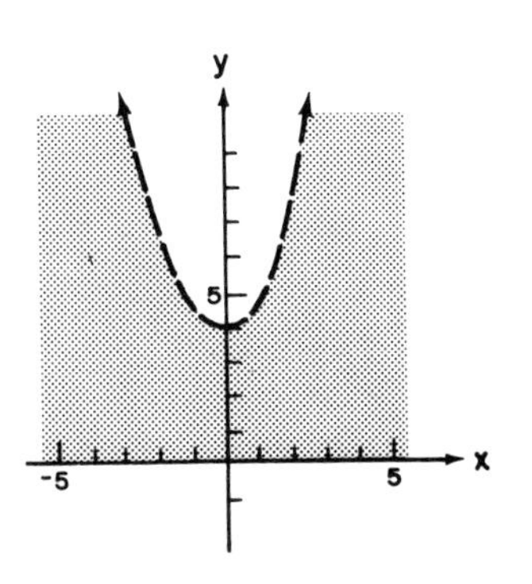

8.

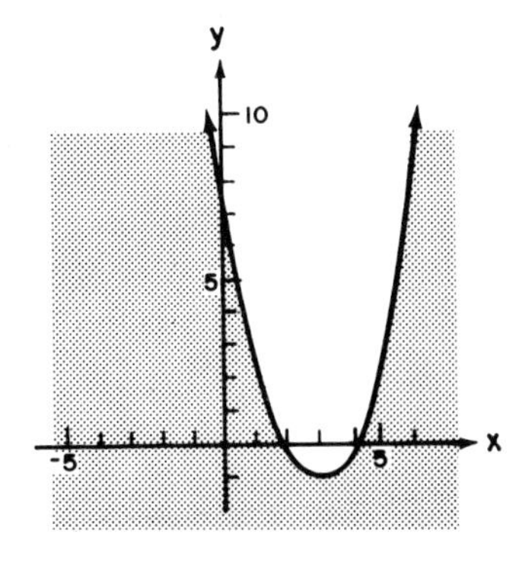

10.

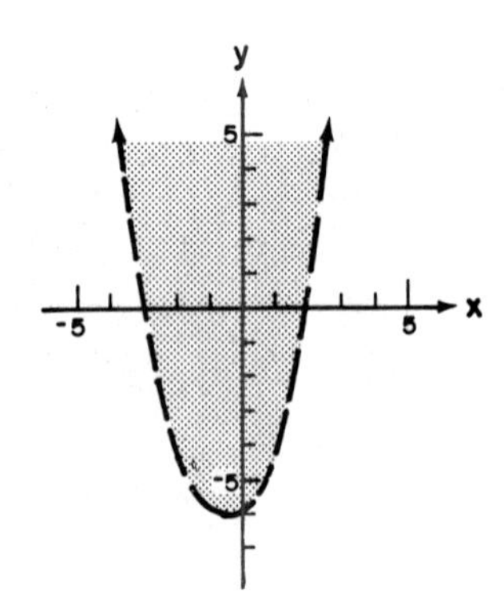

12.

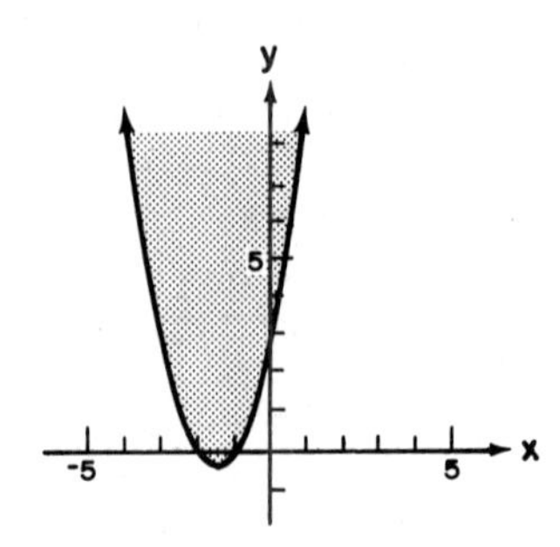

14.

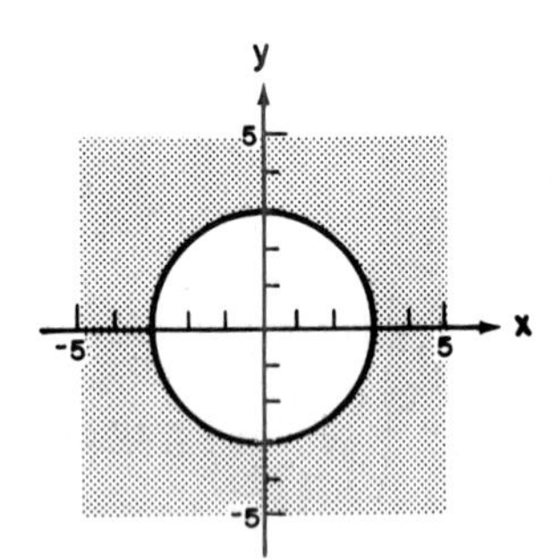

16.

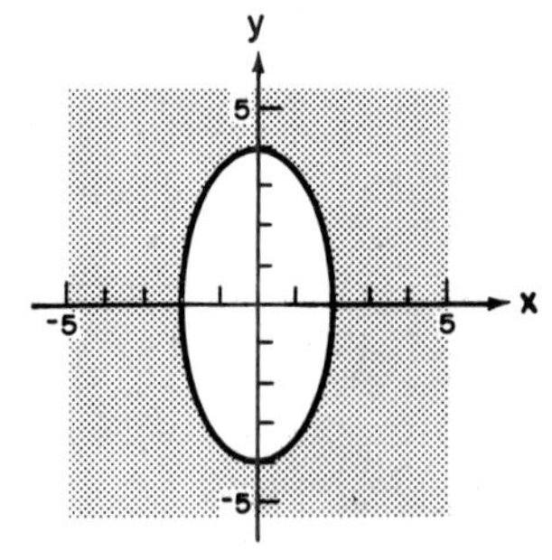

18.

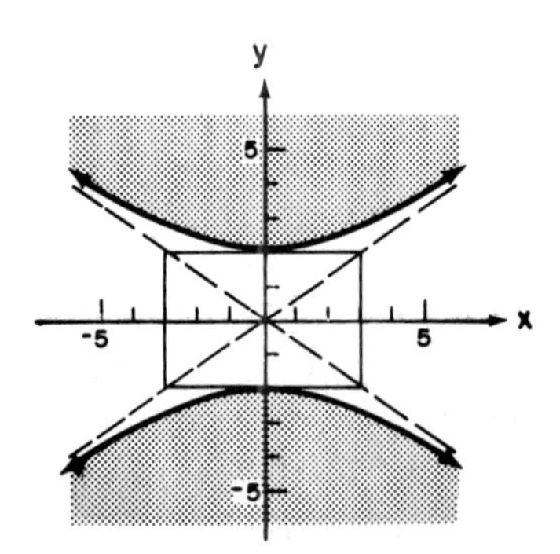

20. First graph the parabola $y = x^2 + 4$ with a dashed line and shade the region below it because (0,0) is a solution of $y < x^2 + 4$. Then graph the straight line $x - y = 4$ with a solid line and shade the region above it because (0,0) is a solution of $x - y \le 4$. The double shaded region is common to both inequalities and is the graph of the solutions of the system of inequalities.

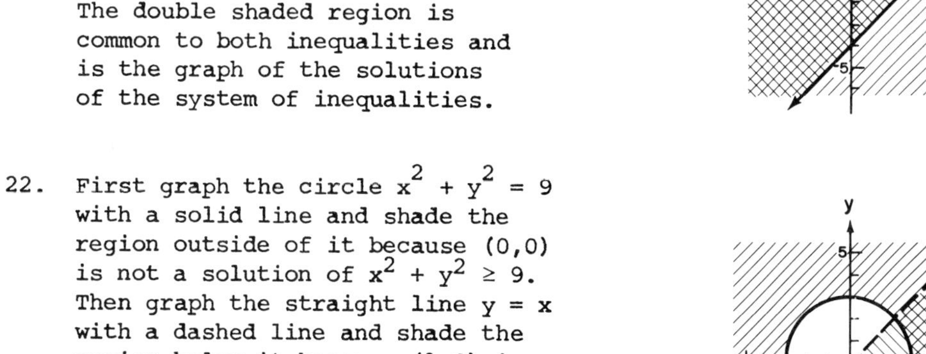

22. First graph the circle $x^2 + y^2 = 9$ with a solid line and shade the region outside of it because (0,0) is not a solution of $x^2 + y^2 \ge 9$. Then graph the straight line $y = x$ with a dashed line and shade the region below it because (1,0) is a solution of $y < x$. The double shaded region is common to both inequalities and is the graph of the solutions of the system of inequalities.

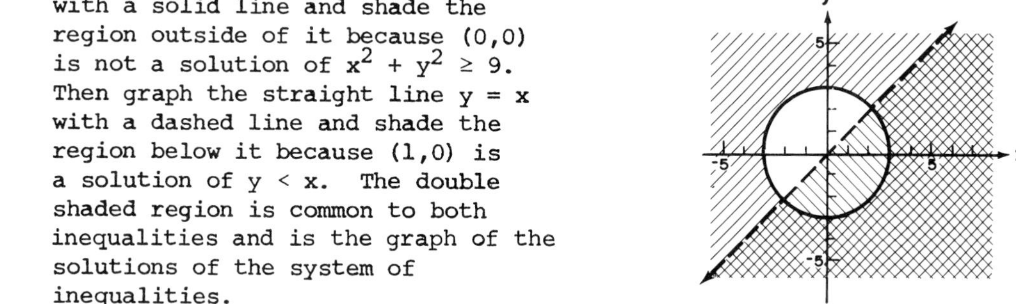

24.

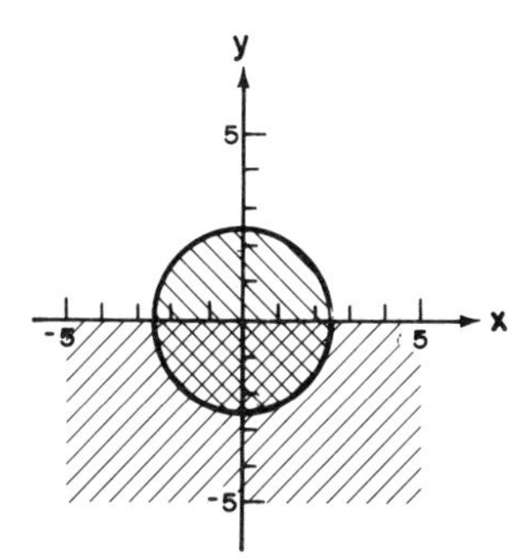

26.

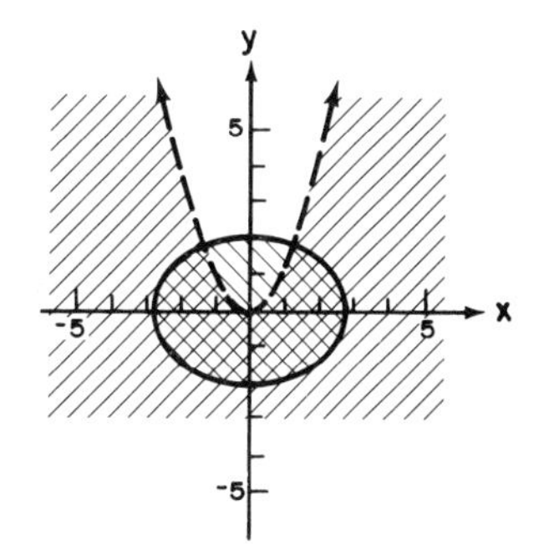

28.

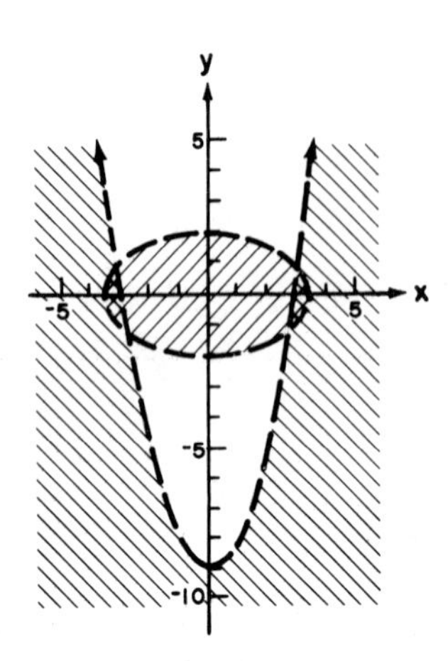

30.

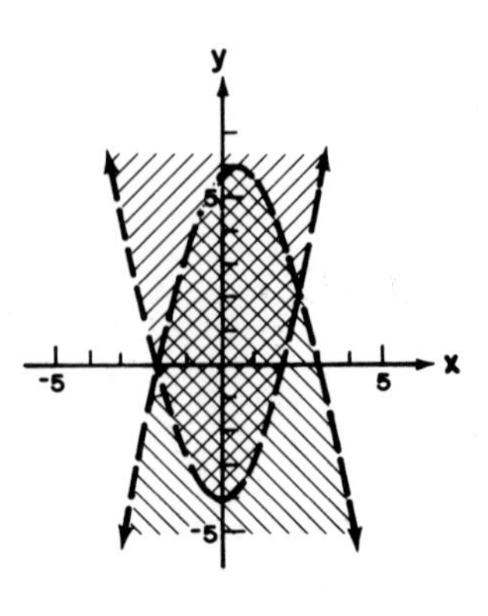

11 Relations and Functions

EXERCISE 11.1

2. a. Domain, the set of first components: $\{-1,0,1,2\}$
   Range, the set of second components: $\{-1,0,1,2\}$

   b. It is a function because no two ordered pairs have the same first component.

4. a. Domain: $\{4\}$
   Range: $\{1,2,3,4\}$

   b. Not a function since two or more pairs have the same first component.

6. a. Domain: $\{-5,2,3,6\}$
   Range: $\{-5,2,3,6\}$

   b. It is a function.

8. a. Domain: $\{2,3,4\}$
   Range: $\{-1,1,2\}$

   b. Not a function since $(3,-1)$ and $(3,1)$ have the same first component.

10. a. $3y = 12 - 2x$

    $y = \dfrac{12 - 2x}{3}$

    b. It is a function.

    c. All real numbers.

12. a. $2y = 3x - 8$

    $y = \dfrac{3x - 8}{2}$

    b. It is a function.

    c. All real numbers.

14. a. $y = \dfrac{6}{x + 3}$

    b. It is a function.

    c. All real numbers except $x = -3$.

16. a. $y(x + 2) = 8$

    $y = \dfrac{8}{x + 2}$

    b. It is a function.

    c. All real numbers except $x = -2$.

18. a. $2y^2 = 8 - x^2$

$$y^2 = \frac{8 - x^2}{2}$$

$$y = \pm\sqrt{\frac{8 - x^2}{2}}$$

b. Not a function because two values of y are paired with each permissible x-value.

c. $8 - x^2 \geq 0$. Hence the domain is all real numbers such that $-\sqrt{8} \leq x \leq \sqrt{8}$.

20. a. $y^2 = 4x^2 + 16$

$$y = \pm\sqrt{4x^2 + 16}$$

$$= \pm 2\sqrt{x^2 + 4}$$

b. Not a function because two values of y are paired with each permissible x-value.

c. $x^2 + 4 \geq 0$. Hence the domain is the set of all real numbers.

22. a. $y \leq 4 - x^2$

b. Not a function because many values of y are paired with each permissible x-value. For example, when $x = 0$, y can be 4,3,2, etc.

24. $g(3) = 2(3)^2 + 3(3) - 1$

$= 26$

26. $f(2) = 3(2) - 1 = 5$
$f(0) = 3(0) - 1 = -1$
Hence,
$f(2) - f(0) = 5 - (-1) = 6.$

28. $f(0) = (0)^2 + 3(0) - 2 = -2$
$f(-2) = (-2)^2 + 3(-2) - 2 = -4$
Hence,
$f(0) - f(-2) = -2 - (-4) = 2.$

30. $f(a) = 3a + 2$
$f(b) = 3b + 2$
$f(a + b) = 3(a + b) + 2$
$= 3a + 3b + 2$

32. $f(a) = a^2 - 2a$
$f(b) = b^2 - 2b$
$f(a + b) = (a + b)^2 - 2(a + b)$
$= a^2 + 2ab + b^2 - 2a - 2b$

34. a. $f(x + h) = 2(x + h) + 5 = 2x + 2h + 5$

b. $f(x + h) - f(x) = (2x + 2h + 5) - (2x + 5) = 2h$

c. $\frac{f(x + h) - f(x)}{h} = \frac{2h}{h} = 2$

36. a. $f(x + h) = (x + h)^2 + 2(x + h) = x^2 + 2hx + h^2 + 2x + 2h$

b. $f(x + h) - f(x) = (x^2 + 2hx + h^2 + 2x + 2h) - (x^2 + 2x)$
$= 2hx + h^2 + 2h$

c. $\frac{f(x + h) - f(x)}{h} = \frac{2hx + h^2 + 2h}{h} = 2x + h + 2$

38. a. $f(x + h) = (x + h)^3 + 3(x + h)^2 - 1$
$= x^3 + 3x^2h + 3xh^2 + h^3 + 3x^2 + 6xh + 3h^2 - 1$

b. $f(x + h) - f(x)$
$= (x^3 + 3x^2h + 3xh^2 + h^3 + 3x^2 + 6xh + 3h^2 - 1) - (x^3 + 3x^2 - 1)$
$= 3x^2h + 3xh^2 + h^3 + 6xh + 3h^2$

c. $\frac{f(x + h) - f(x)}{h} = \frac{3x^2h + 3xh^2 + h^3 + 6xh + 3h^2}{h}$
$= 3x^2 + 3xh + h^2 + 6x + 3h$

40. $f(-3)$ names the ordinate of the point on the graph whose abscissa (first coordinate) is -3. From the graph, $f(-3)$ is approximated to be 7.

Similarly $f(2) \approx 12$ and $f(5) \approx -9$.

The zeros are -4 and 4 because $f(-4) = 0$ and $f(4) = 0$.

42. The zero of
$f(x) = 2x - 8 = 2(x - 4)$
is 4 because $f(4) = 0$.

44. The zeros of
$f(x) = x^2 - 6x + 8 = (x - 2)(x - 4)$
are 2 and 4 because $f(2) = 0$ and $f(4) = 0$.

46. The zeros of
$f(x) = -x^2 + 2x + 15$
$= -(x^2 - 2x - 15)$
$= -(x - 5)(x + 3)$
are -3 and 5 because
$f(-3) = 0$ and $f(5) = 0$.

EXERCISE 11.2

2. $y = \frac{k}{x}$

Since $y = 16$ when $x = 4$, $16 = \frac{k}{4}$ from which $k = 64$.

Hence, $y = \frac{64}{x}$. Then, taking $x = 12$,
$$y = \frac{64}{12} = \frac{16}{3}.$$

4. $y = \frac{kx}{z^2}$

Since $y = 20$ when $x = 4$ and $z = 6$, $20 = \frac{4k}{6^2}$ from which $k = 180$.

Hence, $y = \frac{180x}{z^2}$. Then, taking $x = 12$ and $z = 10$,
$$y = \frac{180(12)}{10^2} = \frac{108}{5}.$$

6. a. $P = kd$. Substitute 40 for P and 10 for d and obtain $40 = k(10)$ from which $k = 4$. Substituting 4 for k in $P = kd$, we have $P = 4d$.

b. Substitute 18 for d in $P = 4d$ and obtain $P = 4(18) = 72$. The pressure is 72 pounds per square foot.

8. a. $R = \frac{k}{d^2}$. Substitute 10 for R and 0.012 for d and obtain

$10 = \frac{k}{(0.012)^2}$ from which $k = 0.00144$. Substituting 0.00144 for k in $R = \frac{k}{d^2}$, we have $R = \frac{0.00144}{d^2}$.

b. Substitute 0.018 for d in $R = \frac{0.00144}{d^2}$ and obtain

$R = \frac{0.00144}{(0.018)^2} \approx 4.44$

The resistance is 4.44 ohms, to the nearest hundredth.

10. $$[\text{Cost}] = \begin{bmatrix}\text{Cost per}\\ \text{Chip}\end{bmatrix} \cdot \begin{bmatrix}\text{Number}\\ \text{of Chips}\end{bmatrix} + \begin{bmatrix}\text{Set-up}\\ \text{Cost}\end{bmatrix}$$

$$C = 24 \cdot n + 12000$$

Substitute 8000 for n and obtain

$$C = 24(8000) + 12000 = 204000$$

The cost is $204,000.

12. The width of the field: w
The length of the field: 2w

$$\begin{bmatrix}\text{Cost of fencing}\\ \text{field}\end{bmatrix} = \begin{bmatrix}\text{Number of feet}\\ \text{of fencing}\end{bmatrix} \times \begin{bmatrix}\text{Cost per foot}\\ \text{of fencing}\end{bmatrix}$$

$$C = [2 \cdot w + 2 \cdot 2w] \times [5.50]$$

$$C = 33w$$

Substitute 140 for w and obtain

$$C = 33(140) = 4620.$$

The cost is $4620.

14. $\left[\text{Cost}\right] = \begin{bmatrix}\text{Cost of}\\ \text{brick walls}\end{bmatrix} + \begin{bmatrix}\text{Cost of}\\ \text{fencing}\end{bmatrix}$

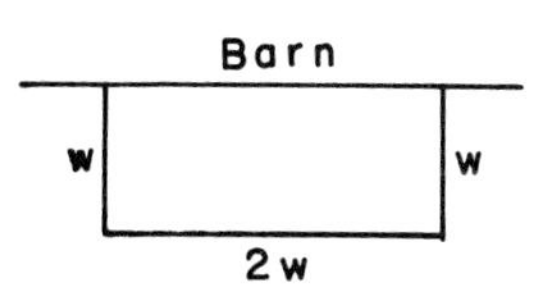

$$C = 15(w + w) + 7.50(2w)$$
$$C = 45w$$

Substitute 30 for w and obtain

$$C = 45(30) = 1350.$$

The cost is \$1350.

16. The area of a circle is given by

$$A = \pi r^2$$

and the circumference is given by

$$C = 2\pi r.$$

Solve $C = 2\pi r$ for r and obtain $r = \frac{C}{2\pi}$.

Substitute $\frac{C}{2\pi}$ for r in $A = \pi r^2$ and obtain

$$A = \pi\left(\frac{C}{2\pi}\right)^2 = \frac{C^2}{4\pi}$$

Substitute 24 for C and obtain

$$A = \frac{(24)^2}{4\pi} \approx 45.84$$

To the nearest hundredth, the area is 45.84 square centimeters.

18. From the diagram,

$$A = x^2$$

and using the Pythagorean theorem,

$$x^2 + x^2 = (2r)^2$$
$$2x^2 = 4r^2 \text{ from which}$$
$$x^2 = 2r^2.$$

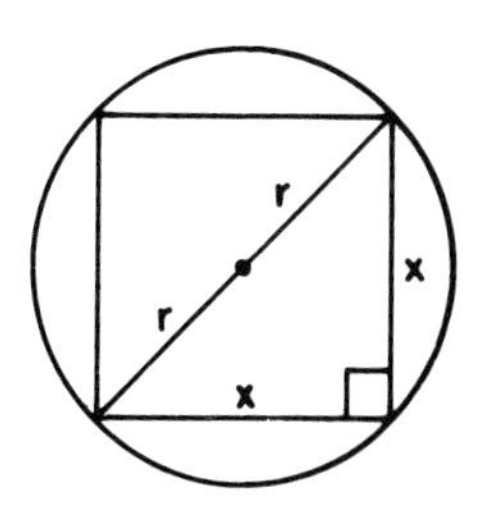

Substitute $2r^2$ for x^2 in $A = x^2$.

Hence, $A = 2r^2$.

Substitute 4 for r and obtain

$$A = 2(4)^2 = 32$$

The area is 32 square inches.

20. From the diagram, the area (A_1) of the square is

$$A_1 = s^2.$$

The area (A_2) of the circle is,

$$A_2 = \pi\left(\frac{1}{2}s\right)^2 = \frac{1}{4}\pi s^2.$$

Hence the area (A) of the waste metal is

$$A = A_1 - A_2$$
$$A = s^2 - \frac{1}{4}\pi s^2.$$

To find the length of the side of the square, the Pythagorean theorem is used to obtain

$$s^2 + s^2 = 8^2$$
$$2s^2 = 64$$
$$s^2 = 32$$
$$s = \sqrt{32} = 4\sqrt{2}.$$

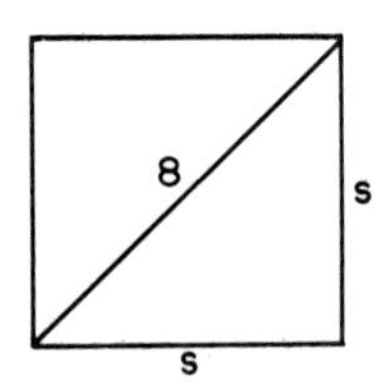

Substitute $4\sqrt{2}$ for s in

$A = s^2 - \frac{1}{4}\pi s^2$ and obtain

$$A = (4\sqrt{2})^2 - \frac{1}{4}\pi(4\sqrt{2})^2 = 32 - 8\pi \approx 6.87.$$

To the nearest hundredth, the area is 6.87 square inches.

22. From the diagram,

$A = (x - 1)(x - 2)$

or

$A = x^2 - 3x + 2$

Substitute 40 for x and obtain

$A = (40 - 1)(40 - 2) = 1482$

The area is 1482 square inches.

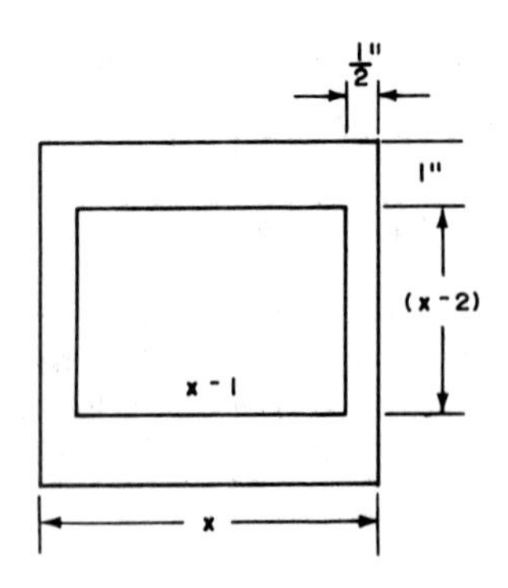

24. From the diagram, one-half the perimeter of the rectangle is given by $\frac{1}{2}(200 - x)$.

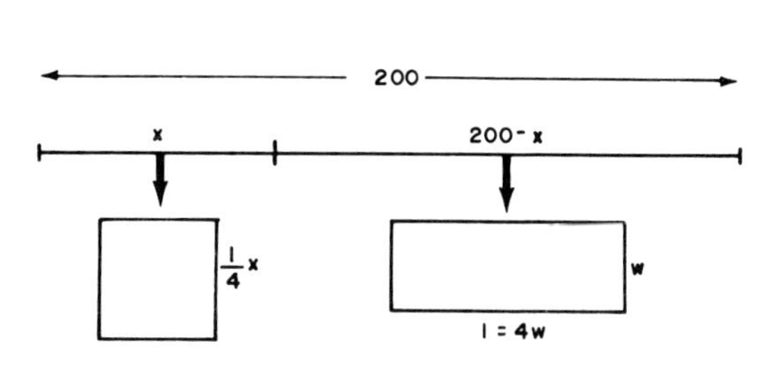

Then,

$$w + 4w = \frac{1}{2}(200 - x)$$

$$5w = \frac{1}{2}(200 - x)$$

$$w = \frac{1}{10}(200 - x)$$

$$\ell = 4w = \frac{4}{10}(200 - x) = \frac{2}{5}(200 - x).$$

The area of the square (A_1) is

$$A_1 = \left(\frac{1}{4}x\right)^2 = \frac{1}{16}x^2$$

The area of the rectangle (A_2) is

$$A_2 = \ell w$$

$$= \frac{2}{5}(200 - x) \cdot \frac{1}{10}(200 - x)$$

$$= \frac{1}{25}(200 - x)^2$$

The total area (A) enclosed is

$$A = A_1 + A_2$$

$$A = \frac{1}{16}x^2 + \frac{1}{25}(200 - x)^2 .$$

26. From the diagram, the area (A_1) of the square is given by

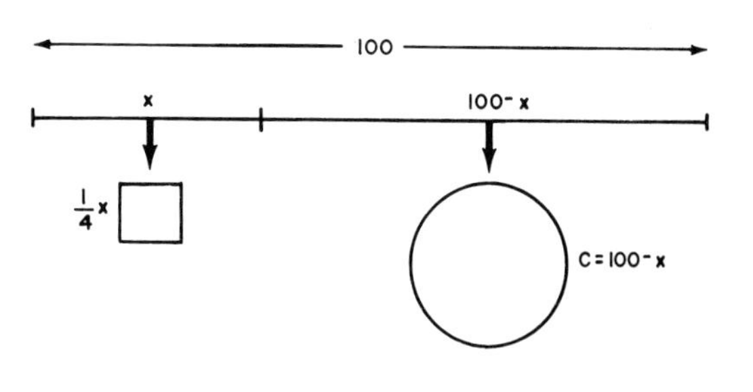

$$A_1 = \frac{1}{16}x^2$$

From the results of Exercise 16 above, the area (A_2) of the circle expressed as a function of the circumference (C) is given by

$$A_2 = \frac{C^2}{4\pi}$$

Substituting (100 - x) for C we have

$$A_2 = \frac{(100 - x)^2}{4}$$

26. cont'd.

The total area (A) is

$$A = A_1 + A_2 = \frac{1}{16}x^2 + \frac{(100 - x)^2}{4\pi}.$$

28. The length (P) fencing needed is given by

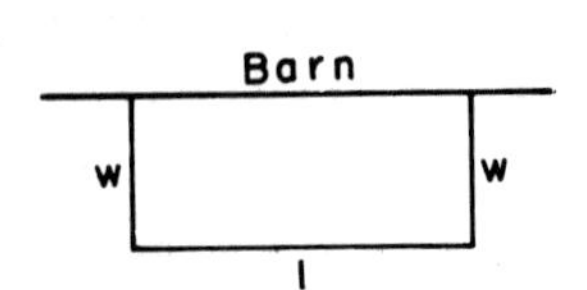

$$P = \ell + 2w.$$

Since the area of the rectangle is given by $\ell \cdot w$, we have

$$\ell w = 125 \text{ from which } \ell = \frac{125}{w}.$$

Hence, $P = \frac{125}{w} + 2w$

The cost (C) of the fencing is given by

$$C = \begin{bmatrix}\text{The number of} \\ \text{feet needed}\end{bmatrix} \cdot \begin{bmatrix}\text{Cost per} \\ \text{foot}\end{bmatrix}$$

$$C = \qquad P \qquad \cdot \qquad 10$$

$$C = 10\left(\frac{125}{w} + 2w\right)$$

or

$$C = \frac{1250}{w} + 20w \quad \text{or} \quad \frac{1250 + 20w^2}{w}$$

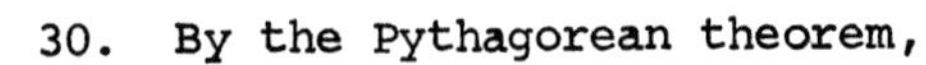

30. By the Pythagorean theorem,

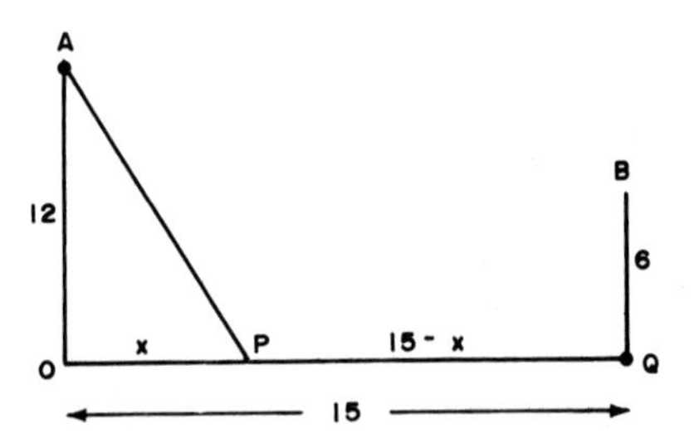

$$AP = \sqrt{x^2 + 144}.$$

The length of cable under water in miles is

$$AP + QB = \sqrt{x^2 + 144} + 6.$$

Since a cost of \$50 per foot is equivalent to \$264,000 per mile, the cost of laying the cable underwater is

$$264{,}000(\sqrt{x^2 + 144} + 6).$$

30. cont'd.

The length of cable underground in miles is $15 - x$ and since a cost of \$30 per foot is equivalent to \$158,400 per mile, the cost of laying the cable underground is $158,400(15 - x)$. Hence, the total cost is given by

$$C = 264,000(\sqrt{x^2 + 144} + 6) + 158,400(15 - x)$$
$$= 264,000\sqrt{x^2 + 144} - 158,400x + 3,960,000.$$

Since $15 - x$ must be a positive number, x is restricted to the interval $(0,15)$.

EXERCISE 11.3

2. $4^{-1/2} = \frac{1}{4^{1/2}} = \frac{1}{2}$

$4^0 = 1$

$4^{1/2} = 2$

4. $5^{-2} = \frac{1}{5^2} = \frac{1}{25}$

$5^0 = 1$

$5^2 = 25$

6. $\left(\frac{1}{3}\right)^{-3} = \frac{1}{\left(\frac{1}{3}\right)^3} = \frac{1}{\frac{1}{27}}$

$= 27$

$\left(\frac{1}{3}\right)^0 = 1$

$\left(\frac{1}{3}\right)^3 = \frac{1}{27}$

8. $10^0 = 1$

$10^1 = 10$

$10^2 = 100$

In Exercises 10 and 12 a scientific calculator was used.

10. $1.8^{-3} \approx 0.17$

$1.8^2 \approx 3.24$

12. $0.3^{0.3} \approx 0.70$

$0.3^{1.6} \approx 0.15$

14.

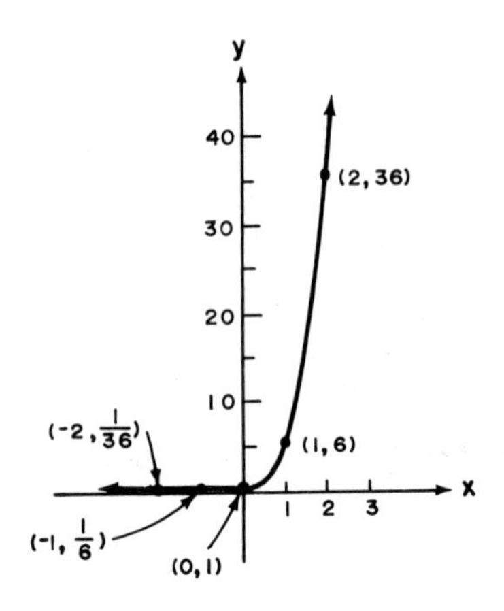

16.

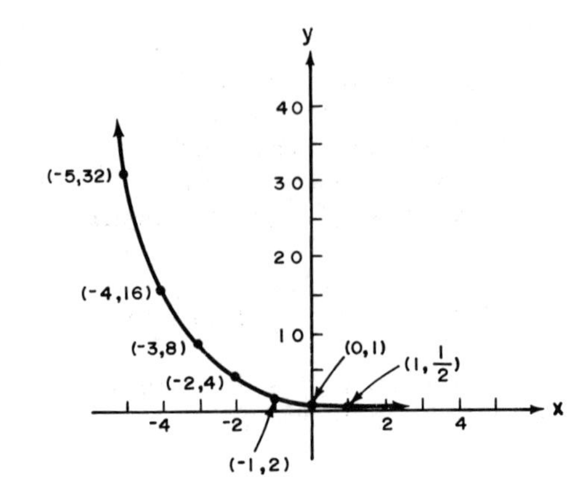

18.

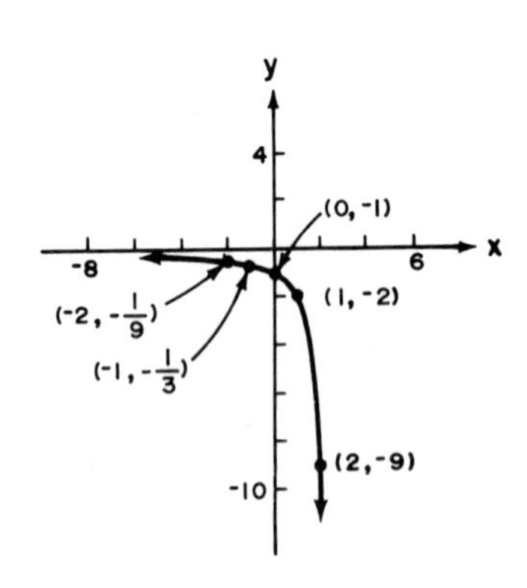

20. $y = \left(\frac{1}{3}\right)^x = \frac{1}{3^x} = 3^{-x}$

Hence, see the answer for Exercise 13.

22. $Y = \left(\frac{1}{10}\right)^x = 10^{-x}$

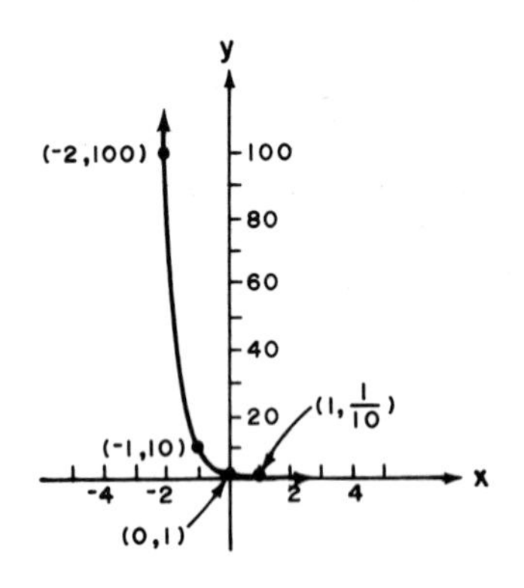

24. $y = \left(\frac{1}{3}\right)^{-x} = \frac{1}{3^{-x}} = 3^x.$

See Example 3 in the text.

26. Using a calculator with -2,-1,0,1,2, and 3 as x-values, we obtain an approximation for each corresponding function value.

(-2,0.2), (-1,0.4), (0,1), (1,2.4), (2,5.8), (3,13.8)

Plot the points and connect with a smooth curve.

28. As was done in Exercise 26, obtain the following ordered pairs and connect with a smooth curve.

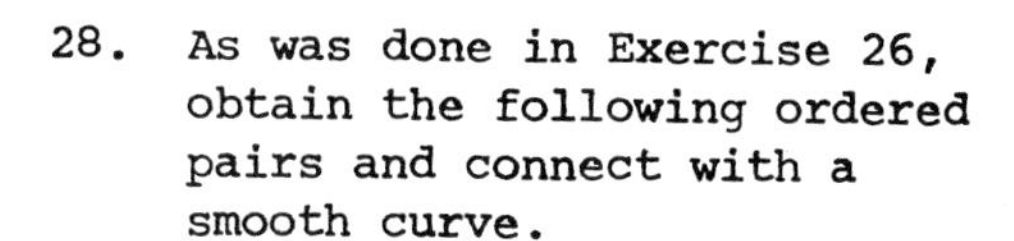

(-2,2.0), (-1,1.4), (0,1), (1,0.7), (2,0.5), (3,0.3)

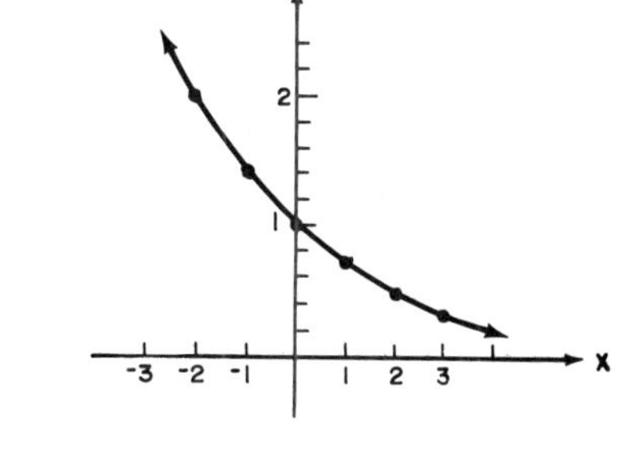

30. Graph $Y_1 = 3^x$ and $y_2 = 4$ on the same axes. The solution is the x-coordinate of the point of intersection; $\left\{\frac{4}{3}\right\}$.

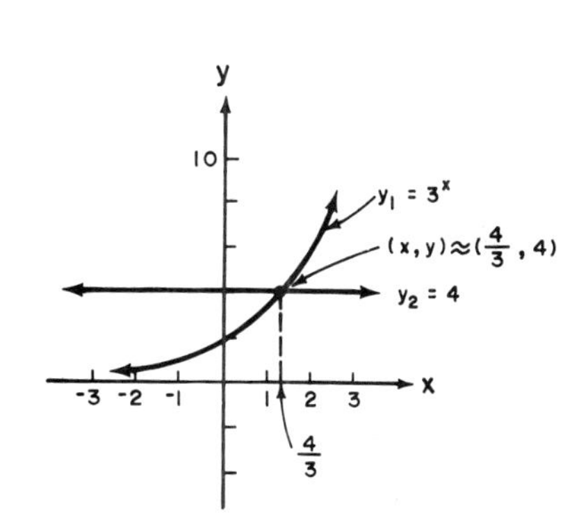

32. $a^{-x} = \frac{1}{a^x} = \left(\frac{1}{a}\right)^x$. Therefore, $y = a^{-x}$, a positive--or, equivalently, $y = \left(\frac{1}{a}\right)^x$, a positive--defines an increasing function, if the base $\frac{1}{a} > 1$. Solving this inequality for a, we have $a < 1$. Similarly, $y = a^{-x}$ defines a decreasing function if $\frac{1}{a} < 1$. Solving this inequality for a, we have $a > 1$.

34.

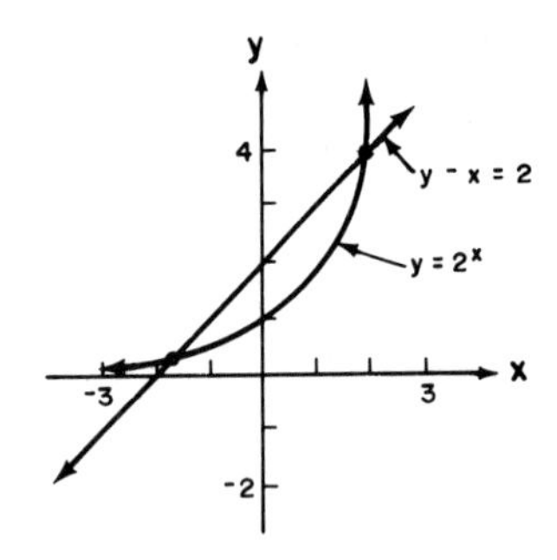

Approximate solutions: (2,4) and (-1.8,0.3)

EXERCISE 11.4

2. Interchanging the components in each ordered pair:

$$f^{-1} = \{(1,-5),\ (2,5)\}.$$

f^{-1} is a function because the ordered pairs do not have the same first component.

4. Interchanging the components in each ordered pair we obtain

$$\{(2,2),\ (3,3),\ (3,4)\},$$

which is not a function because it has ordered pairs with the same first component.

6. $\{(0,-2),\ (0,0),\ (-2,4)\}$ is not a function because it has ordered pairs with the same first component.

8. a. $3y - 2x = 5$

b. 

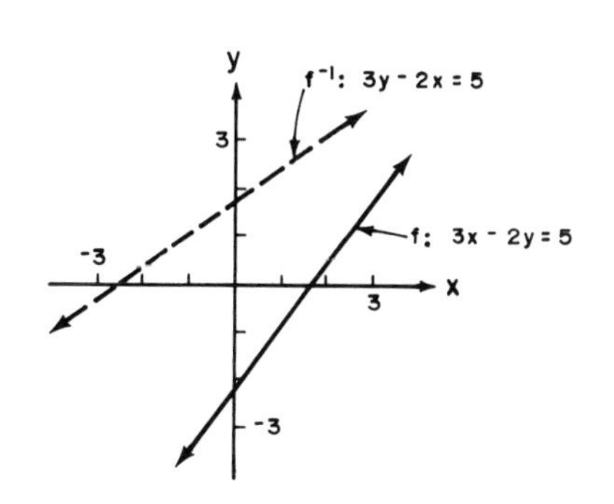

c. A function, because for each x value, there is exactly one y value.

10. a. $4y + x = 4$

b. 

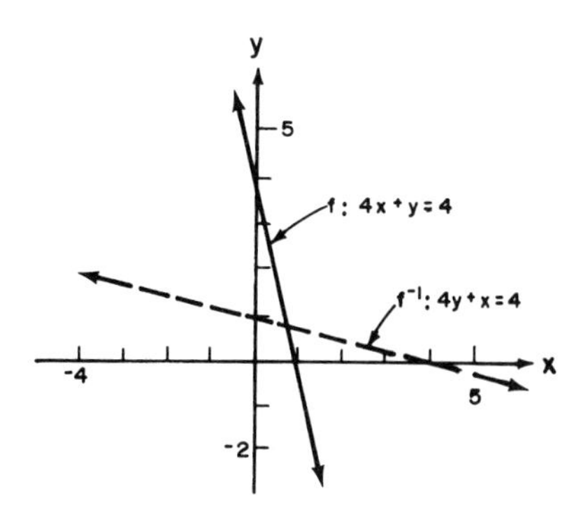

c. A function, because for each x value, there is exactly one y value.

12. a. $x = y^2 - 4$

b. 

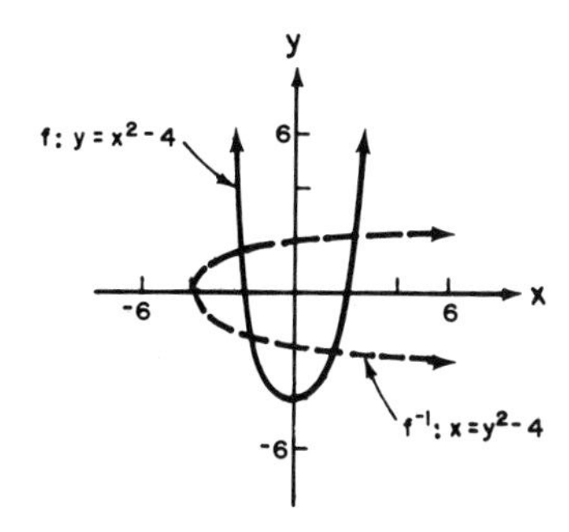

c. Not a function, because for each $x \geq -4$, there are two y values. For example, (-3,1) and (-3,-1) are members of the inverse.

14. a. $x = -\sqrt{y^2 - 4}$

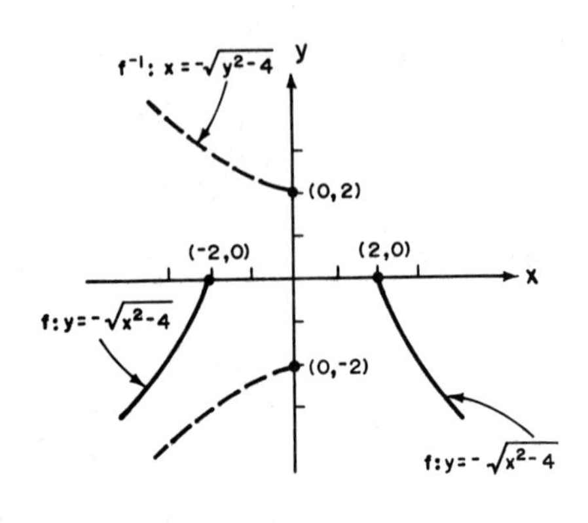

b. To graph $y = -\sqrt{x^2 - 4}$, note that if both members are squared and the terms are rearranged, the equation can be written as $x^2 - y^2 = 4$, which we recognize as a hyperbola. We graph the part of the hyperbola that is on or below the x-axis because $y = -\sqrt{x^2 - 4}$. $x = -\sqrt{y^2 - 4}$ is similarly graphed.

c. Not a function because for each $x < 0$ there are two y values. For example, $(-\sqrt{5},3)$ and $(-\sqrt{5},-3)$ are members of the inverse.

16. a. $x = |y| + 1$

b.

c. Not a function because for every x value there is more than one y value. For example, when $x = 2$, $y = \pm 1$.

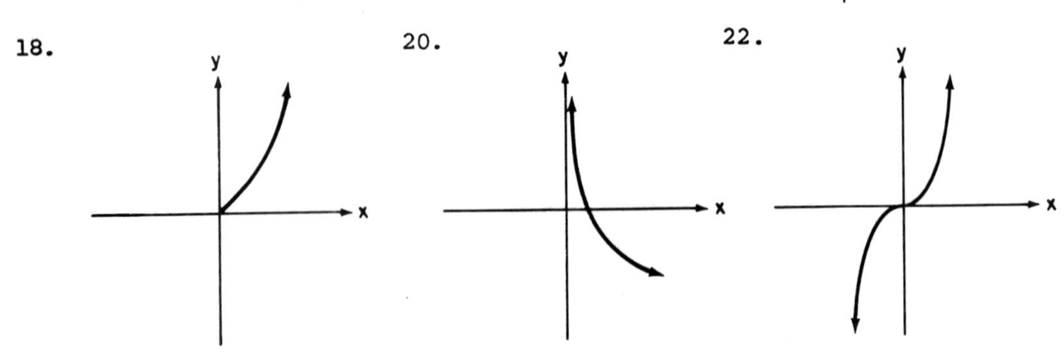

24. $f: y = -x$; hence, $f(x) = -x$.

$f^{-1}: x = -y$ or $y = -x$; hence, $f^{-1}(x) = -x$.

$$f[f^{-1}(x)] = f[-x] = -(-x) = x$$
$$f^{-1}[f(x)] = f^{-1}[-x] = -(-x) = x$$

Thus, $f[f^{-1}(x)] = f^{-1}[f(x)] = x$.

26. $f: 2y = x - 4$, $y = \frac{1}{2}(x - 4)$; hence, $f(x) = \frac{1}{2}(x - 4)$.

$f^{-1}: y - 2x = 4$, $y = 2x + 4$; hence, $f^{-1}(x) = 2x + 4$.

$$f[f^{-1}(x)] = \frac{1}{2}[(2x + 4) - 4] = x$$
$$f^{-1}[f(x)] = 2[\frac{1}{2}(x - 4)] + 4 = x$$

Thus, $f[f^{-1}(x)] = f^{-1}[f(x)] = x$.

28. $f: 4y = -3x + 12$, $y = \frac{1}{4}(-3x + 12)$; hence, $f(x) = \frac{1}{4}(-3x + 12)$.

$f^{-1}: 3y + 4x = 12$, $y = \frac{1}{3}(-4x + 12)$; hence, $f^{-1}(x) = \frac{1}{3}(-4x + 12)$.

$$f[f^{-1}(x)] = \frac{1}{4}[-3 \cdot \frac{1}{3}(-4x + 12) + 12]$$
$$= \frac{1}{4}[-(-4x + 12) + 12] = x$$
$$f^{-1}[f(x)] = \frac{1}{3}[-4 \cdot \frac{1}{4}(-3x + 12) + 12]$$
$$= \frac{1}{3}[-(-3x + 12) + 12] = x$$

Thus, $f[f^{-1}(x)] = f^{-1}[f(x)] = x$.

12 Logarithmic Functions

EXERCISE 12.1

2. $\log_5 125 = 3$

4. $\log_8 64 = 2$

6. $\log_{1/3} \frac{1}{9} = 2$

8. $\log_{64} \frac{1}{2} = \frac{-1}{6}$

10. $\log_{10} 1 = 0$

12. $\log_{10} 0.01 = -2$

14. $5^2 = 25$

16. $16^2 = 256$

18. $\left(\frac{1}{2}\right)^{-3} = 8$

20. $10^0 = 1$

22. $10^{-4} = 0.0001$

24. $\log_2 32 = x$
$2^x = 32$
$x = 5$

26. $\log_3 27 = x$
$3^x = 27$
$x = 3$

28. $\log_5 5^{1/2} = x$
$5^x = 5^{1/2}$
$x = \frac{1}{2}$

30. $\log_3 \frac{1}{3} = x$
$3^x = \frac{1}{3} = 3^{-1}$
$x = -1$

32. $\log_{10} 10 = x$
$10^x = 10$
$x = 1$

34. $\log_{10} 1 = x$
$10^x = 1$
$x = 0$

36. $\log_{10} 0.01 = x$
$10^x = 0.01 = \frac{1}{10^2}$
$10^x = 10^{-2}$
$x = -2$

38. $\log_5 125 = y$
$5^y = 125 = 5^3$
$y = 3$

40. $\log_b 625 = 4$
$b^4 = 625 = 5^4$
$b = 5$

42. $\log_{1/2} x = -5$
$x = \left(\frac{1}{2}\right)^{-5} = 2^5 = 32$

44. $\log_5 \frac{1}{5} = y$
$5^y = \frac{1}{5} = 5^{-1}$
$y = -1$

46. $\log_b 0.1 = -1$
$b^{-1} = 0.1$
$\frac{1}{b} = \frac{1}{10}$; $b = 10$

48. $\log_{10} x = -3$
$x = 10^{-3}$
$= 0.001$

50. $\log_5 (\log_5 5)$

$= \log_5 (1) = 0$

52. $\log_{10}[\log_2(\log_3 9)]$

$= \log_{10}[\log_2(2)]$

$= \log_{10}[1] = 0$

54. $\log_4[\log_2(\log_3 81)]$

$= \log_4[\log_2(4)]$

$= \log_4[2] = \frac{1}{2}$

56. Since $\log_a a^b = b$,

$\log_b(\log_a a^b) = \log_b(b) = 1.$

58. Since $\log_b x$ is defined only if $x > 0$, it follows that $\log_b(x^2 - 4)$ is defined only if $x^2 - 4 > 0$. Hence, $x > 2$ or $x < -2$.

EXERCISE 12.2

2. $\log_b x + \log_b y$

4. $\log_b 4 + \log_b y + \log_b z$

6. $\log_b y - \log_b x$

8. $\log_b x - [\log_b(yz)] = \log_b x - [\log_b y + \log_b z]$

$= \log_b x - \log_b y - \log_b z$

10. $\frac{1}{3} \log_b x$

12. $\log_b y^{1/5} = \frac{1}{5} \log_b y$

14. $\log_b x^{3/2} = \frac{3}{2} \log_b x$

16. $\log_b x^{1/3} + \log_b z^2$

$= \frac{1}{3} \log_b x + 2 \log_b z$

18. $\log_b xy^3 - \log_b z^{1/2} = \log_b x + \log_b y^3 - \log_b z^{1/2}$

$= \log_b x + 3 \log_b y - \frac{1}{2} \log_b z$

20. $\log_{10}\left(\frac{x^2 y}{z^3}\right)^{1/5} = \log_{10} \frac{x^{2/5} y^{1/5}}{z^{3/5}}$

$= \log_{10} x^{2/5} + \log_{10} y^{1/5} - \log_{10} z^{3/5}$

$= \frac{2}{5} \log_{10} x + \frac{1}{5} \log_{10} y - \frac{3}{5} \log_{10} z$

22. $\log_{10} 2y\left(\frac{x}{y}\right)^{1/3} = \log_{10} 2 + \log_{10} y + \log_{10}\left(\frac{x}{y}\right)^{1/3}$

$= \log_{10} 2 + \log_{10} y + \frac{1}{3}\left(\log_{10} \frac{x}{y}\right)$

$= \log_{10} 2 + \log_{10} y + \frac{1}{3}(\log_{10} x - \log_{10} y)$

$= \log_{10} 2 + \log_{10} y + \frac{1}{3} \log_{10} x - \frac{1}{3} \log_{10} y$

24. $\log_{10} \frac{2^{1/2}L^{1/2}}{R} = \frac{1}{2} \log_{10} 2 + \frac{1}{2} \log_{10} L - \log_{10} R$

26. $\log_{10} s(s - a)^{3/2} = \log_{10} s + \frac{3}{2} \log_{10}(s - a)$

28. $\log_b 10 = \log_b(5 \cdot 2)$

$= \log_b 5 + \log_b 2$

$= 0.6990 + 0.3010$

$= 1.0000$

30. $\log_b \frac{3}{2} = \log_b 3 - \log_b 2$

$= 0.4771 - 0.3010$

$= 0.1761$

32. $\log_b 25 = \log_b(5^2)$

$= 2 \log_b 5$

$= 2(0.6990)$

$= 1.3980$

34. $\log_b \frac{6}{5} = \log_b(3 \cdot 2) - \log_b 5$

$= \log_b 3 + \log_b 2 - \log_b 5$

$= 0.4771 + 0.3010 - 0.6990$

$= 0.0791$

36. $\log_{10} 50^{1/2} = \frac{1}{2} \log_{10}(5 \cdot 10) = \frac{1}{2}(\log_{10} 5 + \log_{10} 10)$

$= \frac{1}{2}(0.6990 + 1) = 0.8495$

38. $\log_{10} 0.08 - \log_{10} 15 = \log_{10}(8 \times 10^{-2}) - \log_{10}(3 \cdot 5)$

$= \log_{10}(2^3 \times 10^{-2}) - \log_{10}(3 \cdot 5)$

$= 3 \log_{10} 2 + \log_{10} 10^{-2} - \log_{10} 3 - \log_{10} 5$

$= 3(0.3010) + (-2) - (0.4771) - (0.6990)$

$= -2.2731$

40. $\log_b 5 + \log_b 2 = \log_b 5 \cdot 2$

$= \log_b 10$

42. $\frac{1}{2}(\log_b 6 + 2\log_b 4 - \log_b 2)$

$= \frac{1}{2}(\log_b 6 + \log_b 4^2 - \log_b 2)$

$= \frac{1}{2}(\log_b 6 + \log_b 16 - \log_b 2)$

$= \frac{1}{2}(\log_b 6 \cdot 16 - \log_b 2)$

$= \frac{1}{2}\log_b \frac{96}{2}$

$= \log_b 48^{1/2}$ or $\log_b \sqrt{48}$

or $\log_b 4\sqrt{3}$

44. $\log_b x^{1/4} + \log_b y^{3/4} = \log_b x^{1/4}y^{3/4}$ or $\log_b \sqrt[4]{xy^3}$

46. $\frac{1}{3}(\log_b xy - \log_b z^2) = \frac{1}{3}\log_b\left(\frac{xy}{z^2}\right) = \log_b\left(\frac{xy}{z^2}\right)^{1/3}$ or $\log_b \sqrt[3]{\frac{xy}{z^2}}$

48. $\frac{1}{2}(\log_{10} x - \log_{10} y^3 - \log_{10} z) = \frac{1}{2}(\log_{10} x - [\log_{10} y^3 + \log_{10} z])$

$= \frac{1}{2}\log_{10}\left(\frac{x}{y^3z}\right) = \log_{10}\left(\frac{x}{y^3z}\right)^{1/2}$

or $\log_{10}\sqrt{\frac{x}{y^3z}}$

50. $-1 \cdot \log_b x = \log_b x^{-1}$ or $\log_b\left(\frac{1}{x}\right)$

52. $\log_{10}(x - 1) - \log_{10} 4 = \log_{10}\left(\frac{x-1}{4}\right) = 2$

$\frac{x-1}{4} = 10^2$

$x - 1 = 400$

$= 401;\ \{401\}$

54. $\log_{10}(x + 3) + \log_{10} x = \log_{10} x(x + 3) = 1$

$$x(x + 3) = 10^1$$

$$x^2 + 3x - 10 = 0$$

$$(x + 5)(x - 2) = 0$$

$$x = -5, 2$$

We reject $x = -5$ since we must have $x > 0$ and $x + 3 > 0$. Hence, $x = 2$. $\{2\}$

56. $\log_{10}(x + 3) - \log_{10}(x - 1) = \log_{10}\left(\frac{x + 3}{x - 1}\right) = 1$

$$\frac{x + 3}{x - 1} = 10^1$$

$$x + 3 = 10(x - 1)$$

$$9x = 13$$

$$x = \frac{13}{9}; \left\{\frac{13}{9}\right\}$$

58. Let $x = 1000$ and $y = 100$.

$$\log_{10} \frac{1000}{100} = \log_{10} 10 = 1$$

$$\frac{\log_{10} 1000}{\log_{10} 100} = \frac{3}{2}$$

Since $1 \neq \frac{3}{2}$, $\log_{10} \frac{x}{y}$ is not equivalent to $\frac{\log_{10} x}{\log_{10} y}$.

60. $\log_b 24 - \log_b 2 = \log_b \frac{24}{2} = \log_b 12 = \log_b(3 \cdot 4) = \log_b 3 + \log_b 4$

62. $4 \log_b 3 - 2 \log_b 3 = 2 \log_b 3 = \log_b 3^2 = \log_b 9$

64. $\frac{1}{4} \log_b 8 + \frac{1}{4} \log_b 2 = \frac{1}{4}(\log_b 8 + \log_b 2)$

$$= \frac{1}{4} \log_b(8 \cdot 2) = \log_b 16^{1/4} = \log_b 2$$

EXERCISE 12.3

2. 4

4. -6

6. n

8. $\log_{10}(8.91 \times 10^2) = 2.9499$

10. $\log_{10}(2.14 \times 10^1) = 1.3304$

12. $\log_{10}(2.19 \times 10^2) = 2.3404$

14. $\log_{10}(2.14 \times 10^{-3}) = -3 + 0.3304$
$= 0.3304 - 3$

16. $\log_{10}(4.13 \times 10^{-4})$
$= -4 + 0.6160 = 6.6160 - 10$

18. $-3 + 0.7316 = 0.7316 - 3$

20. $x = \text{antilog}_{10}\ 0.2504$
$= 1.78$

22. $x = \text{antilog}_{10}\ 3.9258$
$= \text{antilog}_{10}(0.9258 + 3)$
$= 8.43 \times 10^3 = 8430$

24. $x = \text{antilog}_{10}(0.9722 - 3)$
$= 9.38 \times 10^{-3}$
$= 0.00938$

26. $x = \text{antilog}_{10}(1.8155 - 4)$
$= \text{antilog}_{10}(0.8155 - 3)$
$= 6.54 \times 10^{-3} = 0.00654$

28. $\log_{10} x = -0.4123 + 1 - 1$
$= 0.5877 - 1$
$x = 0.387$

30. $\log_{10} x = -1.0545 + 2 - 2$
$= 0.9455 - 2$
$x = 0.0882$

32. $\log_{10} x = -2.0670 + 3 - 3$
$= 0.9330 - 3$
$x = 0.00857$

34. $10^{1.6405} = \text{antilog}_{10}\ 1.6405$
$= 4.37 \times 10^1 = 43.7$

36. $10^{4.3766} = \text{antilog}_{10}\ 4.3766$
$= 2.38 \times 10^4$
$= 23800$

38. $10^{-2.0958}$
$= \text{antilog}_{10}(-2.0958 + 3 - 3)$
$= \text{antilog}_{10}(0.9042 - 3)$
$= 8.02 \times 10^{-3} = 0.00802$

40. 1.8590

42. 0.9231

44. 13.464

46. 0.3012

48. 1.8405

50. 4.0073

52. -0.3567

54. $x = e^{0.25}$
$= 1.2840$

56. $x = e^{2.4}$
$= 11.023$

58. $x = e^{6.0}$
$= 403.43$

EXERCISE 12.4

2. 4
4. -6
6. 0
8. -3
10. 1.4456
12. 3.1691
14. -2.2118
16. -1.9830
18. 2.6915
20. 0.048978
22. 1383.6
24. 0.0090929
26. $x = 10^{2.3} = 199.53$
28. $x = 10^{0.8} = 6.3096$
30. $x = 10^{-1.69} = 0.020417$
32. $x = e^{6.3} = 1.8405$
34. 4.0073
36. -0.35668
38. 2.0751
40. 23.571
42. 0.10026
44. 0.53794
46. $x = e^{2.03} = 7.6141$
48. $x = e^{0.59} = 1.8040$
50. $x = e^{-3.4} = 0.033373$

EXERCISE 12.5

2. $x = \log_{10} 8.07 = 0.907$
4. $x = \log_{10} 182.4 = 2.26$
6. $x = \log_{10} 9480.2 = 3.98$
8. $x = \ln 2.1 = 0.742$
10. $x = \ln 60 = 4.09$
12. $x = \ln 0.9 = -0.105$
14.
$$10^{0.2x} = \frac{140}{63.1}$$
$$0.2x = \log_{10} \frac{140}{63.1}$$
$$x = \frac{1}{0.2} \log_{10} \frac{140}{63.1} = 1.73$$
16.
$$10^{-1.3x} = 180 - 64$$
$$-1.3x = \log_{10} 116$$
$$x = -\frac{1}{1.3} \log_{10} 116 = -1.59$$
18.
$$3(10^{0.7x}) = 163 + 49.3$$
$$10^{0.7x} = \frac{212.3}{3}$$
$$0.7x = \log_{10} \frac{212.3}{3}$$
$$x = \frac{1}{0.7} \log_{10} \frac{212.3}{3}$$
$$= 2.64$$
20.
$$4(10^{-0.6x}) = 28.2 - 16.1$$
$$10^{-0.6x} = \frac{12.1}{4}$$
$$-0.6x = \log_{10} \frac{12.1}{4}$$
$$x = -\frac{1}{0.6} \log_{10} \frac{12.1}{4}$$
$$= -0.801$$

22. $e^{0.4x} = \frac{22.26}{5.3}$

$0.4x = \ln \frac{22.26}{5.3}$

$x = \frac{1}{0.4} \ln \frac{22.26}{5.3} = 3.59$

24. $e^{1.4x} = \frac{14.105}{4.03}$

$1.4x = \ln \frac{14.105}{4.03}$

$x = \frac{1}{1.4} \ln \frac{14.105}{4.03} = 0.895$

26. $4e^{2.1x} = 4.5 - 3.3$

$e^{2.1x} = \frac{1.2}{4}$

$2.1x = \ln \frac{1.2}{4}$

$x = \frac{1}{2.1} \ln 0.3 = -0.573$

28. $1.3e^{2.1x} = 1.23 + 17.1$

$e^{2.1x} = \frac{18.33}{1.3}$

$2.1x = \ln \frac{18.33}{1.3}$

$x = \frac{1}{2.1} \ln \frac{18.33}{1.3} = 1.26$

30. $0.6e^{-0.7x} = 55.68 - 23.1$

$e^{-0.7x} = \frac{32.58}{0.6}$

$-0.7x = \ln \frac{32.58}{0.6}$

$x = -\frac{1}{0.7} \ln \frac{32.58}{0.6}$

$= -5.71$

32. $\log_3 24 = \frac{\log_{10} 24}{\log_{10} 3}$

$= \frac{1.38021}{0.477121}$

$= 2.89$

34. $\log_2 14.3 = \frac{\log_{10} 14.3}{\log_{10} 2}$

$= \frac{1.15534}{0.30103}$

$= 3.84$

36. $\log_4 28.1 = \frac{\log_{10} 28.1}{\log_{10} 4}$

$= \frac{1.44871}{0.60206}$

$= 2.41$

38. $\log_{10} 3^x = \log_{10} 4$

$x(\log_{10} 3) = \log_{10} 4$

$\left\{\frac{\log_{10} 4}{\log_{10} 3}\right\}$

In decimal form:

$x = \frac{0.6021}{0.4771} = 1.26.$

40. $\log_{10} 2^{x-1} = \log_{10} 9$

$(x - 1) \log_{10} 2 = \log_{10} 9$

$x - 1 = \frac{\log_{10} 9}{\log_{10} 2}$

$\left\{\frac{\log_{10} 9}{\log_{10} 2} + 1\right\}$

In decimal form:

$x = \frac{0.9542}{0.3010} + 1 = 4.17.$

42. $\log_{10} 3^{x^2} = \log_{10} 21$

$x^2 \log_{10} 3 = \log_{10} 21$

$x^2 = \dfrac{\log_{10} 21}{\log_{10} 3}$

$\left\{\pm\sqrt{\dfrac{\log_{10} 21}{\log_{10} 3}}\right\}$

In decimal form:

$x \approx \pm\sqrt{\dfrac{1.3222}{0.4771}} = 1.66.$

44. $\log_{10} 2.13^{-x} = \log_{10} 8.1$

$-x \log_{10} 2.13 = \log_{10} 8.1$

$-x = \dfrac{\log_{10} 8.1}{\log_{10} 2.13}$

$\left\{\dfrac{-\log_{10} 8.1}{\log_{10} 2.13}\right\}$

In decimal form:

$x \approx \dfrac{-0.9085}{0.3284} = -2.77.$

46. $y = k - ke^{-t}$

$ke^{-t} = k - y$

$e^{-t} = \dfrac{k - y}{k}$

$\ln e^{-t} = \ln\left(\dfrac{k - y}{k}\right)$

$-t(\ln e) = \ln(k - y) - \ln k$

Since $\ln e = 1$,

$-t = \ln(k - y) - \ln k$

$t = -\ln(k - y) + \ln k.$

48. $e^{-t/3} = \dfrac{B - 2}{A + 3}$

$\ln e^{-t/3} = \ln \dfrac{B - 2}{A + 3}$

$-\dfrac{t}{3}(\ln e) = \ln(B - 2) - \ln(A + 3)$

Since $\ln e = 1$,

$-\dfrac{t}{3} = \ln(B - 2) - \ln(A + 3)$

$t = -3 \ln(B - 2) + 3 \ln(A + 3).$

50. $\ln 10k = P - P_0$

$10k = e^{P-P_0}$

$k = \dfrac{1}{10} e^{P-P_0}$

52. $\ln 10 = \dfrac{\log_{10} 10}{\log_{10} e} = \dfrac{1}{\log_{10} e}$

54. Substituting 3 for y and $\frac{1}{2}$ for x in the result of Exercise 53 gives

$$3^{1/2} = 10^{1/2 \log_{10} 3}.$$

EXERCISE 12.6

2. a. $P = 30(10)^{-0.09(0)}$

$\approx 30(10)^{0} = 30$

Pressure at sea level is 30 inches of mercury.

b. 50,000 feet $= \frac{50000}{5280}$ miles

$P = 30(10)^{-0.09(\frac{50000}{5280})}$

$\approx 30(10)^{-0.85} = 4.24$

Hence, the pressure at 50,000 feet is 4.24 inches of mercury.

c. $P = 30(10)^{-0.09(\frac{100,000}{5280})}$

$\approx 30(10^{-1.70}) = 0.599$

Hence, the pressure at 100,000 feet is 0.599 inch of mercury.

4. $16.1 = 30(10)^{-0.09a}$

$\frac{16.1}{30} = 10^{-0.09a}$

$\log_{10} \frac{16.1}{30} = -0.09a$

$a = -\frac{1}{0.09} \log_{10} \frac{16.1}{30}$

$= -\frac{1}{0.09}(-0.270295) = 3.00$

Hence, at a height of 3 miles the pressure is 16.1 inches of mercury.

6. From Exercise 2a, the pressure at sea level is 30 inches of mercury. Hence,

$\frac{30}{4} = 30(10)^{-0.09a}$

$\frac{1}{4} = 10^{-0.09a}$

$\log_{10} \frac{1}{4} = -0.09a$

$a = \frac{1}{-0.09} \log_{10} 0.25$

$= \frac{1}{-0.09}(0.60206) = 6.69$

Hence, at a height of 6.69 miles the pressure is $\frac{1}{4}$ of that at sea level.

8. a. $P = 16{,}782{,}000e^{(0.0083)(10)}$
$= 16{,}782{,}000(1.08654)$
$= 18{,}234{,}000$

The population was 18,234,000 in 1970.

b. $P = 16{,}782{,}000e^{(0.0083)(20)}$
$= 16{,}782{,}000(1.18057)$
$= 19{,}812{,}000$

The population was 19,812,000 in 1980.

$P = 16{,}782{,}000e^{(0.0083)(30)}$
$= 16{,}782{,}000(1.28274)$
$= 21{,}527{,}000$

The population will be 21,527,000 in 1990.

$P = 16{,}782{,}000e^{(0.0083)(40)}$
$= 16{,}782{,}000(1.39375)$
$= 23{,}390{,}000$

The population will be 23,390,000 in 2000.

10. Let $P_0 = 4{,}951{,}600$, $P = 6{,}789{,}400$, and $t = 10$.

$$6{,}789{,}000 = 4{,}951{,}600e^{10r}$$

$$e^{10r} = \frac{6789000}{4951600}$$

$$\approx 1.37107$$

$$10r = \ln 1.37107$$

$$r = \frac{\ln 1.37107}{10}$$

$$\approx 0.0316 = 3.16\%$$

12. Taking $\frac{P}{P_0} = 2$ and $t = 20$,

$2 = e^{20r}$;

$\ln 2 = 20r$;

$r = \frac{\ln 2}{20} = \frac{0.693147}{20}$

$= 0.0347.$

Annual rate is 3.47%.

14. $N = 6000e^{0.04(10)}$
$= 6000(1.49182)$
$= 8950.92$

There are 8950 bacteria 10 hours later.

16. $I = 1000e^{-0.1(0.6)}$
$= 1000(0.941765)$
$= 941.765$

To the nearest tenth, the intensity is 941.8 lumens.

18. Let $y_0 = 40$ and $y = 12$.

$$12 = 40e^{-0.4t}$$
$$e^{-0.4t} = \frac{12}{40} = 0.3$$
$$-0.4t = \ln(0.3)$$
$$t = \frac{\ln(0.3)}{0.4} = 3.00993$$

Hence, to the nearest hundredth, it would take 3.01 seconds.

20. Let $I = 800$.

$$800 = 1000e^{-0.1t}$$
$$\frac{800}{1000} = e^{-0.1t}$$
$$e^{-0.1t} = 0.8$$
$$-0.1t = \ln 0.8$$
$$t = -\frac{\ln 0.8}{0.1} = 2.23144$$

Hence, to the nearest tenth, the thickness would be 2.2 cm.

22. $2000 = 1000(1 + 0.12)^n$

$$2 = 1.12^n$$
$$\log_{10} 2 = n \log_{10} 1.12$$
$$n = \frac{\log_{10} 2}{\log_{10} 1.12}$$
$$= \frac{0.30103}{0.049218} = 6.1162582$$

To the nearest year, 6 years.

24. $60 = 40\left(1 + \frac{r}{4}\right)^{4(3)}$

$$1.5 = \left(1 + \frac{r}{4}\right)^{12}$$
$$\log_{10} 1.5 = 12 \log_{10}\left(1 + \frac{r}{4}\right)$$
$$\log_{10}\left(1 + \frac{r}{4}\right) = \frac{1}{12} \log_{10} 1.5$$
$$= 0.0146742$$
$$1 + \frac{r}{4} = 1.03437$$
$$\frac{r}{4} = 0.03437$$
$$r = 0.13748.$$

To the nearest half percent, the rate of interest is 13.7%.

26. $2 = 1\left(1 + \frac{0.10}{4}\right)^{4n}$

$$\log_{10} 2 = 4n \log_{10}(1 + 0.025)$$
$$4n = \frac{\log_{10} 2}{\log_{10} 1.025}$$
$$n = \frac{1}{4}\left(\frac{\log_{10} 2}{\log_{10} 1.025}\right)$$
$$n = \frac{1}{4}\left(\frac{0.30103}{0.0107239}\right) = 7.017736$$

To the nearest year, 7 years.

28. A's amount after the first 3 months, or 1/4 year, was

$$A = 10{,}000\left(1 + \frac{0.08}{4}\right)^{4(1/4)}$$

$$= 10{,}000(1.02)$$

$$= 10{,}200.$$

Hence, his interest was \$200 for the 3-month period. This interest is withdrawn, and A's principal is once again \$10,000 for the second 3-month period (and for all following quarterly periods, because the procedure is repeated). Since there are four 3-month periods per year, A's interest for the 20 years is

$$20 \text{ years} \cdot 4\,\frac{\text{periods}}{\text{year}} \cdot \$200/\text{period} = \$16{,}000.$$

On the other hand, B's amount after 20 years is

$$B = 10{,}000\left(1 + \frac{0.08}{4}\right)^{4 \cdot 20} = 10{,}000(1.02)^{80}.$$

$$\log_{10} B = \log_{10}[10^4(1.02)^{80}]$$

$$= \log_{10} 10^4 + 80 \log_{10} 1.02$$

$$= 4 + 80(0.0086) = 4.6880$$

$$B = \text{antilog}_{10}\ 4.6880 = 48{,}753$$

Hence, B's interest for the 20 years was \$48,753 - \$10,000 = \$38,753, or \$22,753 more than the \$16,000 A earned.

13 Natural-number Functions

EXERCISE 13.1

2. $s_1 = 2(1) - 3 = -1$ $\quad s_2 = 2(2) - 3 = 1$
   $s_3 = 2(3) - 3 = 3$ $\quad s_4 = 2(4) - 3 = 5$

4. $s_1 = \dfrac{3}{1^2 + 1} = \dfrac{3}{2}$ $\quad s_2 = \dfrac{3}{(2)^2 + 1} = \dfrac{3}{5}$

   $s_3 = \dfrac{3}{(3)^2 + 1} = \dfrac{3}{10}$ $\quad s_4 = \dfrac{3}{(4)^2 + 1} = \dfrac{3}{17}$

6. $s_1 = \dfrac{1}{2(1) - 1} = 1$ $\quad s_2 = \dfrac{2}{2(2) - 1} = \dfrac{2}{3}$

   $s_3 = \dfrac{3}{2(3) - 1} = \dfrac{3}{5}$ $\quad s_4 = \dfrac{4}{2(4) - 1} = \dfrac{4}{7}$

8. $s_1 = \dfrac{5}{1(1 + 1)} = \dfrac{5}{2}$ $\quad s_2 = \dfrac{5}{2(2 + 1)} = \dfrac{5}{6}$

   $s_3 = \dfrac{5}{3(3 + 1)} = \dfrac{5}{12}$ $\quad s_4 = \dfrac{5}{4(4 + 1)} = \dfrac{1}{4}$

10. $s_1 = (-1)^{1+1} = 1$ $\quad s_2 = (-1)^{2+1} = -1$
    $s_3 = (-1)^{3+1} = 1$ $\quad s_4 = (-1)^{4+1} = -1$

12. $s_1 = (-1)^{1-1}3^{1+1} = 9$ $\quad s_2 = (-1)^{2-1}3^{2+1} = -27$
    $s_3 = (-1)^{3-1}3^{3+1} = 81$ $\quad s_4 = (-1)^{4-1}3^{4+1} = -243$

14. $[3(1) - 2] + [3(2) - 2] + [3(3) - 2] = 1 + 4 + 7$

16. $(2^2 + 1) + (3^2 + 1) + (4^2 + 1) + (5^2 + 1) + (6^2 + 1)$
    $= 5 + 10 + 17 + 26 + 37$

18. $\frac{2}{2}(2 + 1) + \frac{3}{2}(3 + 1) + \frac{4}{2}(4 + 1) + \frac{5}{2}(5 + 1) + \frac{6}{2}(6 + 1)$
    $= 3 + 6 + 10 + 15 + 21$

20. $\frac{(-1)^{3+1}}{3-2} + \frac{(-1)^{4+1}}{4-2} + \frac{(-1)^{5+1}}{5-2} = 1 - \frac{1}{2} + \frac{1}{3}$

22. $1 + \frac{1}{2} + \frac{1}{3} + \cdots$

24. $\frac{0}{1+0} + \frac{1}{1+1} + \frac{2}{1+2} + \cdots = 0 + \frac{1}{2} + \frac{2}{3} + \cdots$

26. $\sum_{i=1}^{4} 2i$

28. $\sum_{i=1}^{5} x^{2i+1}$

30. These are the cubes of the first five natural numbers: $\sum_{i=1}^{5} i^3$.

32. Each denominator gives the number of the term, while each numerator is one greater than its denominator: $\sum_{i=1}^{\infty} \frac{i+1}{i}$.

34. Each numerator is two more than the denominator: $\sum_{i=1}^{\infty} \frac{2i+1}{2i-1}$.

36. The numerators are powers of 3. The denominators are multiples of 2: $\sum_{i=1}^{\infty} \frac{3^{i-1}}{2i}$.

EXERCISE 13.2

In Problems 2-12, substitute into the equation $s_n = a + (n - 1)d$.

2. $d = -1 - (-6) = 5$. The sequence is

$$-6, -1, -1 + 5 = 4, 4 + 5 = 9, 9 + 5 = 14;$$

$$s_n = -6 + (n - 1)5 = 5n - 11.$$

4. $d = -20 - (-10) = -10$. The sequence is

$$-10, -20, -30, -40, -50, \cdots;$$

$$s_n = -10 + (n - 1)(-10) = -10n.$$

6. $d = (a + 5) - a = 5$. The sequence is

$$a, a + 5, a + 10, a + 15, a + 20, \cdots;$$

$$s_n = a + (n - 1)5 = a - 5 + 5n.$$

8. $d = y - (y - 2b) = 2b$. The sequence is

$$y - 2b, y, y + 2b, y + 4b, y + 6b, \cdots;$$

$$s_n = (y - 2b) + (n - 1)2b = y - 4b + 2bn.$$

10. $d = (a - 2b) - (a + 2b) = -4b$. The sequence is

$$a + 2b, a - 2b, a - 6b, a - 10b, a - 14b, \cdots;$$

$$s_n = (a + 2b) + (n - 1)(-4b) = a + 6b - 4bn.$$

12. $d = 5a - 3a = 2a$. The sequence is

$$3a, 5a, 7a, 9a, 11a, \cdots;$$

$$s_n = 3a + (n - 1)2a = a + 2an.$$

In Problems 14-18, first find the general term for each progression using the formula $s_n = a + (n - 1)d$, and then replace n by the appropriate natural number.

14. $d = -12 - (-3) = -9$; $a = -3$

$$s_n = -3 + (n - 1)(-9) = 6 - 9n$$

$$s_{10} = 6 - 9(10) = -84$$

16. $d = (-2) - (-5) = 3$; $a = -5$

$$s_n = -5 + (n - 1)3 = 3n - 8$$

$$s_{17} = 3(17) - 8 = 43$$

18. $d = 2 - \frac{3}{4} = \frac{5}{4}$; $a = \frac{3}{4}$

$$s_n = \frac{3}{4} + (n - 1)\frac{5}{4} = \frac{5n}{4} - \frac{1}{2}$$

$$s_{10} = \frac{5(10)}{4} - \frac{1}{2} = 12$$

20. A diagram is helpful.

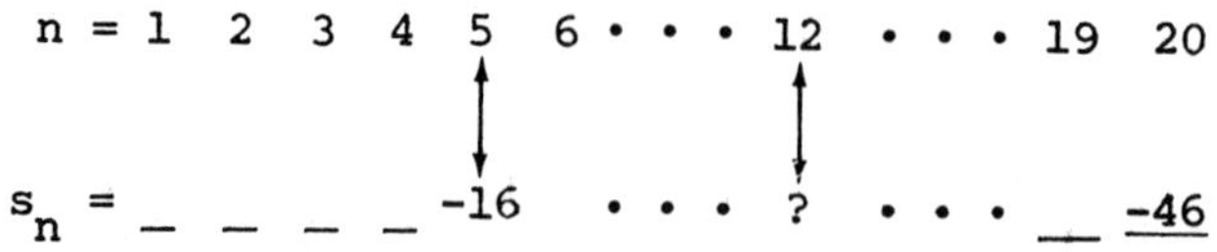

$n =$ 1	2	3	4	5	6 • • •	12	• • •	19	20
				↕		↕			
$s_n =$ ___	___	___	___	-16	• • •	?	• • •	___	-46

Find d: $s_{16} = -46 = -16 + (16 - 1)d$; $-46 = -16 + 15d$; $d = -2$.

Find a: $s_5 = -16 = a + (5 - 1)(-2)$; $a = -8$.

$$s_n = -8 + (n - 1)(-2) = -6 - 2n$$
$$s_{12} = -6 - 2(12) = -30$$

Alternative solution:

Use $s_n = a + (n - 1)d$ twice; once with $s_5 = -16$ and again with $s_{20} = -46$, to obtain the system

$$-16 = a + (5 - 1)d$$
$$-46 = a + (20 - 1)d.$$

Solve the system to obtain $d = -2$ and $a = -8$. Then use $s_n = a + (n - 1)d$ with $n = 12$, $a = -8$, and $d = -2$:

$$s_{12} = -8 + (12 - 1)(-2) = -30.$$

22. $a = 7$, $d = 3 - 7 = -4$

$$s_n = a + (n - 1)d = 7 + (n - 1)(-4)$$
$$= 11 - 4n$$

Let $s_n = -81 = 11 - 4n$; $n = -23$; -81 is the 23rd term.

24. 10 ? ? ? ? 65 (d d d d d)

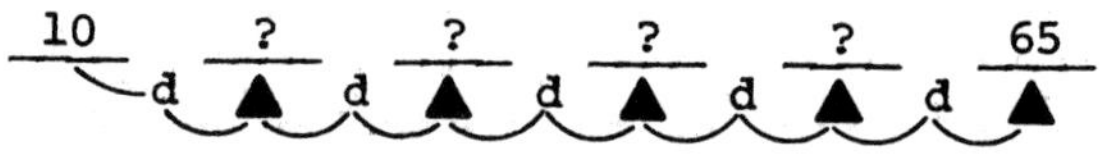

There are 5 differences (d) between 10 and 65. Hence,

$$5d = 65 - 10; \ d = 11.$$

By four successive additions of 11, the required arithmetic means are 21, 32, 43, 54.

26. -11 ? 7

There are 2 differences (d) between -11 and 7. Hence,

$$2d = 7 - (-11); \ d = 9.$$

Adding 9 to -11, the required mean is -2.

28. $\underline{-12}$ d ? d ? d ? d ? d ? d ? d $\underline{23}$

There are 7 differences (d) between -12 and 23. Hence,

$$7d = 23 - (-12);\ d = 5.$$

By six successive additions of 5, the required means are -7, -2, 3, 8, 13, 18.

30. $\sum_{i=1}^{21} (3i - 2) = 1 + 4 + 7 + \cdots;\ d = 3,\ a = 1,\ n = 21.$

Using $S_n = \frac{n}{2}[2a + (n - 1)d]$,

$$S_{21} = \frac{21}{2}[2(1) + (21 - 1)3] = 651.$$

32. $\sum_{j=10}^{20} (2j - 3) = 17 + 19 + 21 + \cdots;\ d = 2,\ a = 17,\ n = 20 - 9 = 11.$

Using $S_n = \frac{n}{2}[2a + (n - 1)d]$,

$$S_{11} = \frac{11}{2}[2(17) + (11 - 1)2] = 297.$$

34. $\sum_{k=1}^{100} k = 1 + 2 + 3 + \cdots;\ d = 1,\ a = 1,\ n = 100.$

Using $S_n = \frac{n}{2}[2a + (n - 1)d]$,

$$S_{100} = \frac{100}{2}[2(1) + (100 - 1)1] = 5050.$$

36. The first term is $14 = 7 \cdot 2$; that is, the 2nd multiple of seven; and the last is $105 = 7 \cdot 15$, the 15th multiple of seven. Thus,

$$S_n = \sum_{i=2}^{15} 7i = 14 + 21 + 28 + \cdots + 105;\ d = 7,\ a = 14,\ n = 15 - 1 = 14.$$

$$S_{14} = \frac{14}{2}[2(14) + (14 - 1)7] = 833.$$

38. The sum of the number of bricks in each row is an arithmetic series with the sum known to be 256.

$$\sum_{j=1}^{n} (2j - 1) = 1 + 3 + \cdots + (2n - 1) = 256; \; a = 1, \; d = 2.$$

$$S_n = 256 = \frac{n}{2}[2(1) + (n - 1)2],$$

from which, $n^2 = 256$; $n = 16$. That is, there are 16 rows; hence, the third row from the bottom is the 14th row. Using $s_n = a + (n - 1)d$ with $a = 1$, $d = 2$, $n = 14$:

$$s_{14} = 1 + (14 - 1)2 = 27.$$

There are 27 bricks in the third row from the bottom.

40.

a: the first number
$a + d$: the second number
$a + 2d$: the third number

$$a + (a + d) + (a + 2d) = 21 \quad \text{or} \quad 3a + 3d = 21 \quad (1)$$
$$a(a + d)(a + 2d) = 231 \quad (2)$$

Solve Equation (1) for d: $d = 7 - a$.

Substitute $7 - a$ for d in Equation (2).

$$\begin{aligned} a[a + (7 - a)][a + 2(7 - a)] &= 231 \\ 7a(-a + 14) &= 231 \\ -7a^2 + 98a - 231 &= 0 \\ a^2 - 14a + 33 &= 0 \\ (a - 11)(a - 3) &= 0 \end{aligned}$$

Therefore, $a = 11$ and $d = 7 - a = -4$, or $a = 3$ and $d = 7 - a = 4$. The required numbers are 11, 7, 3 or 3, 7, 11.

42. $[p + q] + [2p + q] + [3p + q] + [4p + q] = 28$; $10p + 4q = 28$ (1)
$[2p + q] + [3p + q] + [4p + q] + [5p + q] = 44$; $14p + 4q = 44$ (2)

Subtract Equation (1) from Equation (2): $4p = 16$; $p = 4$.

Substitute 4 in Equation (1): $10(4) + 4q = 28$; $q = -3$. Hence, $p = 4$, $q = -3$.

44. $S_n = 2 + 4 + 6 + \cdots + 2n$; $a = 2$, $d = 2$, $n = n$.

$$S_n = \frac{n}{2}[2a + (n - 1)d] = \frac{n}{2}[2(2) + (n - 1)2] = \frac{n}{2}[2n + 2] = n^2 + n.$$

EXERCISE 13.3

2. $r = \frac{8}{4} = 2$; $a = 4$. The next three terms are

$16 \cdot 2 = 32,\ 32 \cdot 2 = 64,\ 64 \cdot 2 = 128$ or $32, 64, 128$.

$s_n = 4(2)^{n-1} = 2^2(2)^{n-1}$

$= (2)^{n-1+2} = 2^{n+1}$

4. $r = \frac{3}{6} = \frac{1}{2}$; $a = 6$. The next three terms are

$$\frac{3}{2} \cdot \frac{1}{2} = \frac{3}{4}, \frac{3}{4} \cdot \frac{1}{2} = \frac{3}{8}, \frac{3}{8} \cdot \frac{1}{2} = \frac{3}{16} \text{ or } \frac{3}{4}, \frac{3}{8}, \frac{3}{16}.$$

$s_n = 6\left(\frac{1}{2}\right)^{n-1}$

6. $r = \frac{-3}{2} \div \frac{1}{2} = -3$; $a = \frac{1}{2}$. The next three terms are $\frac{-27}{2}, \frac{81}{2}, \frac{-243}{2}$.

$s_n = \frac{1}{2}(-3)^{n-1}$

8. $r = \frac{a}{bc} \div \frac{a}{b} = \frac{1}{c}$; the first term is $\frac{a}{b}$. The next three terms are

$$\frac{a}{bc^3}, \frac{a}{bc^4}, \frac{a}{bc^5}.$$

$S_n = \left(\frac{a}{b}\right)\left(\frac{1}{c}\right)^{n-1}$

10. $a = -3$; $r = \frac{3}{2} \div \frac{-3}{1} = \frac{-1}{2}$; $s_n = (-3)\left(\frac{-1}{2}\right)^{n-1}$

The eighth term is $s_8 = (-3)\left(\frac{-1}{2}\right)^7 = \frac{3}{128}$.

12. $a = -81a$; $r = \frac{-27a^2}{-81a} = \frac{1}{3}a$; $s_n = (-81a)\left(\frac{1}{3}a\right)^{n-1}$

The ninth term is $s_9 = (-81a)\left(\frac{1}{3}a\right)^8 = (-3^4a)\left(\frac{1}{3}a\right)^8 = \frac{-1}{81}a^9$.

14. $s_5 = 1$; $r = -\frac{1}{2}$, $n = 5$.

$s_n = ar^{n-1}$; $1 = a\left(-\frac{1}{2}\right)^4$; $a = 16$.

16.  $\underline{-4}, \underline{?}, \underline{?}, \underline{-32}$. There are 3 multiplications by r between -4 and -32. Hence, $r^3 = \frac{-32}{-4} = 8$; $r = 2$. Therefore, from two successive multiplications by 2:

$$-4 \cdot 2 = -8 \quad \text{and} \quad -8 \cdot 2 = -16.$$

The required geometric means are -8 and -16.

18.  $\underline{-12}, \underline{\quad}, \underline{\frac{-1}{12}}$. There are 2 multiplications by r between -12 and $-\frac{1}{12}$.

Hence,

$$r^2 = \frac{\frac{-1}{12}}{-12} = \frac{1}{144}\ ; \ r = \pm\frac{1}{12}.$$

Therefore, the required geometric mean is $-12 \cdot \left(\pm\frac{1}{12}\right) = \pm 1$.

20. 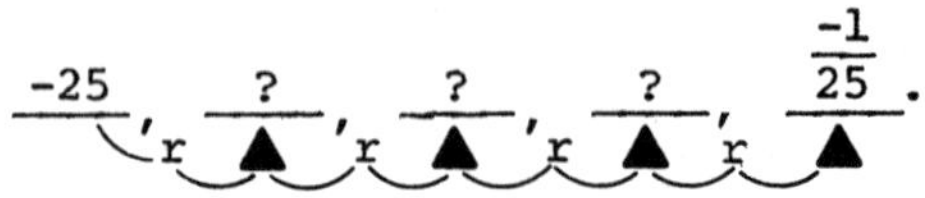 $\underline{-25}, \underline{?}, \underline{?}, \underline{?}, \underline{\frac{-1}{25}}$. There are 4 multiplications by r between -25 and $-\frac{1}{25}$. Hence,

$$r^4 = \frac{\frac{-1}{25}}{-25} = \frac{1}{625}; \quad r = \frac{\pm 1}{5}.$$

With $r = \frac{1}{5}$, the geometric means are -5, -1, and $\frac{-1}{5}$. With $r = \frac{-1}{5}$, the geometric means are 5, -1, and $\frac{1}{5}$.

22. $\sum_{j=1}^{4} (-2)^j = (-2) + (-2)^2 + (-2)^3 + (-2)^4$; $a = -2$, $r = -2$, $n = 4$.

$$S_n = \frac{(-2) - (-2)(-2)^4}{1 - (-2)} = \frac{(-2) + 2(16)}{3} = 10$$

24. $\sum_{i=3}^{12} (2)^{i-5} = 2^{-2} + 2^{-1} + \cdot\cdot\cdot + 2^7$, $a = 2^{-2} = \frac{1}{4}$; $r = 2$, $n = 12 - 2 = 10$.

$$S_{10} = \frac{\frac{1}{4} - \frac{1}{4}(2)^{10}}{1 - 2} = \frac{\frac{1}{4} - 256}{-1} = \frac{4\left(\frac{1}{4} - 256\right)}{4(-1)} = \frac{1023}{4}$$

26. $\sum_{k=1}^{5}\left(\frac{1}{4}\right)^k = \left(\frac{1}{4}\right) + \left(\frac{1}{4}\right)^2 + \cdots + \left(\frac{1}{4}\right)^5$; $a = \frac{1}{4}, r = \frac{1}{4}, n = 5$.

$$S_5 = \frac{\frac{1}{4} - \frac{1}{4}\left(\frac{1}{4}\right)^5}{1 - \frac{1}{4}} = \frac{\frac{4^6}{1}\left(\frac{1}{4} - \frac{1}{4^6}\right)}{\frac{4^6}{1}\left(\frac{3}{4}\right)} = \frac{4^5 - 1}{4^5 \cdot 3}$$

28.

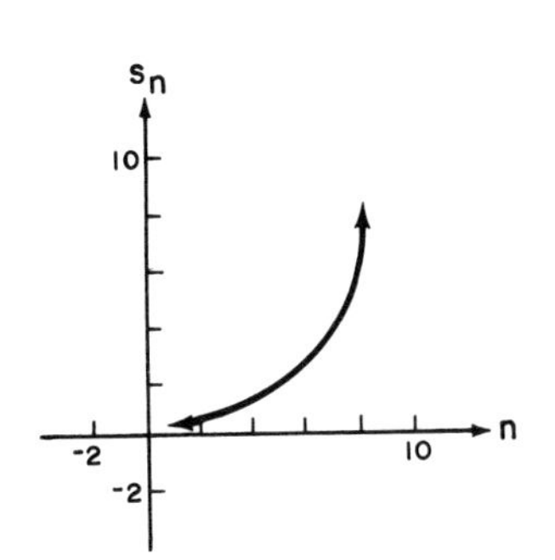

Some ordered pairs are:

$\left(1,\frac{1}{8}\right)$, $\left(2,\frac{1}{4}\right)$, $\left(3,\frac{1}{2}\right)$, (4,1), (5,2), (6,4), (7,8).

30. After 2400 years: 50 grams; after 4800 years: 25 grams; after 7200 years: 12.5 grams; after 9600 years: 6.25 grams; after 2400n years: Note that 50, 25, 12.5, and 6.25 are terms of a geometric progression with $r = \frac{1}{2}$. Hence, $s_n = 50\left(\frac{1}{2}\right)^{n-1}$. Thus, in 2400n years, $50\left(\frac{1}{2}\right)^{n-1}$ grams.

EXERCISE 13.4

In the following problems, we apply the formula for the sum of an infinite geometric progression: $S_\infty = \frac{a}{1 - r}$.

2. $a = 2, r = \frac{1}{2}$;

$|r| < 1$; hence, S_∞ exists.

$$S_\infty = \frac{2}{1 - \frac{1}{2}} = 4$$

4. $a = 1, r = \frac{2}{3}$;

$|r| < 1$; hence, S_∞ exists.

$$S_\infty = \frac{1}{1 - \frac{2}{3}} = \frac{1}{\frac{1}{3}} = 3$$

6. $r = \frac{-1}{8} \div \frac{1}{16} = -2;$

$|r| = 2 > 1$; hence, S_∞ does not exist.

8. $r = \frac{-3}{2} \div \frac{2}{1} = \frac{-3}{4}$

$|r| = \frac{3}{4} < 1$; hence S_∞ exists.

$a = 2;$

$$S_\infty = \frac{2}{1 - \left(\frac{-3}{4}\right)} = \frac{2}{\frac{7}{4}} = \frac{8}{7}$$

10. $\frac{-1}{4} + \left(\frac{-1}{4}\right)^2 + \left(\frac{-1}{4}\right)^3 + \cdots;$

$a = \frac{-1}{4},\ r = \frac{-1}{4};$

$$S_\infty = \frac{\frac{-1}{4}}{1 - \left(\frac{-1}{4}\right)} = \frac{-1}{5}$$

12. $0.666\bar{6} = 0.6 + 0.06 + 0.006 + \cdots$

$= \frac{6}{10} + \frac{6}{10^2} + \frac{6}{10^3} + \cdots;$

$a = \frac{6}{10},\ r = \frac{1}{10};$

$$S_\infty = \frac{\frac{6}{10}}{1 - \frac{1}{10}} = \frac{2}{3}$$

14. $0.4545\overline{45} = 0.45 + 0.0045 + 0.000045 + \cdots$

$= \frac{45}{100} + \frac{45}{10{,}000} + \frac{45}{1{,}000{,}000} + \cdots; \quad a = \frac{45}{100},\ r = \frac{1}{100};$

$$S_\infty = \frac{\frac{45}{100}}{1 - \frac{1}{100}} = \frac{5}{11}$$

16. $3.027\overline{027} = 3 + 0.027 + 0.000027 + \cdots$

$= 3 + \frac{27}{1{,}000} + \frac{27}{1{,}000{,}000} + \cdots; \ a = \frac{27}{1{,}000},\ r = \frac{1}{1{,}000};$

$$S_\infty = \frac{\frac{27}{1{,}000}}{1 - \frac{1}{1{,}000}} = \frac{1}{37}; \text{ hence, } 3.027\overline{027} = 3\,\frac{1}{37}.$$

18. $0.833\bar{3} = 0.8 + 0.03 + 0.003 + 0.0003 + \cdots$

$= \frac{8}{10} + \frac{3}{100} + \frac{3}{1{,}000} + \frac{3}{10{,}000} + \cdots; \ a = \frac{3}{100},\ r = \frac{1}{10};$

$$S_\infty = \frac{\frac{3}{100}}{1 - \frac{1}{10}} = \frac{1}{30}; \text{ hence, } 0.833\bar{3} = \frac{8}{10} + \frac{1}{30} = \frac{5}{6}.$$

20. At 12% semi-annually:

$A = 5000(1 + 0.06)^2 = 5000(1.06)^2$
$= 5618.$

At 11% continuously:

$A = 5000e^{0.11(1)} = 5000e^{0.11}$
$= 5581.39.$

The investment at 12% would produce the greater annual income.

22. In the progression: $12 + 12\left(\frac{9}{10}\right) + 12\left(\frac{9}{10}\right)^2 + \cdots;$

$a = 12,\ r = \frac{9}{10};$

$$S_\infty = \frac{12}{1 - \frac{9}{10}} = 120.$$

The pendulum will swing approximately 120 inches.

24. The figure shows the first few bounces. The sum of the falling segments is

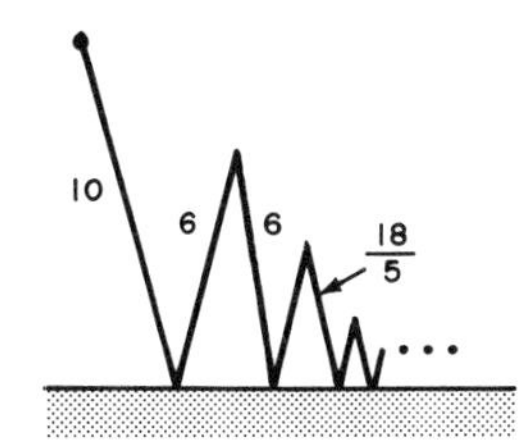

$$10 + \left(\frac{3}{5}\right)(10) + \left(\frac{3}{5}\right)\left(\frac{3}{5}\right)(10) + \cdots$$

or

$10 + 6 + \frac{18}{5} + \cdots$, where $a = 10$ and $r = \frac{3}{5}$. The sum of the rising segments is

$6 + \left(\frac{3}{5}\right)(6) + \left(\frac{3}{5}\right)\left(\frac{3}{5}\right)(6) + \cdots$ or $6 + \frac{18}{5} + \frac{54}{25} + \cdots$, where $a = 6$ and $r = \frac{3}{5}$. Thus, the total distance equals

$\frac{10}{1 - \frac{3}{5}} + \frac{6}{1 - \frac{3}{5}} = \frac{10 + 6}{\frac{2}{5}} = 40.$ The ball travels approximately 40 feet.

26. First compounding: $P + P\left(\frac{r}{2}\right)(1) = P\left(1 + \frac{r}{2}\right)$

Second compounding (end of year 1):

$$\left[P\left(1 + \frac{r}{2}\right)\right] + \left[P\left(1 + \frac{r}{2}\right)\right]\left(\frac{r}{2}\right)(1) = P\left(1 + \frac{r}{2}\right)\left(1 + \frac{r}{2}\right) = P\left(1 + \frac{r}{2}\right)^2;$$

Similarly, the third compounding would yield $P\left(1 + \frac{r}{2}\right)^3$ and the fourth compounding (end of year 2) would yield $P\left(1 + \frac{r}{2}\right)^4$. Thus, we have:

At the end of year 1: $P\left(1 + \frac{r}{2}\right)^2$;

At the end of year 2: $P\left(1 + \frac{r}{2}\right)^4$; and we make the conjecture that at the end of year n: $P\left(1 + \frac{r}{2}\right)^{2n}$.

Similarly, if compounding is quarterly, we have:

At the end of year 1: $P\left(1 + \frac{r}{4}\right)^4$;

At the end of year 2: $P\left(1 + \frac{r}{4}\right)^8$; and we make the conjecture that at the end of year n: $P\left(1 + \frac{r}{4}\right)^{4n}$.

Furthermore, if compounding is t times yearly, we have $P\left(1 + \frac{r}{t}\right)^{tn}$.

EXERCISE 13.5

2. For n = 4,
$(3n)! = 12!$
$= 12 \cdot 11 \cdot 10 \cdot 9 \cdot 8 \cdot 7 \cdot 6 \cdot 5 \cdot 4 \cdot 3 \cdot 2 \cdot 1.$

4. For n = 4,
$3n! = 3 \cdot 4!$
$= 3(4 \cdot 3 \cdot 2 \cdot 1).$

6. For n = 2,
$2n(2n - 1)! = 4 \cdot 3!$
$= 4(3 \cdot 2 \cdot 1).$

8. $7 \cdot 6 \cdot 5 \cdot 4 \cdot 3 \cdot 2 \cdot 1 = 5040$

10. $\frac{12 \cdot 11!}{11!} = 12$

12. $\frac{(12!)(8!)}{16 \cdot 15 \cdot 14 \cdot 13(12!)} = \frac{8 \cdot 7 \cdot 6 \cdot 5 \cdot 4 \cdot 3 \cdot 2 \cdot 1}{16 \cdot 15 \cdot 14 \cdot 13} = \frac{12}{13}$

14. $\frac{10 \cdot 9 \cdot 8 \cdot 7 \cdot 6!}{4!6!} = \frac{10 \cdot 9 \cdot 8 \cdot 7}{4 \cdot 3 \cdot 2 \cdot 1} = 210$

16. $5!$

18. $\frac{(7) \cdot 6!}{6!} = \frac{7!}{6!}$

20. $\frac{(28 \cdot 27 \cdot 26 \cdot 25 \cdot 24) \cdot 23!}{23!} = \frac{28!}{23!}$

22. $(n + 4)(n + 3)(n + 2) \cdot \cdot \cdot \cdot \cdot 3 \cdot 2 \cdot 1$

24. $3 \cdot n \cdot (n - 1) \cdot (n - 2) \cdot \cdot \cdot \cdot \cdot 3 \cdot 2 \cdot 1$

26. $(3n - 2)(3n - 3)(3n - 4) \cdot \cdot \cdot \cdot \cdot 3 \cdot 2 \cdot 1$

28. $(2x)^4 + 4(2x)^3y + \frac{4 \cdot 3}{1 \cdot 2}(2x)^2y^2 + \frac{4 \cdot 3 \cdot 2}{1 \cdot 2 \cdot 3}(2x)y^3 + y^4$

$= 16x^4 + 32x^3y + 24x^2y^2 + 8xy^3 + y^4$

30. $(2x)^5 + 5(2x)^4(-1) + \frac{5 \cdot 4}{1 \cdot 2}(2x)^3(-1)^2 + \frac{5 \cdot 4 \cdot 3}{1 \cdot 2 \cdot 3}(2x)^2(-1)^3$

$+ \frac{5 \cdot 4 \cdot 3 \cdot 2}{1 \cdot 2 \cdot 3 \cdot 4}(2x)(-1)^4 + (-1)^5$

$= 32x^5 - 80x^4 + 80x^3 - 40x^2 + 10x - 1$

32. $\left(\frac{x}{3}\right)^5 + 5\left(\frac{x}{3}\right)^4(3) + \frac{5 \cdot 4}{1 \cdot 2}\left(\frac{x}{3}\right)^3(3)^2 + \frac{5 \cdot 4 \cdot 3}{1 \cdot 2 \cdot 3}\left(\frac{x}{3}\right)^2(3)^3$

$+ \frac{5 \cdot 4 \cdot 3 \cdot 2}{1 \cdot 2 \cdot 3 \cdot 4}\left(\frac{x}{3}\right)(3)^4 + 3^5$

$= \frac{x^5}{243} + \frac{5x^4}{27} + \frac{10x^3}{3} + 30x^2 + 135x + 243$

34. $\left(\frac{2}{3}\right)^4 + 4\left(\frac{2}{3}\right)^3(-a^2) + \frac{4 \cdot 3}{1 \cdot 2}\left(\frac{2}{3}\right)^2(-a^2)^2 + \frac{4 \cdot 3 \cdot 2}{1 \cdot 2 \cdot 3}\left(\frac{2}{3}\right)(-a^2)^3 + (-a^2)^4$

$= \frac{16}{81} - \frac{32a^2}{27} + \frac{8a^4}{3} - \frac{8a^6}{3} + a^8$

36. $x^{15} + 15x^{14}(-y) + \frac{15 \cdot 14}{2!}x^{13}(-y)^2 + \frac{15 \cdot 14 \cdot 13}{3!}x^{12}(-y)^3$

38. $(2a)^{12} + 12(2a)^{11}(-b) + \frac{12 \cdot 11}{2!}(2a)^{10}(-b)^2 + \frac{12 \cdot 11 \cdot 10}{3!}(2a)^9(-b)^3$

40. $\left(\frac{x}{2}\right)^8 + 8\left(\frac{x}{2}\right)^7(2) + \frac{8 \cdot 7}{2!}\left(\frac{x}{2}\right)^6(2)^2 + \frac{8 \cdot 7 \cdot 6}{3!}\left(\frac{x}{2}\right)^5(2)^3$

42. In Formula (2) of this section, substitute 12 for n and 5 for r:

$$\frac{12 \cdot 11 \cdot 10 \cdot 9}{(5 - 1)!}x^{12-5+1}(2)^{5-1} = 7920x^8.$$

44. In Formula (2) of this section, substitute 9 for n and 7 for r:

$$\frac{9 \cdot 8 \cdot 7 \cdot 6 \cdot 5 \cdot 4}{(7 - 1)!}(a^3)^{9-7+1}(-b)^{7-1} = 84a^9b^6.$$

46. In Formula (2) of this section, substitute 8 for n and 4 for r:

$$\frac{8 \cdot 7 \cdot 6}{(4 - 1)!}x^{8-4+1}\left(-\frac{1}{2}\right)^{4-1} = -7x^5.$$

48. In Formula (2) of this section, substitute 10 for n and 8 for r:

$$\frac{10 \cdot 9 \cdot 8 \cdot 7 \cdot 6 \cdot 5 \cdot 4}{(8 - 1)!}\left(\frac{x}{2}\right)^{10-8+1}(4)^{8-1} = 245{,}760x^3.$$

50. a. $(1 + 0.02)^{1/2} = (1)^{1/2} + \frac{1}{2}(1)^{-1/2}(0.02) + \frac{\left(\frac{1}{2}\right)\left(\frac{-1}{2}\right)}{2!}(1)^{-3/2}(0.02)^2 + \cdots$

$= 1 + 0.01 - 0.00005 + \cdots \approx 1.01$

b. $(1 - 0.01)^{1/2} = (1)^{1/2} + \frac{1}{2}(1)^{-1/2}(-0.01) + \frac{\left(\frac{1}{2}\right)\left(\frac{-1}{2}\right)}{2!}(1)^{-3/2}(0.01)^2 + \cdots$

$= 1 - 0.005 - 0.0000125 \approx 0.99$

EXERCISE 13.6

2. $P(7,7) = 7! = 5040$

4. $P(10,10) = 10! = 3{,}628{,}800$

6. $P(7,4) = \frac{7!}{3!}$

$= \frac{7 \cdot 6 \cdot 5 \cdot 4 \cdot \cancel{3 \cdot 2 \cdot 1}}{\cancel{3 \cdot 2 \cdot 1}}$

$= 7 \cdot 6 \cdot 5 \cdot 4 = 840$

8. $P(12,6) = \frac{12!}{6!}$

$= 12 \cdot 11 \cdot 10 \cdot 9 \cdot 8 \cdot 7$

$= 665{,}280$

10. The first digit cannot be zero, and the last digit has only 1 choice, 3. Thus, $6 \cdot 7 \cdot 7 \cdot 1 = 294$.

12. The first digit cannot be zero, and the last digit must be 5. There are 6 choices left for the first digit, 5 choices for the second digit, and 4 for the third digit. Thus, $6 \cdot 5 \cdot 4 \cdot 1 = 120$.

14. $7 \cdot 8 \cdot 1 = 56$ (See Exercise 6.)

16. $7 \cdot 7 \cdot 1 = 49$ (See Exercise 8.)

18. $\underline{26} \cdot \underline{26} \cdot \underline{26} \cdot \underline{26} \cdot \underline{26} = 11{,}881{,}376$

20. $\underline{10} \cdot \underline{26} \cdot \underline{10} \cdot \underline{10} = 26{,}000$

of if zero is not allowable in the first position,

$9 \cdot 26 \cdot 10 \cdot 10 = 23{,}400$.

22. $\frac{6!}{2!} \quad \frac{6 \cdot 5 \cdot 4 \cdot 3 \cdot 2 \cdot 1}{2 \cdot 1} = 360$

24. $\frac{7!}{3!} = \frac{7 \cdot 6 \cdot 5 \cdot 4 \cdot 3 \cdot 2 \cdot 1}{3 \cdot 2 \cdot 1} = 840$

26. $\frac{8!}{3!} = 8 \cdot 7 \cdot 6 \cdot 5 \cdot 4 = 6720$

28. $\frac{11!}{2!} = 11 \cdot 10 \cdot 9 \cdot 8 \cdot 7 \cdot 6 \cdot 5 \cdot 4 \cdot 3 = 19{,}958{,}400$

30. $\frac{7!}{2!2!3!} = \frac{7 \cdot 6 \cdot 5 \cdot 4 \cdot 3 \cdot 2 \cdot 1}{2 \cdot 1 \cdot 2 \cdot 1 \cdot 3 \cdot 2 \cdot 1} = 7 \cdot 6 \cdot 5 = 210$

EXERCISE 13.7

2. $\binom{6}{4} = \frac{6!}{4!2!}$

$= \frac{6 \cdot 5}{2 \cdot 1} = 15$

4. $\binom{10}{4} = \frac{10!}{4!6!}$

$= \frac{10 \cdot 9 \cdot 8 \cdot 7}{4 \cdot 3 \cdot 2 \cdot 1} = 210$

6. $\binom{10}{6} = \frac{10!}{6!4!} = 210$

8. $\binom{9}{5} = \frac{9!}{5!4!} = \frac{9 \cdot 8 \cdot 7 \cdot 6}{4 \cdot 3 \cdot 2 \cdot 1} = 126$

10. 1 coin: $\binom{5}{1} = 5$

2 coins: $\binom{5}{2} = \frac{5 \cdot 4}{2 \cdot 1} = 10$

3 coins: $\binom{5}{3} = \frac{5 \cdot 4 \cdot 3}{3 \cdot 2 \cdot 1} = 10$

4 coins: $\binom{5}{4} = \frac{5 \cdot 4 \cdot 3 \cdot 2}{4 \cdot 3 \cdot 2 \cdot 1} = 5$

5 coins: $\binom{5}{5} = 1$

Therefore, the total number of different amounts is

$$\binom{5}{1} + \binom{5}{2} + \binom{5}{3} + \binom{5}{4} + \binom{5}{5} = 31.$$

12. $\binom{52}{13} = \frac{52!}{13!39!} = 635{,}013{,}559{,}600$

14. An octagon has eight vertices, no three of which are collinear. The number of ways of joining pairs of these eight points is $\binom{8}{2} = 28$. Since eight of these connections will be the sides of the octagon, that leaves $28 - 8 = 20$ diagonals.

16. Two face cards can be drawn from the 12 face cards in a standard deck in $\binom{12}{2}$ ways.

$$\binom{12}{2} = \frac{12!}{2!10!} = \frac{12 \cdot 11 \cdot 10!}{2 \cdot 1 \cdot 10!} = 66$$

18. There are four aces; hence, an ace can be drawn in $\binom{4}{4} = 1$ way. There are forty-eight cards that are not aces; hence, a card that is not an ace can be drawn in $\binom{48}{1} = 48$ ways. Therefore, four aces can be paired with a card that is not an ace to form the specified hand in $1 \cdot 48 = 48$ ways.

20. A committee of eight men can be selected in $\binom{10}{8} = 45$ ways. A committee of eight women can be selected in $\binom{11}{8} = 165$ ways. For each way a committee of eight men can be chosen, there are 165 ways a committee of eight women can be chosen. Hence, a committee of eight men and eight women can be chosen in $45 \cdot 165 = 7425$ ways.

22. $\binom{100}{98} = \binom{100}{2} = \dfrac{100 \cdot 99}{2!} = 4950$

24. Using the results of Problem 21, if r = 7, then

$$n - r = 5;\ n - 7 = 5;\ n = 12.$$

26. The first five terms of $(a + b)^n$ are:

$$a^n + \frac{n}{1!}a^{n-1}b + \frac{n(n-1)}{2!}a^{n-2}b^2 + \frac{n(n-1)(n-2)}{3!}a^{n-3}b^3$$
$$+ \frac{n(n-1)(n-2)(n-3)}{4!}a^{n-4}b^4.$$

Then $\binom{n}{0} = \dfrac{n!}{0!n!} = 1$; $\binom{n}{1} = \dfrac{n!}{1!(n-1)!} = \dfrac{n(n-1)!}{1!(n-1)!} = \dfrac{n}{1!}$;

$$\binom{n}{2} = \frac{n!}{2!(n-2)!} = \frac{n(n-1)(n-2)!}{2!(n-2)!} = \frac{n(n-1)}{2!};$$

$$\binom{n}{3} = \frac{n!}{3!(n-3)!} = \frac{n(n-1)(n-2)(n-3)!}{3!(n-3)!} = \frac{n(n-1)(n-2)}{3!};$$

$$\binom{n}{4} = \frac{n!}{4!(n-4)!} = \frac{n(n-1)(n-2)(n-3)(n-4)!}{4!(n-4)!}$$
$$= \frac{n(n-1)(n-2)(n-3)}{4!}.$$

Therefore, the first five terms of the expansion of $(a + b)^n$ can be written as

$$\binom{n}{0}a^n + \binom{n}{1}a^{n-1}b + \binom{n}{2}a^{n-2}b^2 + \binom{n}{3}a^{n-3}b^3 + \binom{n}{4}a^{n-4}b^4.$$

Appendix A
More about Functions

EXERCISE A.1

2.

1	4	-2	3	0	-5
		4	2	5	5
	4	2	5	5	0

P(1) = 0

-1	4	-2	3	0	-5
		-4	6	-9	9
	4	-6	9	-9	4

P(-1) = 4

4.

-2	1	-10	5	-3	6
		-2	24	-58	122
	1	-12	29	-61	128

P(-2) = 128

3	1	-10	5	-3	6
		3	-21	-48	-153
	1	-7	-16	-51	-147

P(3) = -147

Note for Problems 6-12:

When P(x) is divided by x - a, the remainder term is P(a). Thus, synthetic division is a convenient means of obtaining ordered pairs (a,P(a)), which can be used to approximate the graph of y = P(x). Usually, arbitrary integral values of a are used; however, to better approximate high or low points, fractional values are sometimes selected.

6.

a						P(a)	(a,P(a))
-5	-5	1	5	4	0		
			-5	0	-20		
		1	0	4	-20	-20	(-5,-20)
-4	-4	1	5	4	0		
			-4	-4	0		
		1	1	0	0	0	(-4,0)

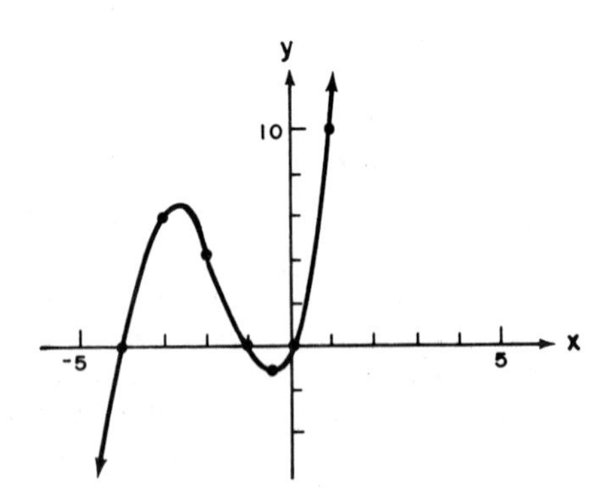

Continuing in the above manner, the following additional ordered pairs are obtained: (-3,6), (-2,4), (-1,0), (-0.5,-0.875), (0,0), (1,10).

8.

a		P(a)	(a,P(a))
-1	-1⌋ 1 -4 3 0		
	-1 5 -8		
	1 -5 8 -8	-8	(-1,-8)
0	0⌋ 1 -4 3 0		
	0 0 0		
	1 -4 3 0	0	(0,0)

Continuing in the above manner, the following additional ordered pairs are obtained: (0.5,0.625), (1,0), (2,-2), (3,0), (4,12).

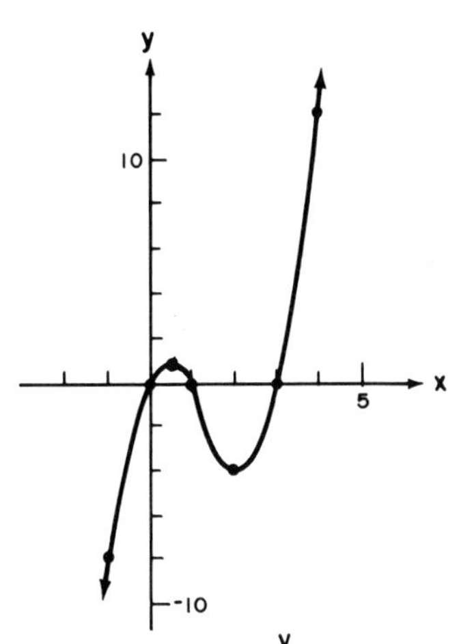

10.

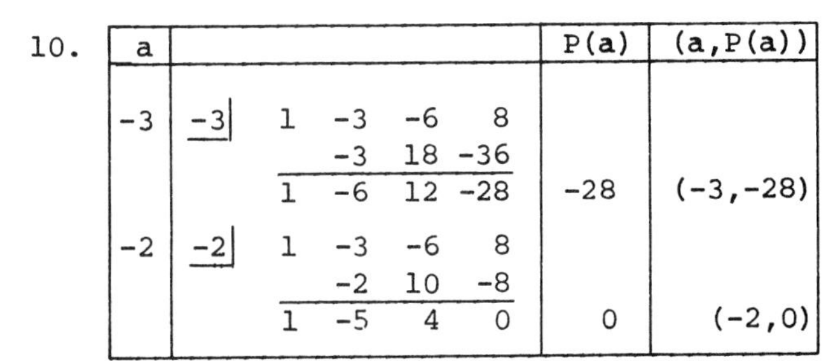

a		P(a)	(a,P(a))
-3	-3⌋ 1 -3 -6 8		
	-3 18 -36		
	1 -6 12 -28	-28	(-3,-28)
-2	-2⌋ 1 -3 -6 8		
	-2 10 -8		
	1 -5 4 0	0	(-2,0)

Continuing in the above manner, the following additional ordered pairs are obtained: (-1,10), (0,8), (1,0), (2,-8), (3,-10), (4,0), (5,28).

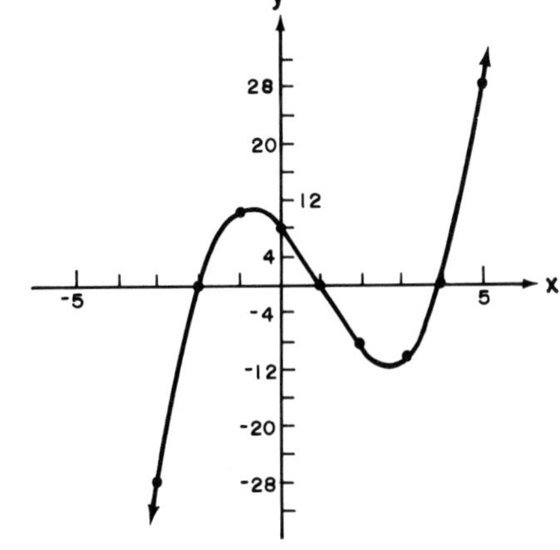

12.

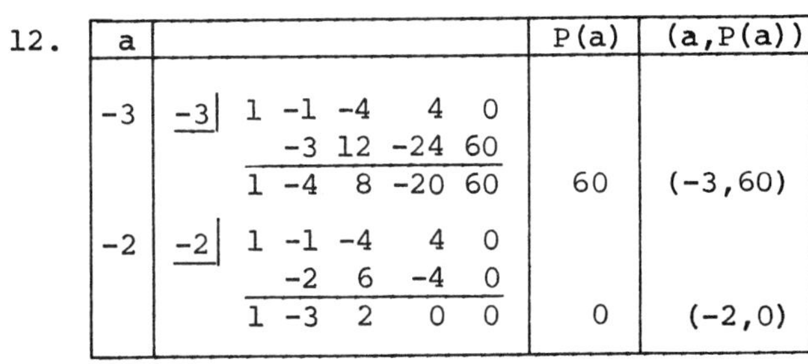

a		P(a)	(a,P(a))
-3	-3⌋ 1 -1 -4 4 0		
	-3 12 -24 60		
	1 -4 8 -20 60	60	(-3,60)
-2	-2⌋ 1 -1 -4 4 0		
	-2 6 -4 0		
	1 -3 2 0 0	0	(-2,0)

Continuing in the above manner, the following additional ordered pairs are obtained: (-1,-6), (0,0), (1,0), (2,0), (3,30).

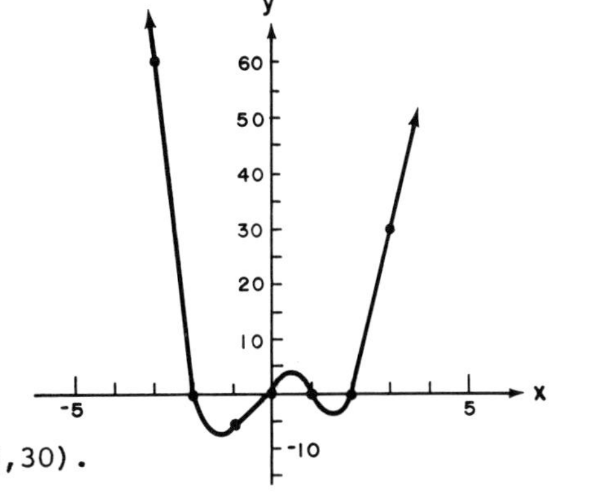

14.

```
1| 2  -5   4  -1
       2  -3   1
   ------------
   2  -3   1   0
```

Since P(1) = 0, by the factor theorem x - 1 is a factor.

16.

```
-1| 2  -5   3    3
       -2   7  -10
   --------------
    2  -7  10   -7
```

Since $P(-1) \neq 0$, by the factor theorem x + 1 is not a factor.

18.

3	1	-6	-1	30
		3	-9	-30
	1	-3	-10	0

$x^3 - 6x^2 - x + 30 = (x - 3)(x^2 - 3x - 10)$

Hence, $x^3 - 6x^2 - x + 30 = 0$ is equivalent to

$$(x - 3)(x^2 - 3x - 10) = 0$$
$$(x - 3)(x - 5)(x + 2) = 0.$$

By inspection, the solution set is $\{3,5,-2\}$.

20.

-5	1	5	-1	-5	0
		-5	0	5	0
	1	0	-1	0	0

$x^4 + 5x^3 - x^2 - 5x = [x - (-5)](x^3 - x)$

Hence, $x^4 + 5x^3 - x^2 - 5x = 0$ is equivalent to

$$(x + 5)(x^3 - x) = 0$$
$$(x + 5)(x)(x^2 - 1) = 0$$
$$(x + 5)(x)(x - 1)(x + 1) = 0.$$

By inspection, the solution set is $\{-5,0,1,-1\}$.

EXERCISE A.2

2. $x - 4 = 0$ if $x = 4$. Thus, $x = 4$ is a vertical asymptote.

4. $(x + 1)(x - 4) = 0$ if $x = -1$ or $x = 4$. Thus, $x = -1$ and $x = 4$ are vertical asymptotes.

6. $x^2 - 3x + 2 = 0$, or $(x - 2)(x - 1) = 0$ if $x = 2$ or $x = 1$. Thus, $x = 2$ and $x = 1$ are vertical asymptotes.

8. $x^2 + 5x + 4 = 0$, or $(x + 4)(x + 1) = 0$, if $x = -4$ or $x = -1$. Thus, $x = -4$ and $x = -1$ are vertical asymptotes. The degree of $2x$ is less than the degree of x^2. Thus, the graph has a horizontal asymptote $y = 0$.

10. $x - 3 = 0$ if $x = 3$. Thus, $x = 3$ is a vertical asymptote. $2x$ and x are of the same degree. Thus, the graph has a horizontal asymptote
$y = \frac{2}{1} = 2$.

12. $x^2 - x - 12 = 0$, or $(x - 4)(x + 3) = 0$ if $x = 4$ or $x = -3$. Thus, $x = 4$ and $x = -3$ are vertical asymptotes. x^2 and x^2 are of the same degree. Thus, the graph has a horizontal asymptote
$y = \frac{1}{1} = 1$.

In the solutions to Exercises 14-26, several ordered pairs which are not shown were used to help obtain the graphs.

14. $x - 3 = 0$ if $x = 3$. Thus, $x = 3$ is a vertical asymptote. Because the degree of the numerator is less than the degree of the denominator, $y = 0$ is a horizontal asymptote. If $x = 0$, $y = \frac{1}{0 - 3} = -\frac{1}{3}$, the y-intercept. If $y = 0$, no value of x exists, and there is no x-intercept.

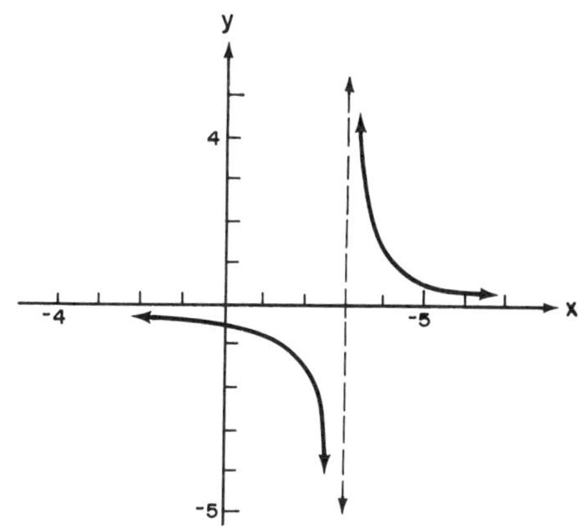

16. $(x + 2)(x - 1) = 0$ if $x = -2$ or $x = 1$. Thus, $x = -2$ and $x = 1$ are vertical asymptotes. Because the degree of the numerator is less than the degree of the denominator, $y = 0$ is a horizontal asymptote. If $x = 0$, $y = \frac{4}{(0 + 2)(0 - 1)}$ $= -2$, the y-intercept. If $y = 0$, no value of x exists, and there is no x=intercept.

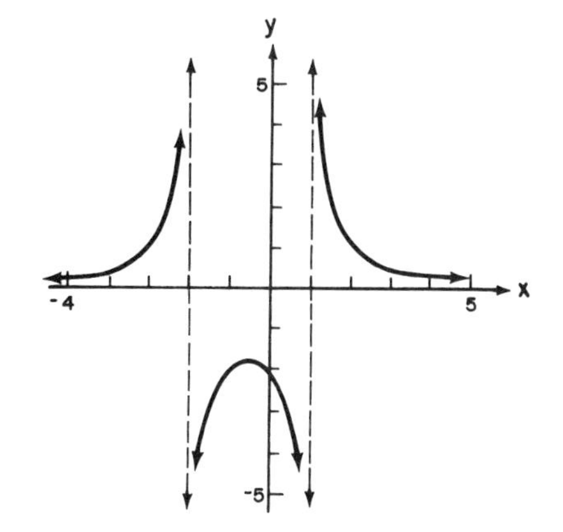

18. $x^2 - x - 6 = 0$, or $(x - 3)(x + 2) = 0$ if $x = 3$ or $x = -2$. Thus, $x = 3$ and $x = -2$ are vertical asymptotes. Because the degree of the numerator is less than the degree of the denominator, $y = 0$ is a horizontal asymptote. If $x = 0$, $y = \frac{4}{0 - 0 - 6} = -\frac{2}{3}$, the y-intercept. If $y = 0$, no value of x exists, and there is no x-intercept.

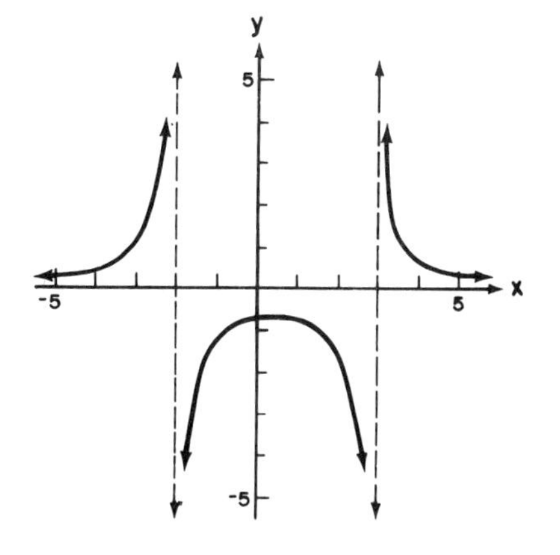

20. Because $x - 2 = 0$ if $x = 2$, a vertical asymptote is $x = 2$. Because the degrees of x and x are the same, a horizontal asymptote is $y = \frac{1}{1} = 1$. If $x = 0$, $y = \frac{0}{0 - 2} = 0$, and the graph passes through the origin.

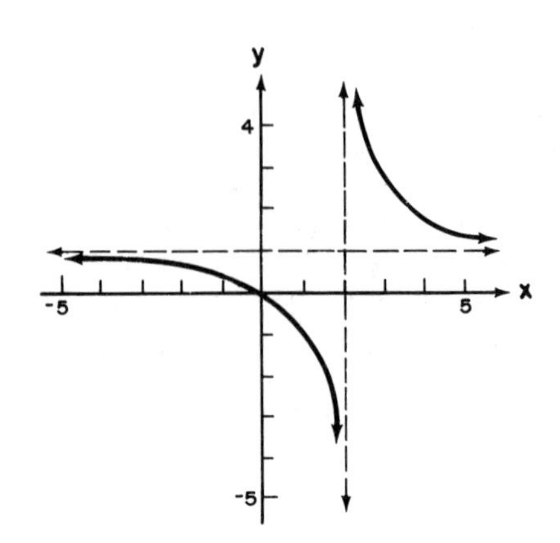

22. Because $x - 3 = 0$ if $x = 3$, a vertical asymptote is $x = 3$. Because the degrees of x and x are the same, a horizontal asymptote is $y = \frac{1}{1} = 1$. If $x = 0$, $y = \frac{0 - 1}{0 - 3} = \frac{1}{3}$, the y-intercept. If $y = 0$, $0 = \frac{x - 1}{x - 3}$, or $x - 1 = 0$ and $x = 1$, the x-intercept.

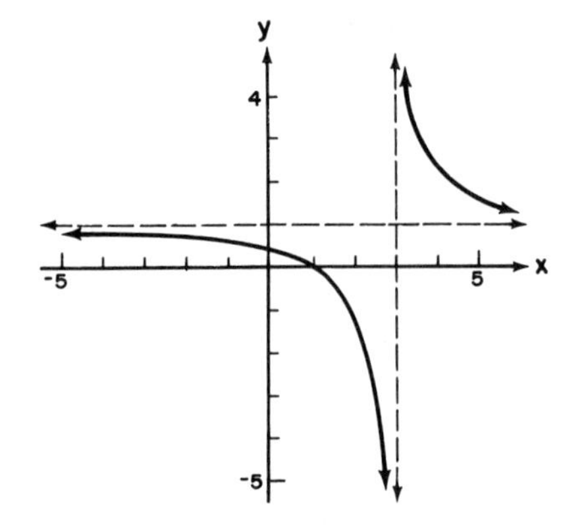

24. Because $x^2 - 9 = 0$, or $(x - 3)(x + 3) = 0$ if $x = 3$ or $x = -3$, vertical asymptotes are $x = 3$ and $x = -3$. Because the degree of x is less than the degree of x^2, the horizontal asymptote is $y = 0$. If $x = 0$, $y = \frac{0}{0 - 9} = 0$, and the graph passes through the origin.

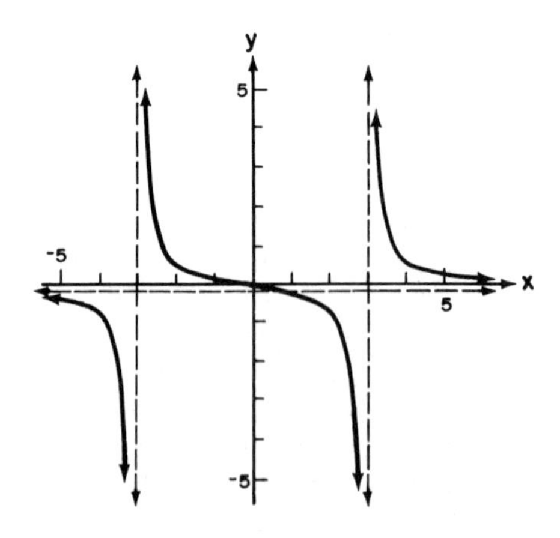

26. Because $x^2 - x - 6 = 0$, or $(x - 3)(x + 2) = 0$ if $x = 3$ or $x = -2$, the vertical asymptotes are $x = 3$ and $x = -2$. Because the degree of x is less than the degree of x^2, the horizontal asymptote is $y = 0$. If $x = 0$, $y = \frac{0 + 1}{0 - 0 - 6} = -\frac{1}{6}$, the y-intercept. If $y = 0$, $0 = \frac{x + 1}{x^2 - x - 6}$, from which $x = -1$, the x-intercept.

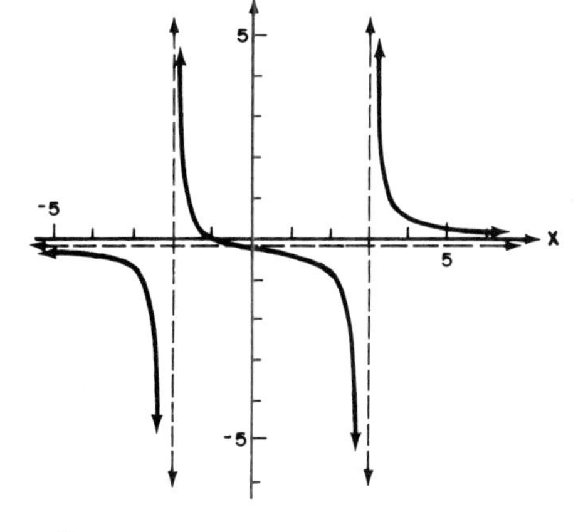

28. $\frac{x^2 - 1}{x + 1} = \frac{\cancel{(x + 1)}(x - 1)}{\cancel{x + 1}}$

$= x - 1,\ x \neq -1.$

Hence, we graph

$y = x - 1,\ x \neq -1.$

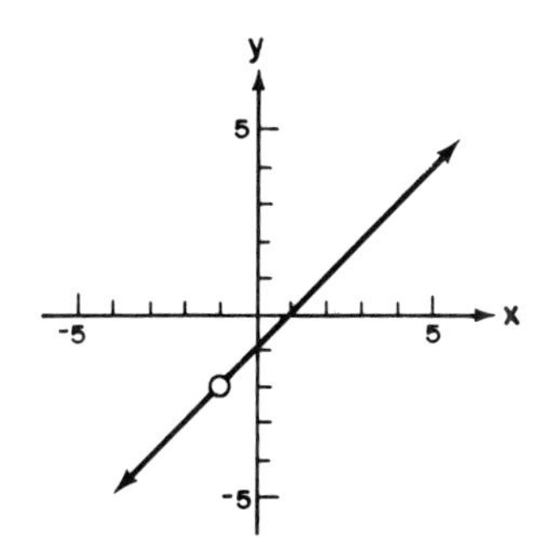

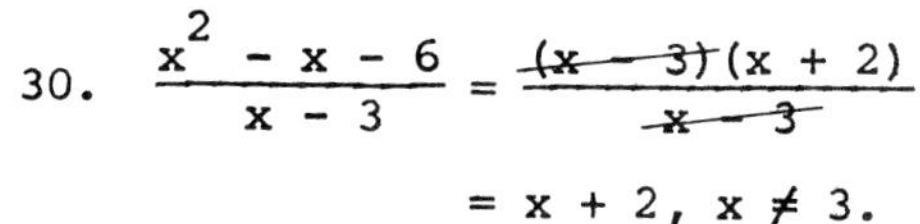

30. $\frac{x^2 - x - 6}{x - 3} = \frac{\cancel{(x - 3)}(x + 2)}{\cancel{x - 3}}$

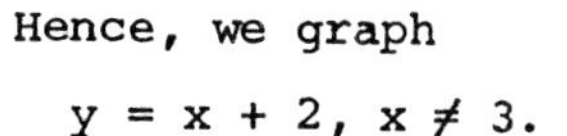

$= x + 2,\ x \neq 3.$

Hence, we graph

$y = x + 2,\ x \neq 3.$

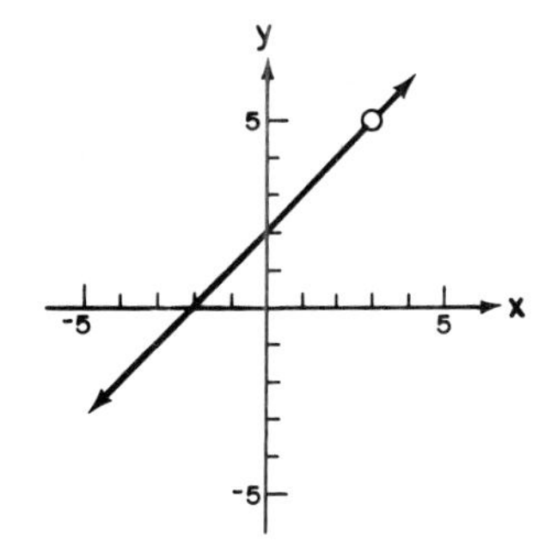

32. We graph

$y = x^2 - 1,\ x \neq 3.$

From the form of $y = x^2 - 1$, its graph is a parabola opening upward with vertex at $(0,-1)$ and x-intercepts at -1 and 1.

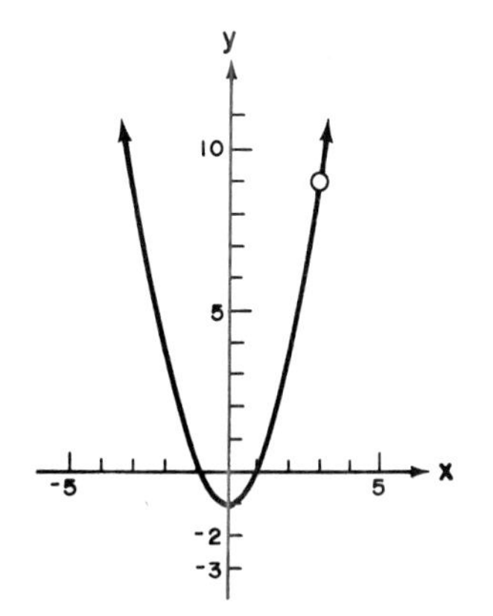

34.

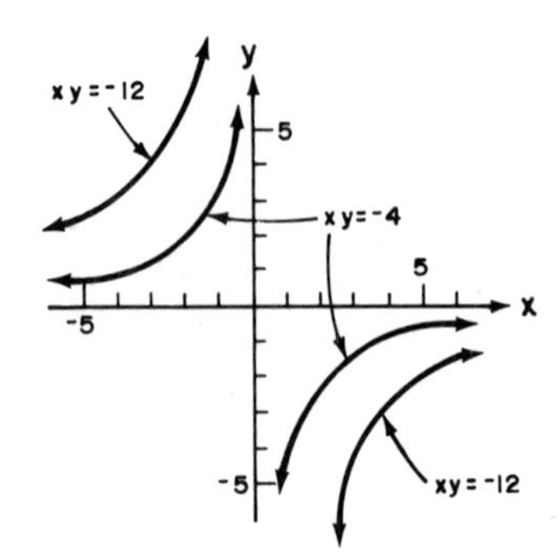

EXERCISE A.3

2. Since $f(x) = 2x + 1$ if $x \geq -1$
$f(1) = 2(1) + 1 = 3.$

4. Since $f(x) = -2x$ if $x \leq -1$,
$f(-1) = -2(-1) = 2.$

6.

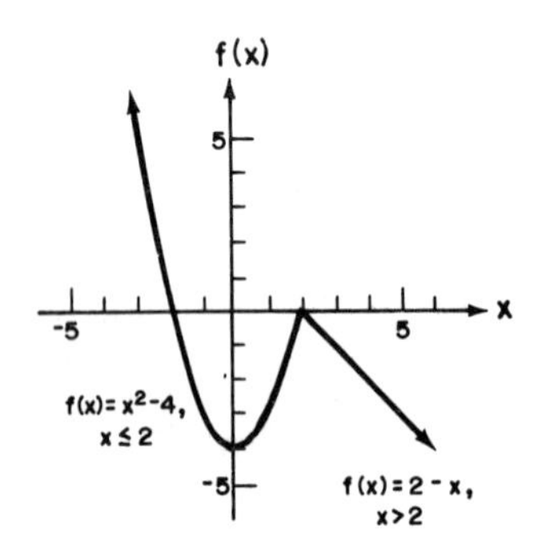

8.

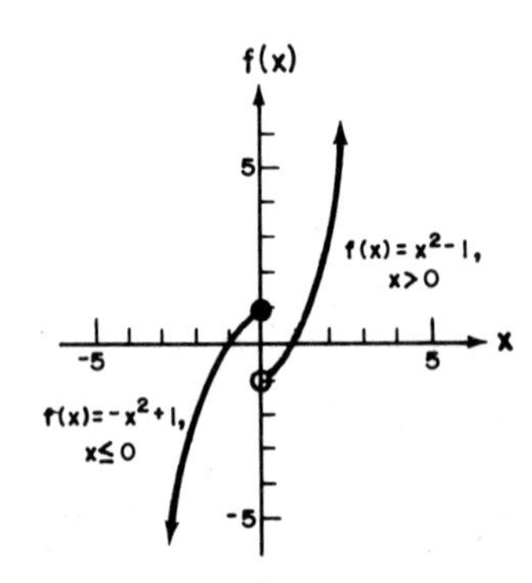

10.

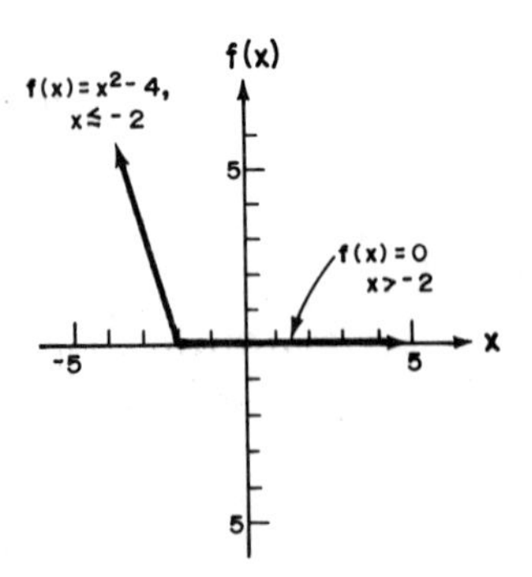

12. The given function is equivalent to $y = \begin{cases} -x + 3 \text{ if } x \geq 0 \\ x + 3 \text{ if } x < 0. \end{cases}$

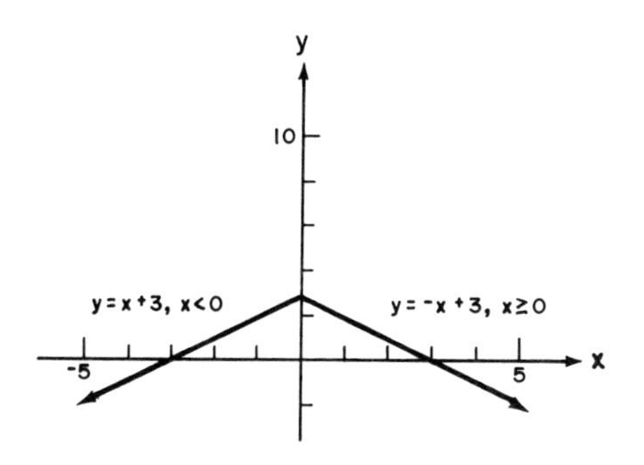

14. $f(x) = x - 2$ if $x \geq 2$
$f(x) = -(x - 2)$
$= -x + 2$ if $x < 2$

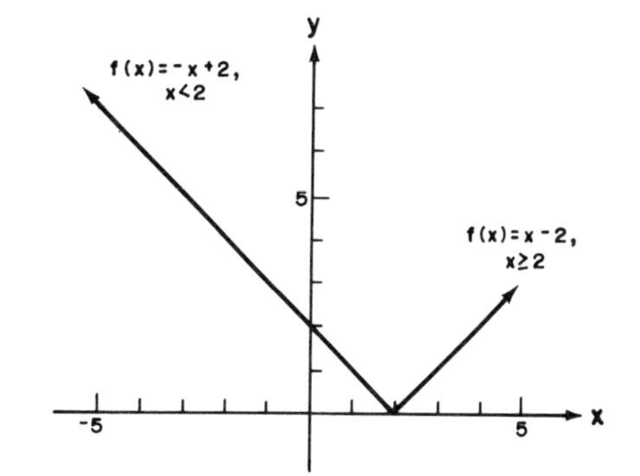

16. $y = -(3x + 2) = -3x - 2$ if $x \geq -\frac{2}{3}$

$y = 3x + 2$ if $x < -\frac{2}{3}$

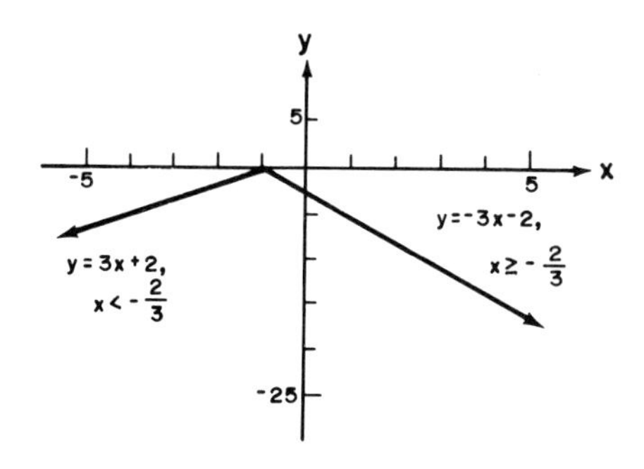

18. $y = 3x + 2$ if $x \geq 0$
$y = -3x + 2$ if $x < 0$

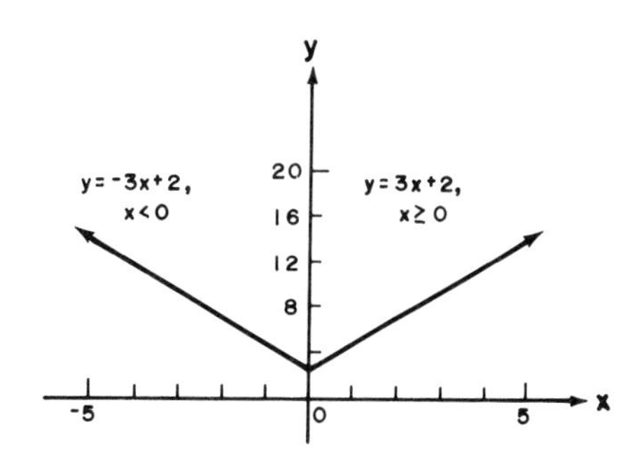

20. First make the table below and then graph the function over each unit interval.

x	[-5,-4)	[-4,-3)	[-3,-2)	[-2,-1)	[-1,0)
2x	[-10,-8)	[-8,-6)	[-6,-4)	[-4,-2)	[-2,0)
[2x]	-10	-8	-6	-4	-2

x	[0,1)	[1,2)	[2,3)	[3,4)	[4,5)	5
2x	[0,2)	[2,4)	[4,6)	[6,8)	[8,10)	10
[2x]	0	2	4	6	8	10

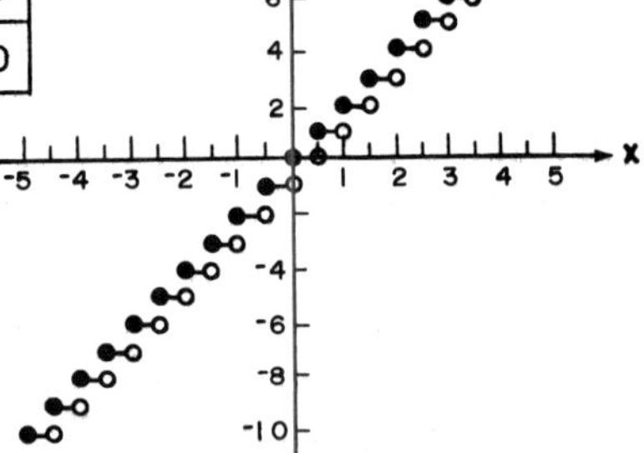

22.

x	[-5,-4)	[-4,-3)	[-3,-2)	[-2,-1)	[-1,0)
[x]	-5	-4	-3	-2	-1
[x] + 1	-4	-3	-2	-1	0

x	[0,1)	[1,2)	[2,3)	[3,4)	[4,5)	5
[x]	0	1	2	3	4	5
[x] + 1	1	2	3	4	5	6

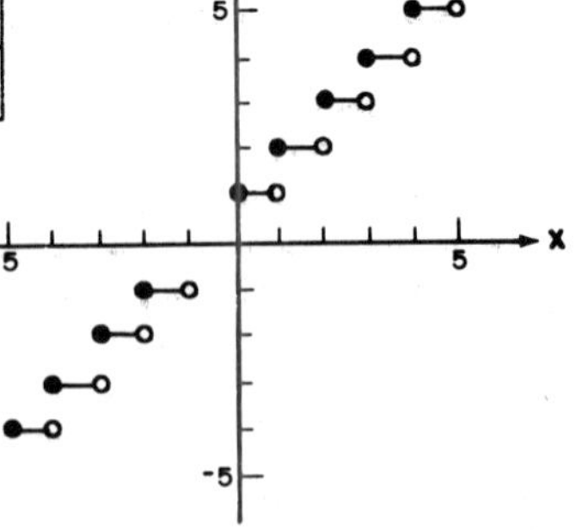

24.

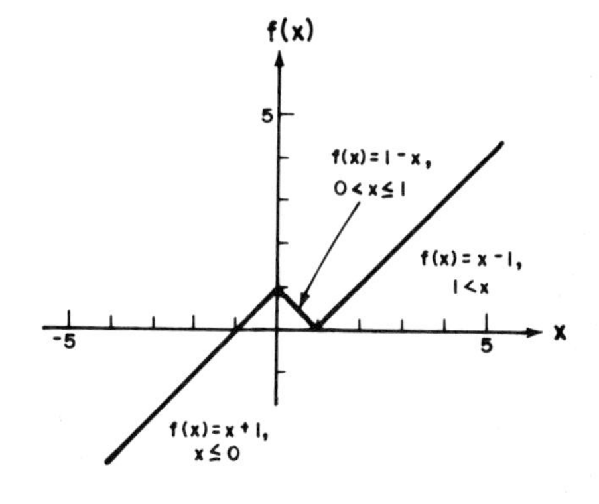

26.

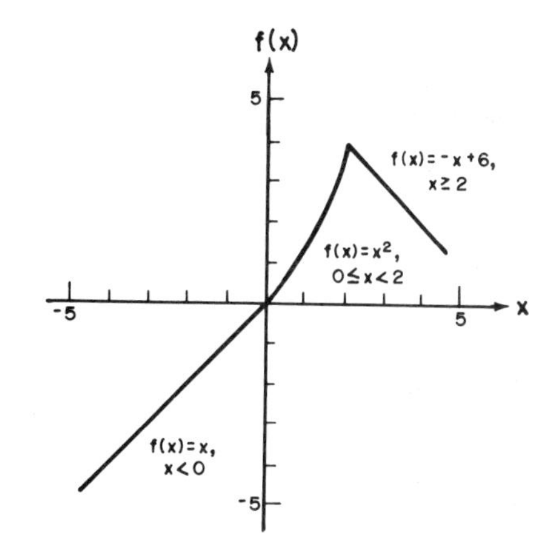

28. The function can be defined equivalently piecewise as follows:

$$f(x) = \begin{cases} 3x - x & \text{if } x \geq 0 \\ -3x - (-x) & \text{if } x < 0. \end{cases}$$

Simplifying the right members,

$$f(x) = \begin{cases} 2x \text{ if } x \geq 0 \\ -2x \text{ if } x < 0. \end{cases}$$

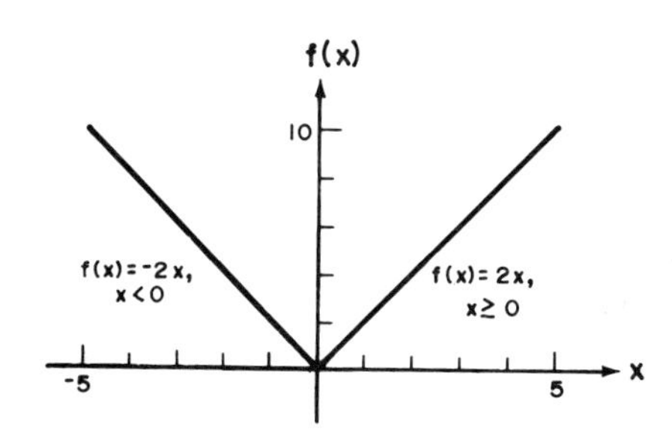

30. The function can be defined piecewise as follows:

$$f(x) = \begin{cases} 3x - x = 2x & \text{if } x \geq 0 \\ 3(-x) - x = -4x & \text{if } x < 0. \end{cases}$$

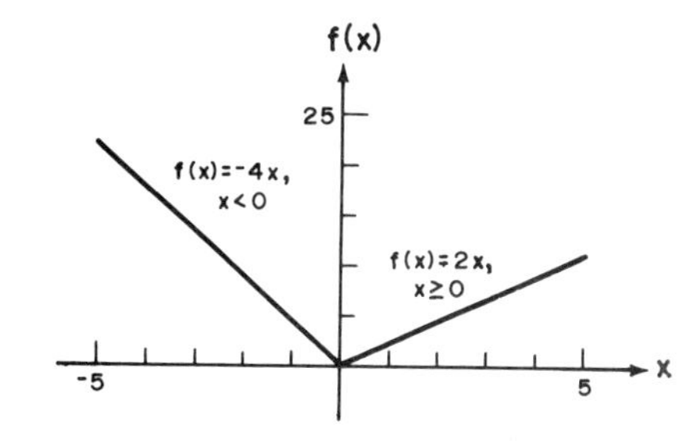

32. Since for all x, $|x^2| = x^2$,

$f(x) = x^2$.

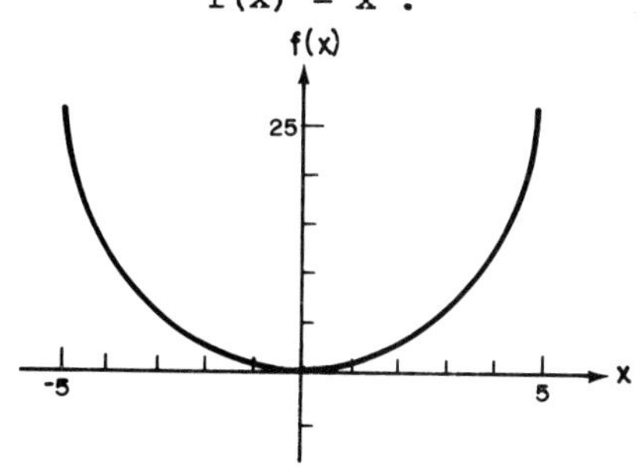

34.

x	[-5,-4)	[-4,-3)	[-3,-2)	[-2,-1)	[-1,0)
[x]	-5	-4	-3	-2	-1
[x] - x	-5 - x	- 4 - x	- 3 - x	- 2 - x	- 1 - x

x	[0,1)	[1,2)	[2,3)	[3,4)	[4,5)	5
[x]	0	1	2	3	4	5
[x] - x	-x	1 - x	2 - x	3 - x	4 - x	0

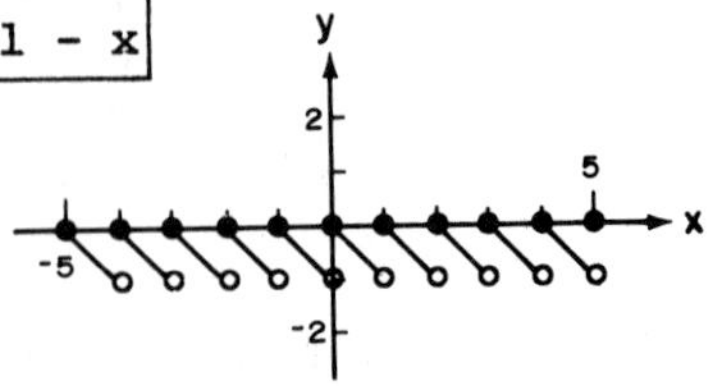

Test Problems

NOTE: These Test Problems are only a sampling of items that instructors might include on a test. Do not assume that they constitute a complete review. For a complete review, see the Chapter Reviews in the text.

CHAPTER 1

1. Let $A = \left\{-7, 6, -\frac{1}{7}, 0, \frac{6}{7}, 7\right\}$.

   a. List the members of $\{x \mid x \text{ is in } A \text{ and } x \text{ is a natural number}\}$.

   b. List the members of $\{x \mid x \text{ is in } A \text{ and } x \text{ is an even number}\}$.

2. Express each of the following relations using appropriate symbols.

   a. x is not less than y.
   b. x is between -2 and 1.

3. Graph each set.

   a. $\{x \mid -1 \le x < 3\}$.

   b. $\{x \mid x \ge -2\}$.

In Problem 4, each of the statements is an application of one of the properties in Section 1.3 of the text. Justify each statement by citing the appropriate property. Assume that all variables represent real numbers.

4. a. $5 + (-5) = 0$

   b. $x(y + z) = (y + z)x$.

   c. $0 + (x + y) = x + y$.

   d. $p + p(a + b) = p + pa + pb$.

In Problem 5, each of the statements is an application of one of the following four properties.

I. Addition property of equality.
II. Multiplication property of equality.
III. Zero-factor property.
IV. $-(-a) = a$.

Justify each statement by citing the appropriate property. Assume that all variables represent real numbers.

5. a. If $x + 2 = 4$, then $-3(x + 2) = -3(4)$.
   b. If $5 = n - 2$, then $2n + 5 = 3n - 2$.
   c. $-[-(x + 2)] = x + 2$.
   d. $0 \cdot (r + p) = 0$.

6. Rewrite without absolute-value notation.

   a. $|-8|$.
   b. $|x - 2|$ if $x - 2 < 0$.

7. Write each product or quotient as a basic numeral if possible.

   a. $(-2)(-2)(2)$.
   b. $\frac{5 \cdot (2)}{0}$.
   c. $\frac{-28}{4}$.
   d. $(-1)(-2)(-3)$.

8. Write each sum or difference as a basic numeral.

   a. $-4 + (4) - (-2)$.
   b. $(8 + 5 - 11) + (2 - 6)$.
   c. $(15 - 8) - (6 - 11)$.
   d. $8 - (-3 + 2 - 7)$.

9. a. Rewrite $\frac{5}{9}$ as a product.
   b. Rewrite $\frac{1}{11}(7)$ as a quotient.

10. a. Write $\frac{6 + 3 \cdot 4}{4 - 5 \cdot 2}$ as a basic numeral.
    b. Write $\frac{4 - \left(\frac{8 + 12}{2 - 7}\right) - 6}{4 \cdot 2 - 2}$ as a basic numeral.

CHAPTER 2

In Problem 1, identify each given polynomial as a monomial, binomial, or trinomial and give its degree.

1. a. $2x^3 - x$
   b. $3y^2 + 5y - 7$

2. If $P(x) = 3x^2 + 2x - 1$, find

   a. $P(0)$
   b. $P(-1)$

3. Find the value of $\frac{3x - 2y^2}{x + y}$ for $x = 2$ and $y = -1$.

In Problems 4-7, simplify each expression.

4. $(2x - 5y) - (5x - 2y)$

5. $3c - [2c - (3c - 1) + 1]$

6. $y - [y^3 - (y^3 - y) + y]$

7. $(-3xy^2)(2x^2y)(xy)$.

In Problems 8-17, write each expression as a polynomial in simple form.

8. $(-xy)^2 + 3x^2(xy) - 2x^2y^2$

9. $-3[(x - 3) - (x + 2)^2]$

10. $5p^3(p^2 - 2p + 1)$

11. $3(3x - 1)(x + 2)$

12. $(y - 1)(y^2 - 2y - 1)$

13. $(z - 2)(z + 2)(z - 1)$

14. $2[x - 3(x - 2) - 2]$

15. $-y\{1 - 3[x + 2(x - 1) + 2x\}$

16. $(2x^n + 1)(2x^n - 1)$

17. $(x^n y^{n-1})^4$

18. If $P(x) = x^2 - 3x$ and $Q(x) = x^4 - 2x^2$, find

a. $P[Q(1)]$.

b. $Q[P(1)]$.

CHAPTER 3

In Problems 1-6, solve for x.

1. $5[2 + 3(x - 2)] + 20 = 0$.

2. $(2x + 1)(x - 3) = (x - 2) + 2x^2$.

3. $-[3(x - 1)^2 - 3x^2] = 0$

4. $0.05x + 0.06(x - 500) = 25$

5. $3x - (x - 2) = 3(x + 1)$

6. $3[2x - (x + 2) + 5] = 10$

7. Solve $5T = 3t + 4p$ for t, given that $T = 4$ and $p = 2$.

8. Solve $2A = h(b + c)$ for c.

In Problems 9-12, solve and graph each solution set on a number line.

9. $2x - 3 \le 10$.

10. $3(x + 2) \ge 4x$

11. $2 \le 3x - 4 \le 8$.

12. $\{x | 2x - 3 < 5\} \cap \{x | 2x - 3 > -5\}$.

13. Solve $|2x + 5| = 2$.

14. Solve $\left|2x + \frac{1}{2}\right| = \frac{1}{4}$.

In Problems 15 and 16, solve and graph each solution set.

15. $|x + 5| > 2$.

16. $|2x + 4| < 6$.

17. Represent $[-3,4)$ using set-builder notation.

18. Write $[-3,5) \cap [-1,6]$ as a single interval.

19. How much water should be added to 2 gallons of pure acid to obtain a 10% solution?

20. A collection of coins containing nickels, dimes, and quarters has a value of $4.55. If there are 5 more dimes than nickels, and the number of quarters is 25 less than twice the number of dimes, how many of each kind of coin is in the collection?

21. The Fahrenheit and Celsius temperatures are related by $C = \frac{5}{9}(F - 32)$. Within what range must the temperature be in Fahrenheit degrees if the temperature in Celsius degrees is between 0° and 100°?

CHAPTER 4

Factor each polynomial if possible.

1. $5m^3 + 2m^2 - m$.
2. $y^{n+3} + 4y^n$.
3. $3x^2 + 3xy - 18y^2$.
4. $8a^3 - b^3$.
5. $3m^2 - 5m + 3mn - 5n$.
6. $9x^4 - 16x^2y^2$.
7. $7y^4z^2 + 14y^3z + 7y^2z^2$
8. $2x^2 - 18$
9. $x^2y^2z^2 - 4x^2z^2$
10. $x^2 + 14x + 49$
11. $2x^2y + 16xy + 32y$
12. $2x^3y^2 + 16x^2y^3 + 32x^2y^2$
13. $3x^2 + 5x + 1$
14. $8x^2 + 14x + 3$
15. $12x^2 - 17x + 6$
16. $12x^2 - 8x - 15$
17. $xy + 3y - x - 3$
18. $xy^2 - 2y^2 + 2xy - 4y$
19. $2x^3 - x^2y - 2xy + y^2$
20. $4x^2 - 4x + 1 - 9z^2$
21. $x^3y^3 - 8$
22. $64x^3 - y^3$
23. $(x + 1)^3 - x^3$
24. $(y^3 - 1)^3 + 27$

25. $x^2y^3 - 8x^2$

26. $x^{2n} - y^{2n}$

Solve for x.

27. $5x^2 - 9x = 0$

28. $2x^2 - 5x - 3 = 0$

29. $2x(4x - 7) = -3$

30. $4x(x + 1) - 24 = 0$

31. $4x^2 - 49 = 0$

32. $6x^2 = 2 - x$

33. Write an equation in standard form for the quadratic equation whose solutions are 3 and -4.

34. The product of two consecutive odd integers is 255. Find the integers.

35. The length of a rectangle is two centimeters more than twice its width. Find the dimensions of the rectangle if its area is 40 square centimeters.

CHAPTER 5

In all problems, assume that no denominator is equal to zero.

In Problem 1, write each fraction in standard form.

1. a. $-\frac{-4}{7}$. b. $\frac{4 - 3x}{-x}$. c. $\frac{-5a^2b^2}{-2c^2}$. d. $\frac{m - 2n}{-n}$.

In Problems 2 and 3, reduce each fraction to lowest terms.

2. $\frac{8r^4 - 12r^3}{4r^2 - 6r}$.

3. $\frac{8a^3 - 27b^3}{4a^2 - 9b^2}$.

In Problems 4 and 5, divide.

4. $\frac{x^2 + 3x - 4}{x - 2}$.

5. $\frac{6x^2 - 2xy + 9x - 3y}{2x + 3}$.

In Problems 6-13, write each expression as a single fraction in lowest terms.

6. $\frac{4}{3x} + \frac{5}{2y} - \frac{2}{xy}$.

7. $\frac{m}{m^2 - 1} - \frac{m}{m^2 - 2m + 1}$.

8. $\frac{3p}{4pq - 6q^2} \cdot \frac{2p - 3q}{12p}$.

9. $\frac{4y^2 + 8y + 3}{2y^2 - 5y + 3} \cdot \frac{6y^2 - 9y}{1 - 4y^2}$.

10. $\frac{m^2 + m - 2}{m^2 + 2m - 3} \div \frac{m^2 + 7m + 10}{m^2 - 2m - 15}$.

11. $\frac{8x^3 - 1}{x - 2} \div \frac{4x^2 + 2x + 1}{x^2 - 4x + 4}$.

12. $\dfrac{\dfrac{m - 2}{m}}{\dfrac{m^2 - 4}{m^2}}$.

13. $\dfrac{\dfrac{a - 1}{a + 1} - \dfrac{a + 1}{a - 1}}{\dfrac{a - 1}{a + 1} + \dfrac{a + 1}{a - 1}}$.

14. Use synthetic division to write each quotient as a polynomial in simple form.

a. $\dfrac{x^3 - 3x^2 + 2x + 6}{x - 3}$

b. $\dfrac{x^5 + 1}{x + 1}$

In Problems 15-17, solve for x.

15. $\dfrac{x}{5} - 9 = \dfrac{x}{2}$.

16. $\dfrac{x}{x + 2} - \dfrac{x^2 + 8}{x^2 - 4} = \dfrac{3}{x - 2}$.

17. $\dfrac{x}{a} - \dfrac{a}{b} = \dfrac{b}{c}$

18. Solve the equation

$S = \dfrac{a}{1 - r}$ for r.

19. Solve $\dfrac{1}{C_n} = \dfrac{1}{C_1} + \dfrac{1}{C_2}$ for C_2 if $C_1 = 15$ and $C_n = 5$.

20. Solve $\dfrac{2x + y}{2} = \dfrac{4x - y}{3}$ for x in terms of y.

21. When a number is divided by 8, it has 4 as a quotient and 5 as a remainder. Find the number.

22. The width of a rectangle is $\frac{3}{4}$ of its length. When the length is increased by 8 feet and the width by 6 feet, the area is increased by 96 square feet. Find the length and width of the original rectangle.

23. A man sailed a boat across a lake and back in 2 1/2 hours. If his rate returning was 2 miles an hour less than his rate going, and if the distance each way was 6 miles, find his rate each way.

CHAPTER 6

Assume that all variables denote positive numbers and that no denominator equals 0.

In Problems 1-4, simplify.

1. a. $(x^3y^2)^3$.

b. $(-2x^3)^2(-2y^2)^2$.

2. a. $\left(\frac{x^3}{y^4}\right)^3$.   b. $\left(\frac{x^2}{y^2}\right)^2\left(-\frac{2xy}{5}\right)^3$.

3. a. $\frac{(-x^3)^2(-x)^3}{(x^2)^3}$.   b. $(x^{2n+1} \cdot x^{n-2})^2$.

4. a. $\left(\frac{x^{2n}x^{3n}}{x^{5n-2}}\right)^3$.   b. $\left(\frac{x^{2n-2}}{x^{2n}}\right)^{2n}$.

In Problems 5 and 6, simplify. Assume m,n > 0.

5. $\left(\frac{x^{3n}y^{4n}}{x^{2n}y^{n}}\right)^{1/3}$.   6. $\left[\left(\frac{x^{4n}}{y^{3n}}\right)^{1/4n}\right]^{1/3}$.

In Problems 7 and 8, simplify.

7. a. $x^{-5}x^2$.   b. $(x^4y^{-2})^{-1/3}$.

8. a. $\frac{x^{-3}y^{-2}z^{-1}}{x^0y^{-4}z^2}$.   b. $\left(\frac{3^0x^{-2}y^{-3}}{x^0y^{-2}}\right)^{-2}$.

9. Simplify $\frac{(2 \times 10^{-4})(4 \times 10^6)(5 \times 10^2)}{8 \times 10^2}$.

10. Simplify $\frac{.009 \times .0008}{.0036}$.

In Problems 11-13, factor as indicated.

11. $x^{1/3} - x = x(? - ?)$.   12. $y^{1/3} - y^{-1/3} = y^{-1/3}(? - ?)$.

13. $y^{2n+2} - y^n = y^n(? - ?)$.

14. a. Express $(2 - y^2)^{-3/4}$ in radical notation.

b. Express $\sqrt[3]{2m^2n}$ in exponential notation.

15. Simplify.

a. $\sqrt[4]{81x^6y^3}$.   b. $\sqrt{\frac{5}{6}}$.   c. $\frac{2m}{\sqrt{2mn}}$.   d. $\frac{\sqrt[3]{x}\sqrt[3]{y^2}}{\sqrt[3]{x^2y}}$.

16. Simplify.

a. $\sqrt{48} + \sqrt{75}$ b. $\sqrt{9m} + 3\sqrt{36m} - 2\sqrt{25m}$.

c. $\frac{2\sqrt{8}}{3} + \frac{3\sqrt{18}}{2}$. d. $\sqrt[3]{8x} - \sqrt[3]{-64x^4}$.

17. Simplify.

a. $\sqrt{5}(\sqrt{6} - \sqrt{5})$. b. $(\sqrt{2} - \sqrt{2x})(\sqrt{2} + \sqrt{2x})$.

c. $\sqrt[4]{4}(\sqrt[4]{4} + \sqrt[4]{64})$.

18. Simplify.

a. $\frac{\sqrt{15} + \sqrt{21}}{\sqrt{3}}$. b. $\frac{m\sqrt{mn^3} - \sqrt{mn}}{\sqrt{mn}}$.

19. Simplify.

a. $\frac{2\sqrt{3} - 2}{2\sqrt{3} + 2}$. b. $\frac{\sqrt{x} + \sqrt{y}}{\sqrt{x} - \sqrt{y}}$.

20. Express each of the following in a + bi or a - bi form:

a. $(2 - 3i) - (4 + 5i)$. b. $(2 - 3i)(4 + 5i)$.

c. $\frac{4 + 5i}{2 - 3i}$.

21. Express each of the following in a + bi or a - bi form:

a. $(3 + \sqrt{-12}) + (1 - \sqrt{-27})$.

b. $(3 - \sqrt{-3})(1 - \sqrt{-3})$.

c. $\frac{1}{2 - \sqrt{-3}}$.

CHAPTER 7

In Problems 1-12, solve for x. Use the quadratic formula in Problems 1 and 2.

1. $5x^2 = 9x$.
2. $2x(x - 2) = x + 3$.
3. $\frac{2x^2}{7} = 8$.
4. $(2x + 1)^2 = 25$.
5. $3x^2 = 5x - 1$.
6. $\frac{x^2 - x}{2} + 1 = 0$.

7. $2x^2 - 3x + 2 = 0$.

8. $\sqrt{x - 3} + 5 = 0$.

9. $\sqrt{3x + 10} = x + 4$.

10. $4\sqrt{x} + \sqrt{16x + 1} = 5$.

11. $x^4 - 2x^2 - 24 = 0$.

12. $x^{2/3} - 2x^{1/3} = 35$.

13. Solve $2x^2 + 3x - 2 = 0$ by completing the square.

14. Determine k so that $x^2 - x + k - 2 = 0$ has imaginary solutions.

In Problems 15 and 16, solve and represent each solution set on a line graph.

15. $x^2 - 5x - 6 \geq 0$.

16. $\frac{3}{x - 6} > 8$.

17. Find two consecutive positive integers the sum of whose squares is 145.

CHAPTER 8

1. Find the missing component in each solution of $2x - 9y = 18$.

   a. (0,?)   b. (?,0)   c. (2,?)

2. In the equation $\frac{2x^4}{3 + 2y^2} = 5$, express y explicitly in terms of x.

3. Find k if $kx - 3y = 9$ has (1,-3) as a solution.

4. Graph $y = 2x - 1$, if x is -2, -1, 0, 1, and 2.

5. Graph $y = -4x + 2$.

6. Graph the equation $3x - 2y = 6$.

7. Given (2,-3) and (-2,-1), find:

   a. the distance between the two points.
   b. the slope of the line segment joining the two points.

8. Find the equation in standard form of the line through (2,-1) with slope equal to $\frac{3}{4}$.

9. Given the equation $3x - 2y = 5$:

   a. Write the equation in slope-intercept form.
   b. Specify the slope of its graph.
   c. Specify the y-intercept of its graph.

10. Write in standard form the equation of the line through (0,5) that is perpendicular to the graph of $3x - 2y + 5 = 0$.

11. Find an equation for the line through the point (2,-3) and is parallel to the graph of $3x - 4y = 12$.

12. Graph the inequality $2x + y > 2$.

13. Graph the inequality $y < 2x - 4$.

14. Graph $x < 2$ and $y < 3$ on the same coordinate system and use double shading to indicate the region common to both (the solution set).

CHAPTER 9

In problems 1-3, solve each system by linear combinations.

1. $x + 4y = -14$
   $3x + 2y = -2.$

2. $\frac{2}{3}x - y = 4$
   $x - \frac{3}{4}y = 6.$

3. $\frac{1}{x} + \frac{2}{y} = \frac{-11}{12}$
   $\frac{1}{x} + \frac{1}{y} = \frac{-7}{12}.$

In Problems 4-6, state whether the equations in each system are dependent, inconsistent, or have a unique solution.

4. $3x + 4y = 4$
   $9x + 12y = 3.$

5. $3x + 4y = 1$
   $9x + 12y = 3.$

6. $3x + 4y = 7$
   $9x + 6y = 7.$

7. Solve by linear combinations:

   $2x + 4y + z = 0$
   $5x + 3y - 2z = 1$
   $4x - 7y - 7z = 6.$

8. Graph the solution set.

   $x + 2y < 8$
   $3x - 2y > 6$

9. The perimeter of a triangle is 155 inches, the side x is 20 inches shorter than the side y, and the side y is 5 inches longer than the side z. Find the lengths of the sides of the triangle.

10. Find the values of a and b so that the graph of $ax + by = 10$ contains the points (-1,2) and (2,-1).

In Problems 11-12, solve each system by row operations on a matrix.

11. $x + 4y = -14$
    $3x + 2y = -2.$

12. $2x + 4y + z = 0$
    $5x + 3y - 2z = 1$
    $4x - 7y - 7z = 6.$

13. Evaluate:

$$\begin{vmatrix} 10 & 3 \\ -10 & -2 \end{vmatrix}.$$

14. Solve by Cramer's rule:

$$\begin{aligned} 3x - 4y &= -2 \\ x - 2y &= 0. \end{aligned}$$

15. Evaluate:

$$\begin{vmatrix} 2 & 3 & 1 \\ 0 & 1 & 0 \\ -4 & 2 & 1 \end{vmatrix}.$$

16. Solve by Cramer's rule:

$$\begin{aligned} 3x - 2y + 5z &= 6 \\ 4x - 4y + 3z &= 0 \\ 5x - 4y + z &= -5. \end{aligned}$$

CHAPTER 10

Graph each conic section.

1. $4x^2 + y^2 = 28$

2. $6y^2 - x^2 = 24$

3. $\dfrac{(x - 2)^2}{9} + \dfrac{(y + 2)^2}{6} = 1$

4. $\dfrac{y^2}{12} - \dfrac{(x + 3)^2}{9} = 1$

5. $x^2 - 6x + y^2 + 4y + 4 = 0$

6. $y = -2x^2 + 4x - 2$

In Problems 7 to 10, solve each system by substitution or linear combinations.

7. $$\begin{aligned} y &= x^2 - 2x + 1 \\ x + y &= 3 \end{aligned}$$

8. $$\begin{aligned} 9x^2 + 16y^2 - 100 &= 0 \\ x^2 + y^2 - 8 &= 0 \end{aligned}$$

9. $$\begin{aligned} 3x^2 - 2xy + 3y^2 &= 34 \\ x^2 + y^2 &= 17 \end{aligned}$$

10. $$\begin{aligned} x^2 - xy + y^2 &= 21 \\ x^2 - 2xy - 8y^2 &= 0 \end{aligned}$$

11. Use double shading to indicate the region common to both inequalities (the solution set).

$$\begin{aligned} x^2 + y^2 &\geq 25 \\ y &> x^2 \end{aligned}$$

12. The area of a rectangle is 240 square feet. If the perimeter is 64 feet, find the dimensions of the rectangle.

13. Find the equation of the parabola $y = ax^2 + bx + c$ whose graph contains the points (1,1), (-1,-5), and (2,10).

CHAPTER 11

1. Specify the domain and range of each relation and state whether or not the relation is a function:

   a. $\{(3,1), (4,1), (5,1), (6,1)\}$.
   b. $\{(1,3), (1,4), (1,5), (1,6)\}$.

2. Given that $h(x) = 2x^2 + 3x - 2$, find:

   a. $h(3)$ b. $h(1) - h(0)$

3. Given that $f(x) = x^2 + 2x$, find:

   a. $f(x + h)$ b. $\dfrac{f(x + h) - f(x)}{h}$

4. Specify the domain of the function defined by $x^2 + y^2 = 49$.

5. If y varies inversely as t^2 and $y = 2$ when $t = 2$,

   a. express y as a function of t; and,
   b. find y when $t = 3$.

6. The resistance (R) of a wire varies directly as the length (l) and inversely as the square of its diameter (d). Twenty-five feet of wire of diameter 0.006 inches has a resistance of 100 ohms.

   a. Express the resistance as a function of the length and diameter of the wire; and,
   b. what is the resistance of 100 feet of the same type of wire whose diameter is 0.012 inches?

7. $2x - 3y = 6$ defines a relation h.

   a. Write an equation defining h^{-1}.
   b. Graph h and h^{-1} on the same axes.
   c. State whether h^{-1} is a function.

8. Graph $y = 3^x$. 9. Graph $y = 3^{-x}$.

10. Find each value.

    a. 10^{-4} b. $\left(\frac{2}{3}\right)^{-2}$

11. The length of a rectangle is three times its width.
    a. Express the area (A) of the rectangle as a function of the length of the diagonal; b. Find the area of the rectangle with a diagonal 10 inches in length.

12. a. Express the volume of a cylinder as a function of its height if the height of the cylinder is four times the radius. [Hint: $V = \pi r^2 h$.]

    b. Find the volume of a cylinder with height 5 centimeters.

13. Approximate the solution of the system $\begin{array}{l} y = 3^x \\ y = \frac{1}{2}x^2 \end{array}$ graphically.

CHAPTER 12

1. Express each of the following in logarithmic notation:

   a. $2^4 = 16$. b. $10^{-2} = 0.01$.

2. Express each of the following in exponential notation:

   a. $\log_{10}(0.00001) = -5$. b. $\log_{1/4}16 = -2$.

3. Find the value of the variable:

   a. $\log_b 64 = 3$. b. $\log_{1/3}x = -3$.

4. Write each expression as a sum or difference of simpler logarithmic quantities. Assume that all variables and expressions represent real numbers.

   a. $\log_b \frac{x^3 y^2}{z}$. b. $\log_{10} \sqrt[4]{\frac{xy^2}{x - y}}$.

5. Write each expression as a single logarithm with coefficient 1:

   a. $3 \log_{10}x + 4 \log_{10}y - \log_{10}z$.

   b. $\frac{1}{2}\left(3 \log_b x + 5 \log_b y - 3 \log_b z\right)$.

6. Simplify: $\log_{10}[\log_2(\log_3 9)]$.

7. Solve $\log_{10}(x + 3) + \log_{10}x = 1$.

In Problems 8-10, use the table of logarithms or a calculator. Round off answer to three significant digits.

8. Find each logarithm:

   a. $\log_{10}21.4$. b. $\log_{10}0.00214$.

9. Find each power.

 a. $10^{3.9258}$ b. $10^{2.8156-4}$

10. Find each power.

 a. $e^{0.47}$ b. $e^{-2.2}$

11. Find the value of ln 20.

12. Solve for x using the base 10.

 a. $5^x = 15$. b. $3^{3-x} = 1000$.

13. Solve $E = 2ke^{1+t}$ for t using the base e.

14. To the nearest table entry, find $[H^+]$, given that $\log_{10} \frac{1}{[H+]} = 6.3$.

15. Given that $N = N_0 10^{0.2t}$, find t to the nearest table entry if N = 180 and $N_0 = 3$.

16. Given that $y = y_0 e^{-0.3t}$, find t if y = 6 and $y_0 = 30$.

CHAPTER 13

1. Find the first four terms in a sequence if the general term is $s_n = \frac{n}{n^2 + 1}$.

2. Write $\sum_{j=0}^{4} \frac{(-1)^j 3^j}{j + 1}$ in expanded form.

3. Write the series 1 - 8 + 27 - 64 + 125 using summation notation.

4. Given that x, x + 3, ••• is an arithmetic progression:

 a. write the next three terms;
 b. find the expression for the general term.

5. Find the seventeenth term in the progression -5, -2, 1, •••.

6. If the twentieth term of an arithmetic progression is -46 and the twelfth term is -30, find the fifth term.

7. Find the sum of the series $\sum_{j=10}^{20} (2j - 3)$.

8. Given that $\frac{x}{a}, -1, \frac{a}{x}, \cdots$ is a geometric progression:

   a. write the next three terms;
   b. find an expression for the general term.

9. Find the ninth term in the geometric progression $-81, -27, -9, \cdots$.

10. Find the second term of a geometric progression if the sixth term is 60 and the ratio is 3.

11. Find the sum $\sum_{j=5}^{8} (2j - 5)$.

12. Insert three geometric means between 3 and 243.

13. Find the sum of the infinite geometric series $2 - \frac{3}{2} + \frac{9}{8} + \cdots$.

14. Write $\frac{(12!)(8!)}{16!}$ in expanded form and simplify.

15. Write the first four terms in the expansion of $(x - 2y)^8$ and simplify.

16. Write the seventh term in the expansion of $(a^3 - b)^9$.

17. Find a fraction equivalent to the repeating decimal $1.027\overline{027}$.

18. In how many ways can 5 people be seated in a row of 5 chairs?

19. In how many ways can 12 boys be assigned to the 11 positions on a football team if there is one center, one fullback, and one quarterback in the group?

20. How many different license plates can be made if each plate consists of 3 letters chosen from the first 8 letters of the alphabet and these are then followed by 3 digits?

21. How many distinguishable permutations are there of the letters of the word "succeeded"?

22. How many ways can a committee of 3 men and 3 women be chosen from a group of 6 men and 8 women?

23. How many different amounts of money can be formed from a dollar bill, a half-dollar, a quarter, and a dime?

24. An urn contains 5 red and 4 black balls. In how many ways can 2 red and 3 black balls be drawn from the urn?

APPENDIX

1. If $P(x) = 2x^3 - 3x^2 + x + 1$, use synthetic division to find P(1) and P(-1).

2. Use synthetic division and the remainder theorem to find solutions to $y = x^3 - x^2 + 3x$ and then graph the equation.

3. a. Is $x + 1$ a factor of $x^3 + x^2 - 4x - 4$?
   b. Is $x - 1$ a factor of $x^3 + x^2 - 4x - 4$?

4. Verify that $x = -1$ is a solution of $x^3 + x^2 - 4x - 4$ and find the other solutions.

5. Graph $y = \frac{x + 3}{x + 2}$ and show all asymptotes.

6. Graph the function defined by $y = |2x - 2|$.

7. Graph the function defined by $y = [x + 1]$ over the domain $-3 \le x < 3$.

8. Graph $f(x) = \begin{cases} 4 & \text{if } x < 2 \\ -x + 2 & \text{if } x \ge 2 \end{cases}$.

Solutions to Test Problems

CHAPTER 1

1. a. 7,6. b. 0,6.

2. a. $x \not< y$ b. $-2 < x < 1$

3. a. 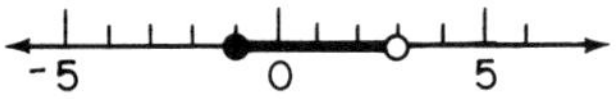 -5 0 5 b. 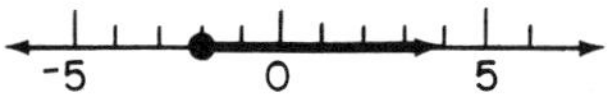 -5 0 5

4. a. Negative or additive-inverse property.
   b. Commutative property for multiplication.
   c. Identity element for addition.
   d. Distributive property.

5. a. II b. I c. IV d. III

6. a. 8 b. $-(x - 2)$

7. a. 8 b. undefined c. -7 d. -6

8. a. 2 b. -2 c. 12 d. 16

9. a. $5\left(\frac{1}{9}\right)$ b. $\frac{7}{11}$

10. a. -3 b. $\frac{1}{3}$

CHAPTER 2

1. a. Binomial b. Trinomial.

2. a. $P(0) = 3(0)^2 + 2(0) - 1 = \boxed{-1}$.

   b. $P(-1) = 3(-1)^2 + 2(-1) - 1 = \boxed{0}$.

3. $\dfrac{3(2) - 2(-1)^2}{2 + (-1)} = \dfrac{6 - 2}{1} = \boxed{4}.$

4. $(2x - 5y) - (5x - 2y) = 2x - 5y - 5x + 2y = (2x - 5x) + (2y - 5y) = \boxed{-3x - 3y}.$

5. $3c - [2c - (3c - 1) + 1] = 3c - [2c - 3c + 1 + 1] = 3c - [-c + 2] = 3c + c - 2 =$
$\boxed{4c - 2}.$

6. $y - [y^3 - (y^3 - y) + y] = y - [\cancel{y^3} - \cancel{y^3} + y + y] = y - 2y = \boxed{-y}.$

7. $(-3xy^2)(2x^2y)(xy) = (-3 \cdot 2)(x \cdot x^2 \cdot x)(y^2 \cdot y \cdot y) = \boxed{-6x^4y^4}.$

8. $(-xy)^2 + 3x^2(xy) - 2x^2y^2 = x^2y^2 + 3x^3y - 2x^2y^2 = \boxed{-x^2y^2 + 3x^3y}.$

9. $-3[(x - 3) - (x + 2)^2] = -3[x - 3 - (x^2 + 4x + 4)] = -3[x - 3 - x^2 - 4x - 4]$
$= -3[(-x^2 - 3x - 7)] = \boxed{3x^2 + 9x + 21}.$

10. $5p^3(p^2 - 2p + 1) = \boxed{5p^5 - 10p^4 + 5p^3}.$

11. $3[(3x - 1)(x + 2)] = 3[3x^2 + 5x - 2] = \boxed{9x^2 + 15x - 6}.$

12. $(y - 1)(y^2 - 2y - 1) = y(y^2 - 2y - 1) - 1(y^2 - 2y - 1)$
$= y^3 - 2y2 - y - y^2 + 2y + 1 = \boxed{y^3 - 3y^2 + y + 1}.$

13. $[(z - 2)(z + 2)](z - 1) = [z^2 - 4](z - 1) = z^2(z - 1) - 4(z - 1)$
$= \boxed{z^3 - z^2 - 4z + 4}.$

14. $2[x - 3(x - 2) - 2] = 2[x - 3x + 6 - 2] = 2[-2x + 4]$
$= \boxed{-4x + 8}.$

15. $-y\{1 - 3[x + 2(x - 1)] + 2x\} = -y\{1 - 3[x + 2x - 2] + 2x\}$
$= -y\{1 - 3[3x - 2] + 2x\}$
$= -y\{1 - 9x + 6 + 2x\}$
$= -y\{7 - 7x\} = \boxed{-7y + 7xy}.$

16. $(2x^n + 1)(2x^n - 1) = \boxed{4x^{2n} - 1}.$

17. $(x^n y^{n-1})^4 = (x^n)^4 (y^{n-1})^4 = \boxed{x^{4n} y^{4n-4}}$.

18. a. $Q(1) = (1)^4 - 2(1)^2 = -1; \quad P[Q(1)] = P(-1) = (-1)^2 - 3(-1) = \boxed{4}$.

b. $P(1) = (1)^2 - 3(1) = -2; \quad Q[P(1)] = Q(-2) = (-2)^4 - 2(-2)^2 = \boxed{8}$.

CHAPTER 3

1. $$\begin{aligned} 5[2 + 3(x - 2)] + 20 &= 0 \\ 5[2 + 6x - 6] + 20 &= 0 \\ 5[6x - 4] + 20 &= 0 \\ 30x - 20 + 20 &= 0 \\ 30x &= 0 \\ x &= 0. \end{aligned}$$

$\boxed{\{0\}}$

2. $$\begin{aligned} (2x + 1)(x - 3) &= (x - 2) + 2x^2 \\ \cancel{2x^2} - 5x - 3 &= x - 2 + \cancel{2x^2} \\ -6x &= 1 \\ x &= -\frac{1}{6}. \end{aligned}$$

$\boxed{\left\{-\frac{1}{6}\right\}}$

3. $$\begin{aligned} -[3(x - 1)^2 - 3x^2] &= 0 \\ -[3(x^2 - 2x + 1) - 3x^2] &= 0 \\ -[\cancel{3x^2} - 6x + 3 - \cancel{3x^2}] &= 0 \\ 6x - 3 &= 0 \\ 6x &= 3 \\ x &= \frac{3}{6} = \frac{1}{2}. \end{aligned}$$

$\boxed{\left\{\frac{1}{2}\right\}}$

4. $$\begin{aligned} 0.05x + 0.06(x - 500) &= 25 \\ 5x + 6(x - 500) &= 2500 \\ 5x + 6x - 3000 &= 2500 \\ 11x - 3000 &= 2500 \\ 11x &= 5500 \\ x &= 50. \end{aligned}$$

$\boxed{\{50\}}$

5. $$\begin{aligned} 3x - (x - 2) &= 3(x + 1) \\ 3x - x + 2 &= 3x + 3 \\ 2x + 2 &= 3x + 3 \\ -x &= 1 \\ x &= -1 \end{aligned}$$

$\boxed{\{-1\}}$

6. $$\begin{aligned} 3[2x - (x + 2) + 5] &= 10 \\ 3[2x - x - 2 + 5] &= 10 \\ 3[x + 3] &= 10 \\ 3x + 9 &= 10 \\ 3x &= 1 \\ x &= \frac{1}{3}. \end{aligned}$$

$\boxed{\left\{\frac{1}{3}\right\}}$

7. $5(4) = 3t + 4(2)$

$20 = 3t + 8$

$12 = 3t$

$4 = t$

$\boxed{\{4\}}$

8. $2A = h(b + c)$

$2A = hb + hc$

$2A - hb = hc$

$\boxed{\dfrac{2A - hb}{h} = c}$

9. $\dfrac{2x - 3}{2} \le 5$

$2x - 3 \le 10$

$2x \le 13$

$x \le \dfrac{13}{2}$;

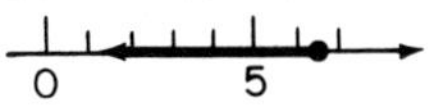

10. $3(x + 2) \ge 4x$

$3x + 6 \ge 4x$

$-x \ge -6$

$x \le 6$;

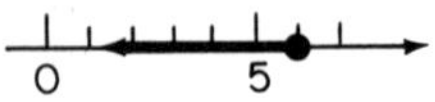

11. $2 \le 3x - 4 \le 8$

$6 \le 3x \le 12$

$2 \le x \le 4$;

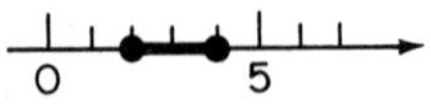

12. $\{x|2x - 3 < 5\} \cap \{x|2x - 3 > -5\}$

$\{x|2x < 8\} \cap \{x|2x > -2\}$

$\{x|x < 4\} \cap \{x|x > -1\}$

$\{x|x < 4\}$ and $\{x > -1\}$ = $\{x|-1 < x < 4\}$;

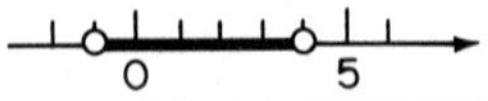

13. $|2x + 5| = 2$

$2x + 5 = 2$ or $-(2x + 5) = 2$

$2x = -3$ or $-2x = 7$

$\boxed{x = -\dfrac{3}{2} \quad \text{or} \quad x = -\dfrac{7}{2}}$.

14. $\left|2x + \frac{1}{2}\right| = \frac{1}{4}$

$$2x + \frac{1}{2} = \frac{1}{4} \quad \text{or} \quad -\left(2x + \frac{1}{2}\right) = \frac{1}{4}$$

$$8x + 2 = 1 \quad \text{or} \quad -8x - 2 = 1$$

$$8x = -1 \quad \text{or} \quad -8x = 3$$

$$\boxed{x = -\frac{1}{8} \quad \text{or} \quad x = -\frac{3}{8}}\ .$$

15. $|x + 5| > 2$

$$x + 5 > 2 \quad \text{or} \quad -(x + 5) > 2$$

$$x > -3 \quad \text{or} \quad -x - 5 > 2$$

$$x > -3 \quad \text{or} \quad -x > 7$$

$$x > -3 \quad \text{or} \quad x < -7$$

$\boxed{\{x|x > -3\} \cup \{x|x < -7\};}$ 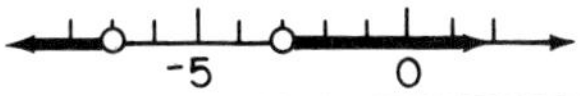

16. $|2x + 4| < 6$

$$-6 < 2x + 4 < 6$$

$$-10 < 2x < 2$$

$$-5 < x < 1$$

$\boxed{\{x|-5 < x < 1\};}$ 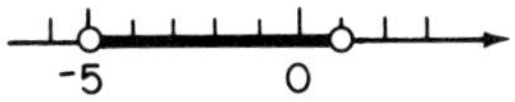

17. $\boxed{\{x|-3 \le x < 4\}}$

18. $[-3,5) \cap [-1,6] = \boxed{[-1,5)}$.

19. The number of gallons of water to be added: x

[amount of acid in original 2 gallons] = [amount of acid in the final 10% solution.]

$$2 = .1(2 + x)$$

$$2 = .2 + .1x$$

$$20 = 2 + x$$

$$x = 18.$$

$\boxed{\text{18 gallons should be added}}$.

20. The number of nickels: x
The number of dimes: $x + 5$
The number of quarters: $2(x + 5) - 25$

$$\begin{aligned} 5x + 10(x + 5) + 25[2(x + 5) - 25] &= 455 \\ 5x + 10x + 50 + 50x + 250 - 625 &= 455 \\ 65x - 325 &= 455 \\ 65x &= 780 \\ x &= 12 \\ x + 5 &= 17 \\ 2(x + 5) - 25 &= 9 \end{aligned}$$

Hence, there are $\boxed{\text{12 nickels, 17 dimes, and 9 quarters}}$.

21.
$$\begin{aligned} 0 < \tfrac{5}{9}(F - 32) &< 100 \\ 0 < 5(F - 32) &< 900 \\ 0 < 5F - 160 &< 900 \\ 160 < 5F &< 1060 \\ 32 < F &< 212 \end{aligned}$$

In Fahrenheit degrees the temperature will have to range $\boxed{\text{between 32 and 212}}$.

CHAPTER 4

1. $5m^3 + 2m^2 - m = \boxed{m(5m^2 + 2m - 1)}$.

2. $y^{n+3} + 4y^n = y^3 \cdot y^n + 4y^n = \boxed{y^n(y^3 + 4)}$.

3. $3(x^2 + xy - 6y^2) = \boxed{3(x + 3y)(x - 2y)}$.

4. $8a^3 - b^3 = (2a)^3 - b^3 = \boxed{(2a - b)(4a^2 + 2ab + b^2)}$.

5. $3m^2 - 5m + 3mn - 5n = m(3m - 5) + n(3m - 5) = \boxed{(3m - 5)(m + n)}$.

6. $9x^4 - 16x^2y^2 = x^2(9x^2 - 16y^2) = \boxed{x^2(3x - 4y)(3x + 4y)}$.

7. $7y^4z^2 + 14y^3z + 7y^2z^2 = \boxed{7y^2z(y^2z + 2y + z)}$.

8. $2x^2 - 18 = 2(x^2 - 9) = 2(x^2 - 3^2) = \boxed{2(x - 3)(x + 3)}$.

9. $x^2y^2z^2 - 4x^2z^2 = x^2z^2(y^2 - 4) = \boxed{x^2z^2(y + 2)(y - 2)}$.

10. $x^2 + 14x + 49 = \boxed{(x + 7)^2}$.

11. $2x^2y + 16xy + 32y = 2y(x^2 + 8x + 16) = \boxed{2y(x + 4)^2}$.

12. $2x^3y^2 + 16x^2y^3 + 32x^2y^2 = \boxed{2x^2y^2(x + 8y + 16)}$.

13. Trial and error leads to the conclusion that $3x^2 + 5x + 1$ is $\boxed{\text{prime}}$.

14. $8x^2 + 14x + 3 = \boxed{(4x + 1)(2x + 3)}$.

15. $12x^2 - 17x + 6 = \boxed{(4x - 3)(3x - 2)}$.

16. $12x^2 - 8x - 15 = \boxed{(6x + 5)(2x - 3)}$.

17. $xy + 3y - x - 3 = y(x + 3) - 1 \cdot (x + 3) = \boxed{(y - 1)(x + 3)}$.

Alternate procedure: $xy + 3y - x - 3 = (xy - x) + (3y - 3)$

$$= x(y - 1) + 3(y - 1)$$

$$= \boxed{(x + 3)(y - 1)} .$$

18. $xy^2 - 2y^2 + 2xy - 4y = (x - 2)y^2 + 2y(x - 2)$

$$= (x - 2)(y^2 + 2y)$$

$$= \boxed{(x - 2)y(y + 2)} .$$

19. $2x^3 - x^2y - 2xy + y^2 = x^2(2x - y) - y(2x - y)$

$$= \boxed{(2x - y)(x^2 - y)} .$$

20. $(4x^2 - 4x + 1) - 9z^2 = (2x - 1)^2 - (3z)^2$

$$= \boxed{(2x - 1 - 3z)(2x - 1 + 3z)} .$$

21. $x^3y^3 - 8 = (xy)^3 - 2^3$

$$= ((xy) - 2)((xy)^2 + (xy) \cdot 2 + 2^2)$$

$$= \boxed{(xy - 2)(x^2y^2 + 2xy + 4)} .$$

22. $64x^3 - y^3 = (4x)^3 - y^3 = (4x - y)[(4x)^2 + 4xy + y^2]$

$$= \boxed{(4x - y)(16x^2 + 4xy + y^2)}.$$

23. $(x + 1)^3 - x^3 = [(x + 1) - x][(x + 1)^2 + (x + 1) \cdot x + x^2]$

$$= 1[(x^2 + 2x + 1 + x^2 + x + x^2)]$$

$$= \boxed{3x^2 + 3x + 1} .$$

24. $(y^3 - 1)^3 + 27 = (y^3 - 1)^3 + 3^3$

$= [(y^3 - 1) + 3][(y^3 - 1)^2 - 3(y^3 - 1) + 3^2]$

$= [y^3 + 2][y^6 - 2y^3 + 1 - 3y^3 + 3 + 9]$

$= \boxed{[y^3 + 2][y^6 - 5y^3 + 13]}$.

25. $x^2y^3 - 8x^2 = x^2(y^3 - 2^3) = \boxed{x^2(y - 2)(y^2 + 2y + 4)}$.

26. $x^{2n} - y^{2n} = \boxed{(x^n - y^n)(x^n + y^n)}$.

27. $5x^2 - 9x = 0$

$x(5x - 9) = 0$

$x = 0$ or $x = \frac{9}{5}$.

$\boxed{\left\{0, \frac{9}{5}\right\}}$

28. $2x^2 - 5x - 3 = 0$

$(2x + 1)(x - 3) = 0$

$2x + 1 = 0$ or $x - 3 = 0$

$x = -\frac{1}{2}$ or $x = 3$.

$\boxed{\left\{-\frac{1}{2}, 3\right\}}$

29. $2x(4x - 7) = -3$

$8x^2 - 14x + 3 = 0$

$(4x - 1)(2x - 3) = 0$

$4x - 1 = 0$ or $2x - 3 = 0$

$x = \frac{1}{4}$ or $x = \frac{3}{2}$.

$\boxed{\left\{\frac{1}{4}, \frac{3}{2}\right\}}$

30. $4x(x + 1) - 24 = 0$

$4x^2 + 4x - 24 = 0$

$4(x^2 + x - 6) = 0$

$4(x + 3)(x - 2) = 0$

$x = -3$ or $x = 2$.

$\boxed{\{-3, 2\}}$.

31. $4x^2 - 49 = 0$

$(2x + 7)(2x - 7) = 0$

$2x + 7 = 0$ or $2x - 7 = 0$

$x = \frac{-7}{2}$ or $x = \frac{7}{2}$.

$\boxed{\left\{-\frac{7}{2}, \frac{7}{2}\right\}}$

32. $6x^2 = 2 - x$

$6x^2 + x - 2 = 0$

$(3x + 2)(2x - 1) = 0$

$3x + 2 = 0$ or $2x - 1 = 0$

$x = \frac{-2}{3}$ or $x = \frac{1}{2}$.

$\boxed{\left\{-\frac{2}{3}, \frac{1}{2}\right\}}$

33. $(x - 3)(x - (-4)) = 0$

$(x - 3)(x + 4) = 0$

$\boxed{x^2 + x - 12 = 0}$.

34. The first positive odd integer: n
The next positive odd integer: $n + 2$

$$n(n + 2) = 255$$
$$n^2 + 2n - 255 = 0$$
$$(n - 15)(n + 17) = 0, \{-17, 15\}$$

Hence the positive consecutive odd integers are $\boxed{15 \text{ and } 17}$.

35. Width of rectangle: x
Length of rectangle: $2x + 2$

$$x(2x + 2) = 40$$
$$2x^2 + 2x - 40 = 0$$
$$2(x^2 + x - 20) = 0$$
$$2(x + 5)(x - 4) = 0, \quad \{4, -5\}$$

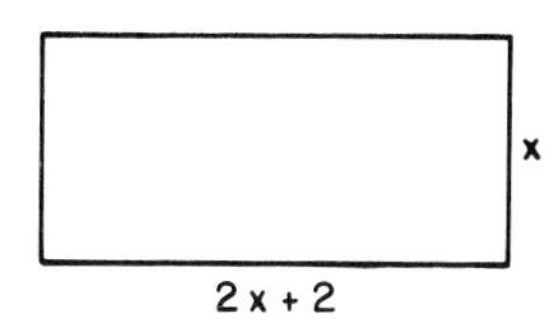

Since a dimension of a geometric figure cannot be negative, -5 is rejected. $\boxed{\text{Hence the width is 4 cm. and the length is } [2(4) + 2]\text{cm} = 10\text{cm.}}$

CHAPTER 5

1. a. $-\frac{-4}{7} = \boxed{\frac{4}{7}}$.

b. $\frac{4 - 3x}{-x} \quad \frac{-1(4 - 3x)}{-1(-x)} = \boxed{\frac{3x - 4}{x}}$.

c. $\frac{-5a^2b^2}{-2c^2} = \boxed{\frac{5a^2b^2}{2c^2}}$.

d. $-\frac{m - 2n}{-n} = \boxed{\frac{m - 2n}{n}}$.

2. $\frac{8r^4 - 12r^3}{4r^2 - 6r} = \frac{4r^3(2r - 3)}{2r(2r - 3)} = \frac{\cancel{2r(2r - 3)} \cdot 2r^2}{\cancel{2r(2r - 3)} \cdot 1} = \boxed{2r^2}$.

3. $\frac{8a^3 - 27b^3}{4a^2 - 9b^2} = \frac{\cancel{(2a - 3b)}(4a^2 + 6ab + 9b^2)}{\cancel{(2a - 3b)}(2a + 3b)} = \boxed{\frac{4a^2 + 6ab + 9b^2}{2a + 3b}}$.

4.
```
              x +  5
x - 2 | x² + 3x -  4
        x² - 2x
             5x -  4
             5x - 10
                   6
```

Therefore, $\frac{x^2 + 3x - 4}{x - 2} = \boxed{x + 5 + \frac{6}{x - 2}}$.

5. $$\frac{6x^2 - 2xy + 9x - 3y}{2x + 3} = \frac{2x(3x - y) + 3(3x - y)}{2x + 3} = \frac{(2x + 3)(3x - y)}{2x + 3} = 3x - y.$$

Therefore, $\dfrac{6x^2 - 2xy + 9x - 3y}{2x + 3} = \boxed{3x - y}$.

6. $$\frac{4}{3x} + \frac{5}{2y} - \frac{2}{xy} = \frac{8y}{2 \cdot 3xy} + \frac{15x}{2 \cdot 3xy} - \frac{12}{2 \cdot 3xy} = \boxed{\frac{8y + 15x - 12}{6xy}}.$$

7. $$\frac{m}{m^2 - 1} - \frac{m}{m^2 - 2m + 1} = \frac{m}{(m + 1)(m - 1)} - \frac{m}{(m - 1)(m - 1)}$$
$$= \frac{m(m - 1)}{(m + 1)(m - 1)^2} - \frac{m(m + 1)}{(m + 1)(m - 1)^2}$$
$$= \frac{m(m - 1) - m(m + 1)}{(m + 1)(m - 1)^2} = \frac{m^2 - m - m^2 - m}{(m + 1)(m - 1)^2} = \boxed{\frac{-2m}{(m + 1)(m - 1)^2}}.$$

8. $$\frac{3p}{4pq - 6q^2} \cdot \frac{2p - 3q}{12p} = \frac{\cancel{3p}}{2q\cancel{(2p - 3q)}} \cdot \frac{\cancel{2p - 3q}}{\cancel{3p} \cdot 4} = \boxed{\frac{1}{8q}}.$$

9. $$\frac{4y^2 + 8y + 3}{2y^2 - 5y + 3} \cdot \frac{6y^2 - 9y}{1 - 4y^2} = \frac{(2y + 3)\cancel{(2y + 1)}}{\cancel{(2y - 3)}(y - 1)} \cdot \frac{3y\cancel{(2y - 3)}}{\cancel{(1 + 2y)}(1 - 2y)} = \boxed{\frac{3y(2y + 3)}{(y - 1)(1 - 2y)}}.$$

10. $$\frac{m^2 + m - 2}{m^2 + 2m - 3} \div \frac{m^2 + 7m + 10}{m^2 - 2m - 15} = \frac{\cancel{(m + 2)}\cancel{(m - 1)}}{\cancel{(m + 3)}\cancel{(m - 1)}} \cdot \frac{(m - 5)\cancel{(m + 3)}}{(m + 5)\cancel{(m + 2)}} = \boxed{\frac{m - 5}{m + 5}}.$$

11. $$\frac{8x^3 - 1}{x - 2} \div \frac{4x^2 + 2x + 1}{x^2 - 4x + 4} = \frac{(2x - 1)\cancel{(4x^2 + 2x + 1)}}{\cancel{x - 2}} \cdot \frac{\cancel{(x - 2)}(x - 2)}{\cancel{4x^2 + 2x + 1}} = \boxed{(2x - 1)(x - 2)}.$$

12. $$\frac{\frac{m - 2}{m}}{\frac{m^2 - 4}{m^2}} = \frac{\frac{m - 2}{m}(m^2)}{\frac{m^2 - 4}{m^2}(m^2)} = \frac{m(m - 2)}{m^2 - 4} = \frac{m\cancel{(m - 2)}}{\cancel{(m - 2)}(m + 2)} = \boxed{\frac{m}{m + 2}}.$$

13. $$\frac{\frac{a-1}{a+1}-\frac{a+1}{a-1}}{\frac{a-1}{a+1}+\frac{a+1}{a-1}} = \frac{\left[\left(\frac{a-1}{a+1}\right)-\left(\frac{a+1}{a-1}\right)\right]}{\left[\left(\frac{a-1}{a+1}\right)+\left(\frac{a+1}{a-1}\right)\right]}\frac{(a+1)(a-1)}{(a+1)(a-1)} = \frac{(a-1)(a-1)-(a+1)(a+1)}{(a-1)(a-1)+(a+1)(a+1)}$$

$$= \frac{a^2-2a+1-a^2-2a-1}{a^2-2a+1+a^2+2a+1} = \frac{-4a}{2a^2+2}$$

$$= \frac{\cancel{2}(-2a)}{\cancel{2}(a^2+1)} = \boxed{\frac{-2a}{a^2+1}} .$$

14. a.

$$\begin{array}{r|rrrr} 3 & 1 & -3 & 2 & 6 \\ & & 3 & 0 & 6 \\ \hline & 1 & 0 & 2 & 0 \end{array}$$

Therefore, we have

$$\boxed{x^2+2} .$$

b.

$$\begin{array}{r|rrrrrr} -1 & 1 & 0 & 0 & 0 & 0 & 1 \\ & & -1 & 1 & -1 & 1 & -1 \\ \hline & 1 & -1 & 1 & -1 & 1 & 0 \end{array}$$

Therefore, we have

$$\boxed{x^4-x^3+x^2-x+1} .$$

15.

$$\frac{x}{5} - 9 = \frac{x}{2}$$

$$10\left(\frac{x}{5}-9\right) = 10\left(\frac{x}{2}\right)$$

$$2x - 90 = 5x$$

$$-90 = 3x$$

$$\boxed{x = -30} .$$

16.

$$\frac{x}{x+2} - \frac{x^2+8}{x^2-4} = \frac{3}{x-2}$$

$$(x+2)(x-2)\left[\frac{x}{x+2} - \frac{x^2+8}{x^2-4}\right] = (x+2)(x-2)\left[\frac{3}{x-2}\right]$$

$$x(x-2) - (x^2+8) = 3(x+2)$$

$$x^2 - 2x - x^2 - 8 = 3x + 6$$

$$-2x - 8 = 3x + 6$$

$$-5x = 14$$

$$\boxed{x = \frac{-14}{5}} .$$

17.
$$\frac{x}{a} - \frac{a}{b} = \frac{b}{c}$$
$$abc\left[\frac{x}{a} - \frac{a}{b}\right] = abc\left[\frac{b}{c}\right]$$
$$bcx - a^2c = ab^2$$
$$bcx = a^2c + ab^2$$
$$\boxed{x = \frac{a^2c + ab^2}{bc}}\ .$$

18.
$$S = \frac{a}{1 - r}$$
$$S(1 - r) = a$$
$$S - Sr = a$$
$$-Sr = a - S$$
$$r = \frac{a - S}{-S} = \boxed{\frac{S - a}{S}}\ .$$

19.
$$\frac{1}{C_n} = \frac{1}{C_1} + \frac{1}{C_2}$$
$$\frac{1}{5} = \frac{1}{15} + \frac{1}{C_2}$$
$$(15C_2)\frac{1}{5} = 15C_2\left(\frac{1}{15} + \frac{1}{C_2}\right)$$
$$3C_2 = C_2 + 15$$
$$2C_2 = 15$$
$$\boxed{C_2 = 7.5}\ .$$

20.
$$\frac{2x + y}{2} = \frac{4x - y}{3}$$
$$6\left(\frac{2x + y}{2}\right) = 6\left(\frac{4x - y}{3}\right)$$
$$3(2x + y) = 2(4x - y)$$
$$6x + 3y = 8x - 2y$$
$$-2x + 3y = -2y$$
$$-2x = -5y$$
$$x = \frac{-5y}{-2}$$

In standard form $\boxed{x = \frac{5y}{2}}$.

21. The number: x

$$\frac{x}{8} = 4 + \frac{5}{8}$$
$$x = 32 + 5$$
$$x = 37.$$

The number is $\boxed{37}$.

22. The length: x

The width: $\frac{3}{4}x$

New area = original area + 96

$$(x + 8)\left(\frac{3}{4}x + 6\right) = x\left(\frac{3}{4}x\right) + 96$$
$$\cancel{\frac{3}{4}x^2} + 12x + 48 = \cancel{\frac{3}{4}x^2} + 96$$
$$12x = 48$$
$$x = 4,\ \frac{3}{4}x = 3.$$

The dimensions are $\boxed{4 \text{ feet} \times 3 \text{ feet}}$.

23. The rate going: x
The rate returning: $x - 2$

	d	r	t = d/r
going	6	x	$\frac{6}{x}$
returning	6	$x - 2$	$\frac{6}{x-2}$

$$\begin{bmatrix}\text{time} \\ \text{going}\end{bmatrix} + \begin{bmatrix}\text{time} \\ \text{returning}\end{bmatrix} = \frac{5}{2}$$

$$\frac{6}{x} + \frac{6}{x-2} = \frac{5}{2}$$

$$2x(x-2)\left(\frac{6}{x} + \frac{6}{x-2}\right) = 2x(x-2)\frac{5}{2}$$

$$12(x-2) + 12x = 5x(x-2)$$

$$12x - 24 + 12x = 5x^2 - 10x$$

$$5x^2 - 34x + 24 = 0$$

$$(5x - 4)(x - 6) = 0$$

$$x = 6 \quad \text{or} \quad \frac{4}{5}.$$

When $x = \frac{4}{5}$, $x - 2 = \frac{4}{5} - 2 = \frac{4}{5} - \frac{10}{5} = \frac{-6}{5}$, which makes no sense for the stated problem. So we reject $x = \frac{4}{5}$. When $x = 6$, $x - 2 = 4$. Therefore, rate going = $\boxed{6 \text{ mph}}$; rate returning = $\boxed{4 \text{ mph}}$.

CHAPTER 6

1. a. $(x^3y^2)^3 = (x^3)^3(y^2)^3 = \boxed{x^9y^6}$.

   b. $(-2x^3)^2(-2y^2)^2 = (-2)^2(x^3)^2(-2)^2(y^2)^2 = (-2)^2(-2)^2(x^3)^2(y^2)^2 = \boxed{16x^6y^4}$.

2. a. $\left(\frac{x^3}{y^4}\right)^3 = \frac{(x^3)^3}{(y^4)^3} = \boxed{\frac{x^9}{y^{12}}}$.

   b. $\left(\frac{x^2}{y^2}\right)^2\left(-\frac{2xy}{5}\right)^3 = \frac{(x^2)^2}{(y^2)^2} \cdot \frac{(-2)^3x^3y^3}{5^3} = \frac{x^4}{y^4} \cdot \frac{-8x^3y^3}{125} = \frac{-8x^7y^3}{125y^4} = \boxed{\frac{-8x^7}{125y}}$.

3. a. $\dfrac{(-x^3)^2(-x)^3}{(x^2)^3} = \dfrac{x^6(-x^3)}{x^6} = \boxed{-x^3}$.

b. $(x^{2n+1} \cdot x^{n-2})^2 = (x^{2n+1+n-2})^2 = (x^{3n-1})^2 = x^{2(3n-1)} = \boxed{x^{6n-2}}$.

4. a. $\left(\dfrac{x^{2n}x^{3n}}{x^{5n-2}}\right)^3 = \left(\dfrac{x^{5n}}{x^{5n-2}}\right)^3 = (x^{5n-(5n-2)})^3 = (x^2)^3 = \boxed{x^6}$.

b. $\left(\dfrac{x^{2n-2}}{x^{2n}}\right)^{2n} = (x^{2n-2-2n})^{2n} = (x^{-2})^{2n} = x^{-4n} = \boxed{\dfrac{1}{x^{4n}}}$.

5. $\left(\dfrac{x^{3n}y^{4n}}{x^{2n}y^{n}}\right)^{1/3} = (x^{3n-2n}y^{4n-n})^{1/3} = (x^n y^{3n})^{1/3} = x^{n(1/3)}y^{3n(1/3)} = \boxed{x^{n/3}y^n}$.

6. $\left[\left(\dfrac{x^{4n}}{y^{3n}}\right)^{1/4n}\right]^{1/3} = \left[\dfrac{(x^{4n})^{1/4n}}{(y^{3n})^{1/4n}}\right]^{1/3} = \left[\dfrac{x^{4n/4n}}{y^{3n/4n}}\right]^{1/3} = \left[\dfrac{x}{y^{3/4}}\right]^{1/3} = \dfrac{x^{1/3}}{(y^{3/4})^{1/3}} = \boxed{\dfrac{x^{1/3}}{y^{1/4}}}$.

7. a. $x^{-5}x^2 = x^{-5+2} = x^{-3} = \boxed{\dfrac{1}{x^3}}$.

b. $(x^4y^{-2})^{-1/3} = (x^4)^{-1/3}(y^{-2})^{-1/3} = x^{-4/3}y^{2/3} = \boxed{\dfrac{y^{2/3}}{x^{4/3}}}$.

8. a. $\dfrac{x^{-3}y^{-2}z^{-1}}{x^0y^{-4}z^2} = x^{-3}y^{-2-(-4)}z^{-1-2} = x^{-3}y^2z^{-3} = \dfrac{1}{x^3} \cdot y^2 \cdot \dfrac{1}{z^3} = \boxed{\dfrac{y^2}{x^3z^3}}$.

b. $\left(\dfrac{3^0x^{-2}y^{-3}}{x^0y^{-2}}\right)^{-2} = (1 \cdot x^{-2-0}y^{-3-(-2)})^{-2} = (x^{-2}y^{-1})^{-2} = (x^{-2})^{-2}(y^{-1})^{-2} = \boxed{x^4y^2}$.

9. $\dfrac{(2 \times 10^{-4})(4 \times 10^6)(5 \times 10^2)}{8 \times 10^2} = \dfrac{2 \times 4 \times 5}{8} \times \dfrac{10^{-4} \times 10^6 \times 10^2}{10^2} = 5 \times 10^2 = \boxed{500}$.

10. $\dfrac{.009 \times .0008}{.0036} = \dfrac{9 \times 10^{-3} \times 8 \times 10^{-4}}{36 \times 10^{-4}} = \dfrac{9 \times 8}{36} \times \dfrac{10^{-3} \times 10^{-4}}{10^{-4}} = 2 \times 10^{-3} = \boxed{.002}$.

11. $x^{1/3} - x = \boxed{x(x^{-2/3} - 1)}$.

12. $y^{1/3} - y^{-1/3} = \boxed{y^{-1/3}(y^{2/3} - 1)}$.

13. $y^{2n+2} - y^{n} = \boxed{y^{n}(y^{n+2} - 1)}$.

14. a. $(2 - y^{2})^{-3/4} = \dfrac{1}{(2 - y^{2})^{3/4}} = \boxed{\dfrac{1}{\sqrt[4]{(2 - y^{2})^{3}}}}$.

b. $\sqrt[3]{2m^{2}n} = (2m^{2}n)^{1/3} = (2n)^{1/3}(m^{2})^{1/3} = \boxed{(2n)^{1/3}m^{2/3}}$.

15. a. $\sqrt[4]{81x^{6}y^{3}} = \sqrt[4]{(3^{4}x^{4})x^{2}y^{3}} = \sqrt[4]{(3x)^{4}}\ \sqrt[4]{x^{2}y^{3}} = \boxed{3x\sqrt[4]{x^{2}y^{3}}}$.

b. $\sqrt{\dfrac{5}{6}} = \sqrt{\dfrac{5 \cdot 6}{6 \cdot 6}} = \dfrac{\sqrt{30}}{\sqrt{36}} = \boxed{\dfrac{\sqrt{30}}{6}}$.

c. $\dfrac{2m}{\sqrt{2mn}} = \dfrac{2m\sqrt{2mn}}{\sqrt{2mn}\sqrt{2mn}} = \dfrac{2m\sqrt{2mn}}{2mn} = \boxed{\dfrac{\sqrt{2mn}}{n}}$.

d. $\dfrac{\sqrt[3]{x}\ \sqrt[3]{y^{2}}}{\sqrt[3]{x^{2}y}} = \dfrac{\sqrt[3]{x}\ \sqrt[3]{y^{2}}\ \sqrt[3]{xy^{2}}}{\sqrt[3]{x^{2}y}\ \sqrt[3]{xy^{2}}} = \dfrac{\sqrt[3]{x^{2}y^{4}}}{xy}$

$= \dfrac{y\ \sqrt[3]{x^{2}y}}{xy} = \boxed{\dfrac{\sqrt[3]{x^{2}y}}{x}}$.

16. a. $\sqrt{48} + \sqrt{75} = \sqrt{16}\sqrt{3} + \sqrt{25}\sqrt{3} = 4\sqrt{3} + 5\sqrt{3} = \boxed{9\sqrt{3}}$.

b. $\sqrt{9m} + 3\sqrt{36m} - 2\sqrt{25m} = \sqrt{9}\sqrt{m} + 3\sqrt{36}\sqrt{m} - 2\sqrt{25}\sqrt{m} = 3\sqrt{m} + 3(6)\sqrt{m} - 2(5)\sqrt{m}$

$= 3\sqrt{m} + 18\sqrt{m} - 10\sqrt{m} = \boxed{11\sqrt{m}}$.

c. $\dfrac{2\sqrt{8}}{3} + \dfrac{3\sqrt{18}}{2} = \dfrac{2\sqrt{4}\sqrt{2}}{3} + \dfrac{3\sqrt{9}\sqrt{2}}{2} = \dfrac{2(2)\sqrt{2}}{3} + \dfrac{3(3)\sqrt{2}}{2} = \dfrac{4\sqrt{2}}{3} + \dfrac{9\sqrt{2}}{2} = \dfrac{8\sqrt{2}}{6} + \dfrac{27\sqrt{2}}{6}$

$= \boxed{\dfrac{35\sqrt{2}}{6}}$.

d. $\sqrt[3]{8x} - \sqrt[3]{-64x^{4}} = \sqrt[3]{8}\sqrt[3]{x} - \sqrt[3]{-64x^{3}}\ \sqrt[3]{x} = 2(\sqrt[3]{x}) + 4x(\sqrt[3]{x}) = \boxed{(2 + 4x)(\sqrt[3]{x})}$.

17. a. $\sqrt{5}(\sqrt{6} - \sqrt{5}) = \sqrt{5}\sqrt{6} - \sqrt{5}\sqrt{5} = \boxed{\sqrt{30} - 5}$.

b. $(\sqrt{2} - \sqrt{2x})(\sqrt{2} + \sqrt{2x}) = (\sqrt{2})^2 - (\sqrt{2x})^2 = \boxed{2 - 2x}$.

c. $\sqrt[4]{4}(\sqrt[4]{4} + \sqrt[4]{64}) = (\sqrt[4]{4})(\sqrt[4]{4}) + (\sqrt[4]{4})(\sqrt[4]{64}) = \sqrt[4]{16} + \sqrt[4]{256} = \sqrt[4]{2^4} + \sqrt[4]{2^8}$

$$= \sqrt[4]{2^4} + \sqrt[4]{(2^2)^4} = 2 + 2^2 = \boxed{6}.$$

18. a. $\dfrac{\sqrt{15} + \sqrt{21}}{\sqrt{3}} = \dfrac{\sqrt{3}\sqrt{5} + \sqrt{3}\sqrt{7}}{\sqrt{3}} = \dfrac{\sqrt{3}(\sqrt{5} + \sqrt{7})}{\sqrt{3}} = \boxed{\sqrt{5} + \sqrt{7}}$.

b. $\dfrac{m\sqrt{mn^3} - \sqrt{mn}}{\sqrt{mn}} = \dfrac{m\sqrt{n^2}\sqrt{mn} - \sqrt{mn}}{\sqrt{mn}} = \dfrac{mn\sqrt{mn} - \sqrt{mn}}{\sqrt{mn}} = \dfrac{\sqrt{mn}(mn - 1)}{\sqrt{mn}} = \boxed{mn - 1}$.

19. a. $\dfrac{2\sqrt{3} - 2}{2\sqrt{3} + 2} = \dfrac{(2\sqrt{3} - 2)(2\sqrt{3} - 2)}{(2\sqrt{3} + 2)(2\sqrt{3} - 2)} = \dfrac{4(3) - 8\sqrt{3} + 4}{4(3) - 4} = \dfrac{16 - 8\sqrt{3}}{8} = \dfrac{8(2 - \sqrt{3})}{8}$

$$= \boxed{2 - \sqrt{3}}.$$

b. $\dfrac{\sqrt{x} + \sqrt{y}}{\sqrt{x} - \sqrt{y}} = \dfrac{(\sqrt{x} + \sqrt{y})(\sqrt{x} + \sqrt{y})}{(\sqrt{x} - \sqrt{y})(\sqrt{x} + \sqrt{y})} = \boxed{\dfrac{x + 2\sqrt{xy} + y}{x - y}}$.

20. a. $(2 - 3i) - (4 + 5i) = (2 - 4) + (-3 - 5)i = \boxed{-2 - 8i}$.

b. $(2 - 3i)(4 + 5i) = 8 - 12i + 10i - 15i^2 = 8 - 2i - 15(-1)$

$$= 8 - 2i + 15 = \boxed{23 - 2i}.$$

c. $\dfrac{4 + 5i}{2 - 3i} = \dfrac{(4 + 5i)(2 + 3i)}{(2 - 3i)(2 + 3i)} = \dfrac{8 + 22i + 15i^2}{4 - 9i^2} = \dfrac{8 + 22i - 15}{4 - (-9)} = \dfrac{-7 + 22i}{13}$

$$= \boxed{\frac{-7}{13} + \frac{22}{13}i}.$$

21. a. $(3 + \sqrt{-12}) + (1 - \sqrt{-27}) = (3 + i\sqrt{12}) + (1 - i\sqrt{27}) = (3 + i\sqrt{4}\sqrt{3}) + (1 - i\sqrt{9}\sqrt{3})$

$$= (3 + 2i\sqrt{3}) + (1 - 3i\sqrt{3}) = (3 + 1) + (2\sqrt{3} - 3\sqrt{3})i$$

$$= \boxed{4 - i\sqrt{3}}.$$

b. $(3 - \sqrt{-3})(1 - \sqrt{-3}) = (3 - i\sqrt{3})(1 - i\sqrt{3}) = 3 - 4i\sqrt{3} + i^2(3)$

$$= 3 - 4i\sqrt{3} - 3 = \boxed{-4i\sqrt{3}}.$$

21. cont'd.

c. $\frac{1}{2 - \sqrt{-3}} = \frac{1}{2 - i\sqrt{3}} = \frac{2 + i\sqrt{3}}{(2 - i\sqrt{3})(2 + i\sqrt{3})} = \frac{2 + i\sqrt{3}}{4 - i^2(3)} = \frac{2 + i\sqrt{3}}{4 + 3} = \frac{2 + i\sqrt{3}}{7}$

$= \boxed{\frac{2}{7} + \frac{\sqrt{3}}{7} i}$.

CHAPTER 7

1. $5x^2 - 9x = 0$

$a = 5,\ b = -9,\ c = 0$

$x = \frac{-(-9) \pm \sqrt{(-9)^2 - 4(5)(0)}}{2(5)}$

$= \frac{9 \pm \sqrt{81}}{10} = \frac{9 \pm 9}{10}$

$x = \frac{9 - 9}{10} = 0$ or $x = \frac{9 + 9}{10} = \frac{9}{5}$.

$\boxed{\left\{0, \frac{9}{5}\right\}}$

2. $2x^2 - 4x = x + 3$

$2x^2 - 5x - 3 = 0$

$a = 2,\ b = -5,\ c = -3$

$x = \frac{-(-5) \pm \sqrt{(-5)^2 - 4(2)(-3)}}{2(2)}$

$= \frac{5 \pm \sqrt{49}}{4} = \frac{5 \pm 7}{4}$

$= \frac{5 - 7}{4} = -\frac{1}{2}$ or $x = \frac{5 + 7}{4} = 3.$

$\boxed{\left\{-\frac{1}{2}, 3\right\}}$

3. $\frac{2x^2}{7} = 8$

$2x^2 = 56$

$x^2 = 28$

$x^2 = \pm\sqrt{28} = \pm 2\sqrt{7}.$

$\boxed{\{2\sqrt{7}, -2\sqrt{7}\}}$.

4. $(2x + 1)^2 = 25$

$2x + 1 = \pm 5$

$2x = -1 \pm 5$

$x = \frac{-1 \pm 5}{2}$.

$\boxed{\{-3, 2\}}$.

5. $3x^2 = 5x - 1$

$3x^2 - 5x + 1 = 0$. $a = 3,\ b = -5,\ c = 1.$

$$x = \frac{-(-5) \pm \sqrt{(-5)^2 - 4(3)(1)}}{2(3)} = \frac{5 \pm \sqrt{25 - 12}}{6} = \frac{5 \pm \sqrt{13}}{6} .$$

$\boxed{\left\{\frac{5 + \sqrt{13}}{6}, \frac{5 - \sqrt{13}}{6}\right\}}$.

6. $\frac{x^2 - x}{2} + 1 = 0$

$x^2 - x + 2 = 0.$ $a = 1,\ b = -1,\ c = 2.$

$$x = \frac{-(-1) \pm \sqrt{(-1)^2 - 4(1)(2)}}{2(1)} = \frac{1 \pm \sqrt{1 - 8}}{2} = \frac{1 \pm \sqrt{-7}}{2} = \frac{1 \pm i\sqrt{7}}{2}$$

$\boxed{\left\{\frac{1 + i\sqrt{7}}{2}, \frac{1 - i\sqrt{7}}{2}\right\}}$.

7. $2x^2 - 3x + 2y = 0$. $a = 2,\ b = -3,\ c = 2y$

$$x = \frac{-(-3) \pm \sqrt{(-3)^2 - 4(2)(2y)}}{2(2)} = \frac{3 \pm \sqrt{9 - 16y}}{4} .$$

$\boxed{\left\{\frac{3 - \sqrt{9 - 16y}}{4}, \frac{3 + \sqrt{9 - 16y}}{4}\right\}}$.

8. $\sqrt{x - 3} + 5 = 0$

$\sqrt{x - 3} = -5.$ For every x, $\sqrt{x - 3}$ is nonnegative. Therefore, the solution set is $\boxed{\emptyset}$.

9. $$\sqrt{3x + 10} = x + 4$$
$$(\sqrt{3x + 10})^2 = (x + 4)^2$$
$$3x + 10 = x^2 + 8x + 16$$
$$x^2 + 5x + 6 = 0$$
$$(x + 2)(x + 3) = 0$$
$x = -2$ or -3. Neither of these is extraneous; therefore

$\boxed{\{-2,-3\}}$.

10. $$4\sqrt{x} + \sqrt{16x + 1} = 5$$
$$\sqrt{16x + 1} = 5 - 4\sqrt{x}$$
$$(\sqrt{16x + 1})^2 = (5 - 4\sqrt{x})^2$$
$$16x + 1 = 25 - 40\sqrt{x} + 16x$$
$$40\sqrt{x} = 24$$
$$5\sqrt{x} = 3$$
$$(5\sqrt{x})^2 = 3^2$$
$$25x = 9$$
$x = \frac{9}{25}$, which is not extraneous. Therefore, $\boxed{\left\{\frac{9}{25}\right\}}$.

11. $$x^4 - 2x^2 - 24 = 0$$
$$(x^2 - 6)(x^2 + 4) = 0$$
$x^2 - 6 = 0$ or $x^2 + 4 = 0$

$x = \pm\sqrt{6}$ $\quad$ $x = \pm\sqrt{-4} = \pm 2i$. $\boxed{\{\sqrt{6},-\sqrt{6},2i,-2i\}}$.

12. $x^{2/3} - 2x^{1/3} = 35$. Let $u = x^{1/3}$.
$$u^2 - 2u - 35 = 0$$
$$(u - 7)(u + 5) = 0$$
$u = 7$ or $u = -5$.

Then $x^{1/3} = 7$ or $x^{1/3} = -5$

$x = 7^3$ or $x = (-5)^3$

$x = 343$ or $x = -125$. $\boxed{\{343,-125\}}$.

13. $$2x^2 + 3x - 2 = 0$$
$$x^2 + \frac{3}{2}x = 1$$
$$x^2 + \frac{3}{2}x + \frac{9}{16} = 1 + \frac{9}{16}$$
$$\left(x + \frac{3}{4}\right)^2 = \frac{25}{16}$$
$$x + \frac{3}{4} = \pm\frac{5}{4}$$
$$x = \frac{-3}{4} \pm \frac{5}{4}. \qquad \boxed{\left\{-2, \frac{1}{2}\right\}}.$$

14. For imaginary solutions, $b^2 - 4ac < 0$. $a = 1$, $b = -1$, $c = k - 2$.

$b^2 - 4ac = (-1)^2 - 4(1)(k - 2) = 1 - 4k + 8 < 0$. $-4k < -9$. $\boxed{k > \frac{9}{4}}$.

15. $$x^2 - 5x - 6 \ge 0$$
$$(x - 6)(x + 1) \ge 0$$
$(x - 6 \ge 0$ and $x + 1 \ge 0)$ or $(x - 6 \le 0$ and $x + 1 \le 0)$

$(x \ge 6$ and $x \ge -1)$ or $(x \le 6$ and $x \le -1)$

$(x \ge 6)$ or $(x \le -1)$.

$\boxed{\{x \mid x \ge 6 \text{ or } x \le -1\}}$; 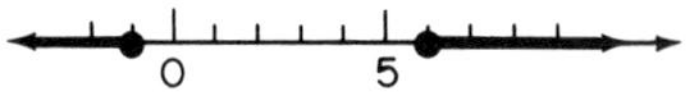

16. $$\frac{3}{x - 6} > 8$$
$$(x - 6)^2\left(\frac{3}{x - 6}\right) > (x - 6)^2(8)$$
$$3(x - 6) > 8(x - 6)^2$$
$$3(x - 6) - 8(x - 6)^2 > 0$$
$$(x - 6)[3 - 8(x - 6)] > 0$$
$$(x - 6)(-8x + 51) > 0$$
$(x - 6 > 0$ and $-8x + 51 > 0)$ or $(x - 6 < 0$ and $-8x + 51 < 0)$

$\left(x > 6 \text{ and } x < \frac{51}{8}\right)$ or $\left(x < 6 \text{ and } x > \frac{51}{8}\right)$

$\left(6 < x < \frac{51}{8}\right)$ or $(\emptyset)$

$6 < x < \frac{51}{8}$. $\boxed{\left\{x \mid 6 < x < \frac{51}{8}\right\}}$; 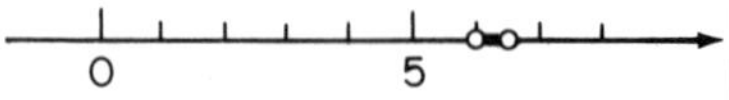

17. The first positive integer: x
The next positive integer: $x + 1$

$$\begin{aligned} x^2 + (x+1)^2 &= 145 \\ x^2 + x^2 + 2x + 1 &= 145 \\ 2x^2 + 2x - 144 &= 0 \\ 2(x+9)(x-8) &= 0. \end{aligned}$$

The only positive solution is $x = 8$. Therefore, $x + 1 = 9$.

$\boxed{8 \text{ and } 9}$

CHAPTER 8

1. a. $2(0) - 9y = 18$. $y = -2$. $\boxed{-2}$.

b. $2x - 9(0) = 18$. $x = 9$. $\boxed{9}$.

c. $2(2) - 9y = 18$. $4 - 9y = 18$. $-9y = 14$. $y = \frac{-14}{9}$. $\boxed{\frac{-14}{9}}$.

2. $$\frac{2x^4}{3 + 2y^2} = 5$$

$$\begin{aligned} 2x^4 &= 15 + 10y^2 \\ 10y^2 &= 2x^4 - 15 \\ y^2 &= \frac{2x^4 - 15}{10} \end{aligned}$$

$$y = \pm\sqrt{\frac{2x^4 - 15}{10}} = \pm\sqrt{\frac{10(2x^4 - 15)}{10^2}} = \pm\frac{1}{10}\sqrt{10(2x^4 - 15)}.$$

$$\boxed{y = \pm\frac{1}{10}\sqrt{10(2x^4 - 15)}}.$$

3. $kx - 3y = 9$. $k(1) - 3(-3) = 9$. $k + 9 = 9$. $\boxed{k = 0}$.

4. $\boxed{\{(-2,-5),\ (-1,-3),\ (0,-1),\ (1,1),\ (2,3)\}}$.

The graph is:

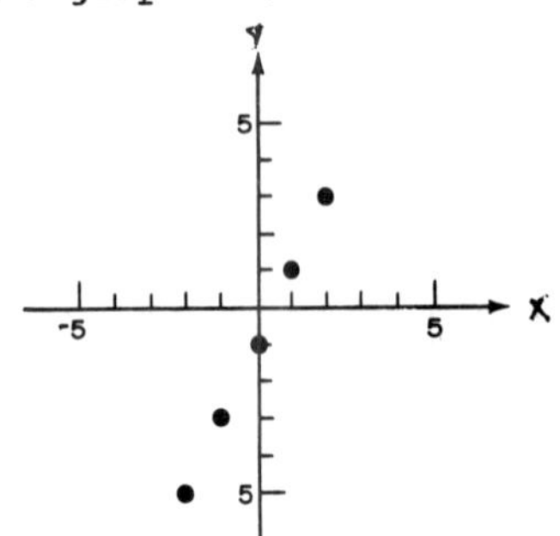

5. x-intercept is $\boxed{\frac{1}{2}}$.

y-intercept is $\boxed{2}$.

The graph is

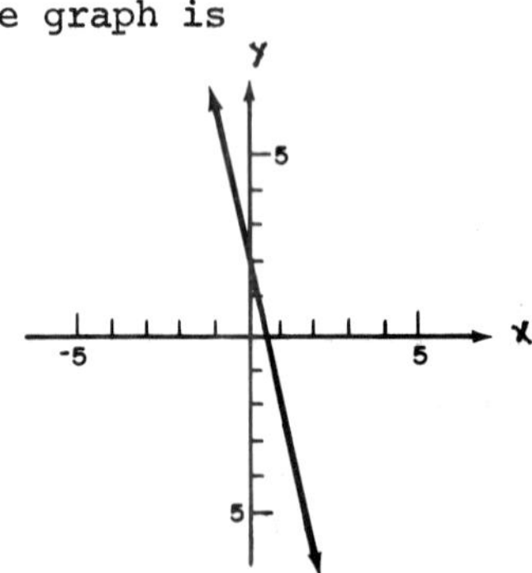

6. x-intercept is $\boxed{2}$.

y-intercept is $\boxed{-3}$.

The graph is

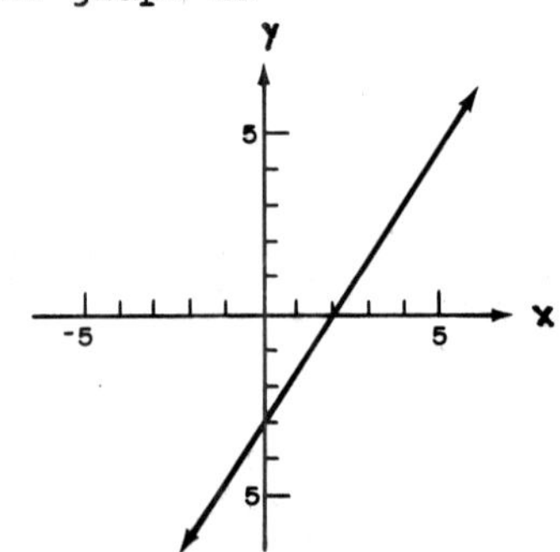

7. a. $d = \sqrt{(2 - (-2))^2 + (-3 - (-1))^2} = \sqrt{4^2 + (-2)^2} = \sqrt{20} = \boxed{2\sqrt{5}}$.

b. $m = \dfrac{-3 - (-1)}{2 - (-2)} = \dfrac{-2}{4} = \boxed{-\dfrac{1}{2}}$.

8. $y - y_1 = m(x - x_1)$

$(x_1, y_1) = (2,-1)$, and $m = \dfrac{3}{4}$

$y - (-1) = \dfrac{3}{4}(x - 2)$

$y + 1 = \dfrac{3}{4}(x - 2)$

$4(y + 1) = 3(x - 2)$

$4y + 4 = 3x - 6$

$\boxed{3x - 4y - 10 = 0}$.

9. a. $3x - 2y = 5$

$-2y = -3x + 5$

$\boxed{y = \dfrac{3}{2}x - \dfrac{5}{2}}$.

b. Slope is $\boxed{\dfrac{3}{2}}$.

c. y-intercept is $\boxed{-\dfrac{5}{2}}$.

10. $3x - 2y + 5 = 0$

$$-2y = -3x - 5$$

$$y = \frac{3}{2}x + \frac{5}{2}.$$

Slope of graph of $3x - 2y + 5 = 0$ is $\frac{3}{2}$.

Slope of perpendicular is $-\frac{2}{3}$.

In point-slope form, we have

$$y - 5 = -\frac{2}{3}(x - 0)$$

$$3y - 15 = -2x$$

$\boxed{2x + 3y - 15 = 0}$.

11. $3x - 4y = 12$

$$-4y = -3x + 12$$

$y = \frac{3}{4}x - 3$. So, $m = \frac{3}{4}$. Then,

$$y - (-3) = \frac{3}{4}(x - 2)$$

$$4x + 12 = 3x - 6$$

$\boxed{-3x + 4y + 18 = 0 \quad \text{or} \quad 3x - 4y - 18 = 0}$

12.

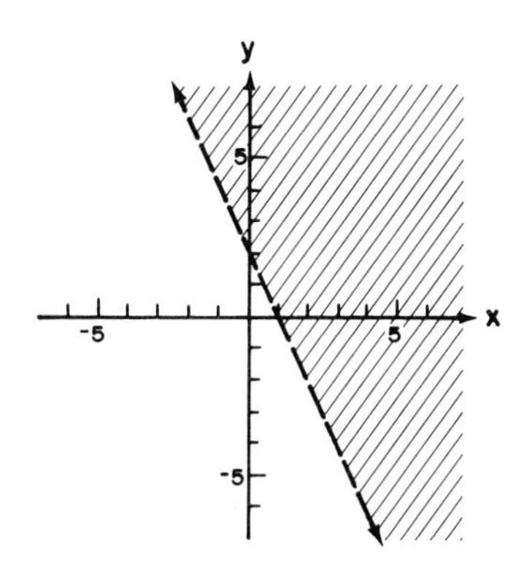

13. The graph is

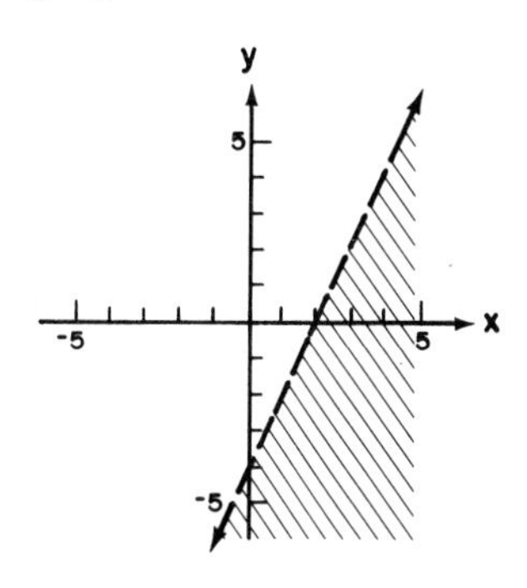

14. The graph is the double-shaded region:

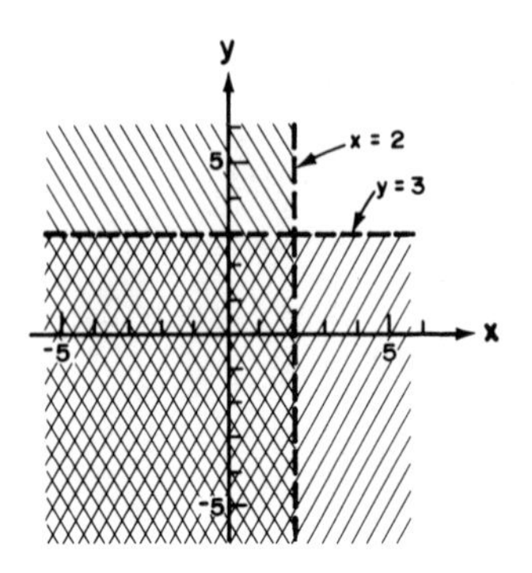

CHAPTER 9

1. $\begin{aligned} x + 4y &= -14 \\ 3x + 2y &= -2. \end{aligned}$ $\quad \begin{aligned} x + 4y &= -14 \\ -6x - 4y &= 4. \end{aligned}$ $\quad -5x = -10. \quad x = 2.$

$\begin{aligned} 2 + 4y &= -14 \\ y &= -4. \end{aligned}$ $\boxed{\{(2,-4)\}}$.

2. $\begin{aligned} \frac{2}{3}x - y &= 4 \\ x - \frac{3}{4}y &= 6. \end{aligned}$ $\quad \begin{aligned} 2x - 3y &= 12 \\ -4x + 3y &= -24. \end{aligned}$ $\quad -2x = -12. \quad x = 6.$

$\begin{aligned} 2(6) - 3y &= 12 \\ -3y &= 0 \\ y &= 0. \end{aligned}$ $\boxed{\{(6,0)\}}$.

3. Let $u = \frac{1}{x}$ and $v = \frac{1}{y}$. Then the system becomes

$\begin{aligned} u + 2v &= -\frac{11}{12} \\ u + v &= -\frac{7}{12}. \end{aligned}$ $\quad \begin{aligned} 12u + 24v &= -11 \\ -12u - 12v &= 7. \end{aligned}$ $\quad 12v = -4. \quad v = -\frac{4}{12} = -\frac{1}{3}.$

$\begin{aligned} u + v &= -\frac{7}{12} \\ u + \left(-\frac{4}{12}\right) &= -\frac{7}{12} \\ u &= -\frac{3}{12} = -\frac{1}{4}. \end{aligned}$

Since $u = \frac{1}{x}$, we have $-\frac{1}{4} = \frac{1}{x}$; $x = -4.$

Since $v = \frac{1}{y}$, we have $-\frac{1}{3} = \frac{1}{y}$; $y = -3.$

$\boxed{\{(-4,-3)\}}$.

4. Since $\frac{3}{9} = \frac{4}{12} \neq \frac{4}{3}$, there are no solutions; the equations are $\boxed{\text{inconsistent}}$.

5. Since $\frac{3}{9} = \frac{4}{12} = \frac{1}{3}$, the equations are $\boxed{\text{dependent}}$.

6. Since $\frac{3}{9} \neq \frac{4}{6}$, the equations have a unique solution.

7. $2x + 4y + z = 0$ | $4x + 8y + 2z = 0$ | $9x + 11y = 1.$
$5x + 3y - 2z = 1.$ | $5x + 3y - 2z = 1.$

$2x + 4y + z = 0$ | $14x + 28y + 7z = 0$ | $18x + 21y = 6.$
$4x - 7y - 7z = 6.$ | $4x - 7y - 7z = 6.$

$9x + 11y = 1$ | $-18x - 22y = -2$ | $-y = 4.$ | $y = -4.$
$18x + 21y = 6.$ | $18x + 21y = 6.$

$9x + 11y = 1$
$9x + 11(-4) = 1$
$9x = 45$
$x = 5.$

$2x + 4y + z = 0$
$2(5) + 4(-4) + z = 0$
$z = 6.$

$\boxed{\{(5,-4,6)\}}$.

8.

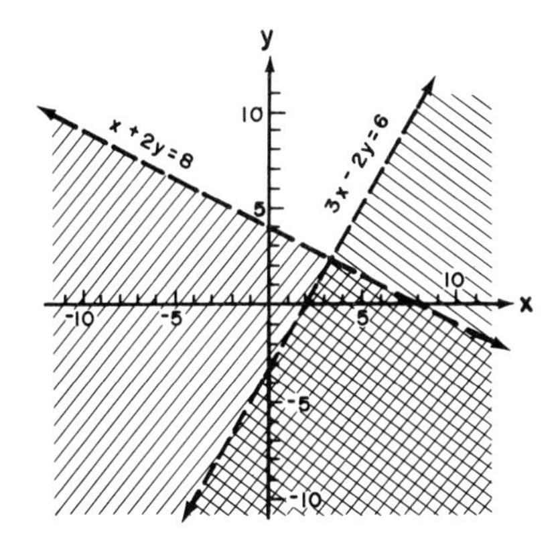

9. $x + y + z = 155$ | $x + y + z = 155$ | $x + y + z = 155$ | $2y + z = 175.$
$x = y - 20$ | $x - y = -20$ | $-x + y = 20.$
$y = 5 + z.$ | $y - z = 5.$

$2y + z = 175$ | $3y = 180$ | $x = 60 - 20$ | $60 - z = 5$
$y - z = 5.$ | $y = 60.$ | $x = 40.$ | $z = 55.$

$$\boxed{\begin{array}{l} x = 40 \text{ inches} \\ y = 60 \text{ inches} \\ z = 55 \text{ inches} \end{array}}$$

10. Substitute (-1,2) and (2,-1) into $ax + by = 10$.

$$\begin{aligned} -a + 2b &= 10 \\ 2a + b &= 10. \end{aligned}$$

The solution of this system yields $\boxed{a = 2 \text{ and } b = 6}$.

11. $\left[\begin{array}{cc|c} 1 & 4 & -14 \\ 3 & 2 & -2 \end{array}\right] \sim \left[\begin{array}{cc|c} 1 & 4 & -14 \\ 0 & -10 & 40 \end{array}\right]$. The last matrix corresponds to the system $\begin{aligned} x + 4y &= -14 \\ -10y &= 40 \end{aligned}$.

From the second equation, $y = -4$. When -4 is substituted for y in the first equation, we find $x = 2$. $\boxed{\{(2,-4)\}}$.

12. We start with the augmented matrix $\left[\begin{array}{ccc|c} 2 & 4 & 1 & 0 \\ 5 & 3 & -2 & 1 \\ 4 & -7 & -7 & 6 \end{array}\right]$. To obtain a row-equivalent matrix with 0 as the first entry in rows 2 and 3 we proceed as follows: multiply each entry of row 1 by $\frac{-5}{2}$ and add each product to the corresponding entry of row 2 to obtain a new row 2; multiply each entry of row 1 by -2 and add each product to the corresponding entry of row 3 to obtain a new row 3. Thus we have $\left[\begin{array}{ccc|c} 2 & 4 & 1 & 0 \\ 5 & 3 & -2 & 1 \\ 4 & -7 & -7 & 6 \end{array}\right] \sim \left[\begin{array}{ccc|c} 2 & 4 & 1 & 0 \\ 0 & -7 & \frac{-9}{2} & 1 \\ 0 & -15 & -9 & 6 \end{array}\right]$.

Then, to obtain a matrix which is row-equivalent to this last one but with a 0 as the second entry in row 3 (without disturbing the already existing zeros), we multiply each entry of row 2 by $\frac{-15}{7}$ and add each product to the corresponding entry of row 3 to obtain a new row 3. Thus, we have

$\left[\begin{array}{ccc|c} 2 & 4 & 1 & 0 \\ 0 & -7 & \frac{-9}{2} & 1 \\ 0 & -15 & -9 & 6 \end{array}\right] \sim \left[\begin{array}{ccc|c} 2 & 4 & 1 & 0 \\ 0 & -7 & \frac{-9}{2} & 1 \\ 0 & 0 & \frac{9}{14} & \frac{27}{7} \end{array}\right]$. This last matrix

corresponds to
$$\begin{aligned} 2x + 4y + z &= 0 \quad (1) \\ -7y - \frac{9}{2}z &= 1 \quad (2) \\ \frac{9}{14}z &= \frac{27}{7} \quad (3) \end{aligned}$$

From Equation (3), $z = 6$. In Equation (2) substitute 6 for z and solve for y to obtain $y = -4$. Then, in Equation (1) substitute 6 for z and -4 for y and solve for x to obtain $x = 5$. The solution set is $\boxed{\{(5,-4,6)\}}$.

13. $\begin{vmatrix} 10 & 3 \\ -10 & -2 \end{vmatrix} = (10)(-2) - (-10)(3) = -20 - (-30) = \boxed{10}$.

14. $D = \begin{vmatrix} 3 & -4 \\ 1 & -2 \end{vmatrix} = -6 - (-4) = -2.$ $\quad D_x = \begin{vmatrix} -2 & -4 \\ 0 & -2 \end{vmatrix} = 4 - 0 = 4.$

$D_y = \begin{vmatrix} 3 & -2 \\ 1 & 0 \end{vmatrix} = 0 - (-2) = 2.$ $\quad x = \frac{D_x}{D} = \frac{4}{-2} = -2.$

$y = \frac{D_y}{D} = \frac{2}{-2} = -1.$ $\boxed{\{(-2,-1)\}}$.

15. Expanding about the 2nd row, we have

$-0\begin{vmatrix} 3 & 1 \\ 2 & 1 \end{vmatrix} + 1\begin{vmatrix} 2 & 1 \\ -4 & 1 \end{vmatrix} - 0\begin{vmatrix} 2 & 3 \\ -4 & 2 \end{vmatrix};$ $\begin{vmatrix} 2 & 1 \\ -4 & 1 \end{vmatrix} = 2 - (-4) = \boxed{6}$.

16. $D = \begin{vmatrix} 3 & -2 & 5 \\ 4 & -4 & 3 \\ 5 & -4 & 1 \end{vmatrix} = \begin{vmatrix} -22 & 18 & 0 \\ -11 & 8 & 0 \\ 5 & 4 & 1 \end{vmatrix} = \begin{vmatrix} -22 & 18 \\ -11 & 8 \end{vmatrix} = -11\begin{vmatrix} 2 & 18 \\ 1 & 8 \end{vmatrix} = -11(16-18) = 22.$

$D_x = \begin{vmatrix} 6 & -2 & 5 \\ 0 & -4 & 3 \\ -5 & -4 & 1 \end{vmatrix} = \begin{vmatrix} 31 & 18 & 0 \\ 15 & 8 & 0 \\ 5 & 4 & 1 \end{vmatrix} = \begin{vmatrix} 31 & 18 \\ 15 & 8 \end{vmatrix} = 248 - 270 = -22.$

$D_y = \begin{vmatrix} 3 & 6 & 5 \\ 4 & 0 & 3 \\ 5 & -5 & 1 \end{vmatrix} = \begin{vmatrix} -22 & 31 & 0 \\ -11 & 15 & 0 \\ 5 & -1 & 1 \end{vmatrix} = \begin{vmatrix} -22 & 31 \\ -11 & 15 \end{vmatrix} = -11\begin{vmatrix} 2 & 31 \\ 1 & 15 \end{vmatrix} = -11(30-31) = 11.$

$D_z = \begin{vmatrix} 3 & -2 & 6 \\ 4 & -4 & 0 \\ 5 & -4 & -5 \end{vmatrix} = 4\begin{vmatrix} 3 & -2 & 6 \\ 1 & -1 & 0 \\ 5 & -4 & -5 \end{vmatrix} = 4\begin{vmatrix} 3 & 1 & 6 \\ 1 & 0 & 0 \\ 5 & 1 & -5 \end{vmatrix} = 4(-1)\begin{vmatrix} 1 & 6 \\ 1 & -5 \end{vmatrix} = -4(-5-6) = 44.$

$x = \frac{D_x}{D} = \frac{-22}{22} = -1.$ $\quad y = \frac{D_y}{D} = \frac{11}{22} = \frac{1}{2}.$ $\quad z = \frac{D_z}{D} = \frac{44}{22} = 2.$ $\quad \boxed{\left\{\left(-1,\frac{1}{2},2\right)\right\}}$.

CHAPTER 10

1. $\frac{x^2}{7} + \frac{y^2}{28} = 1$

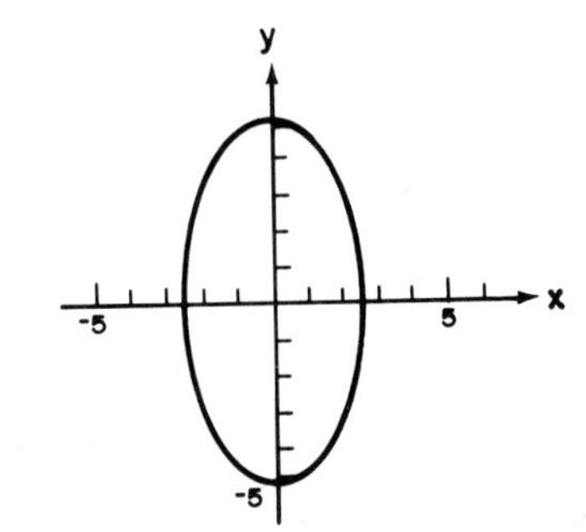

2. $\frac{y^2}{4} - \frac{x^2}{24} = 1$

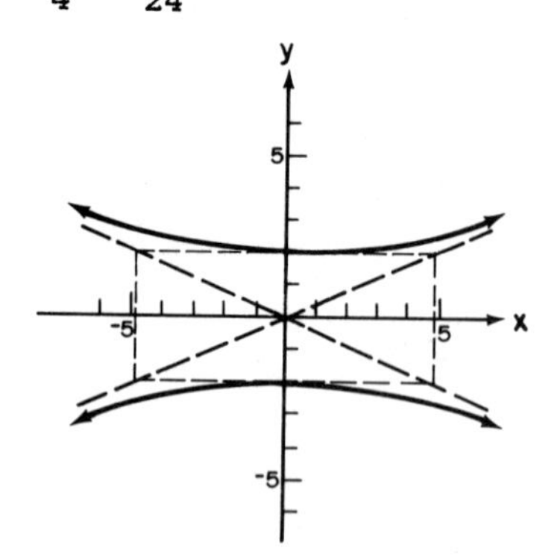

3.

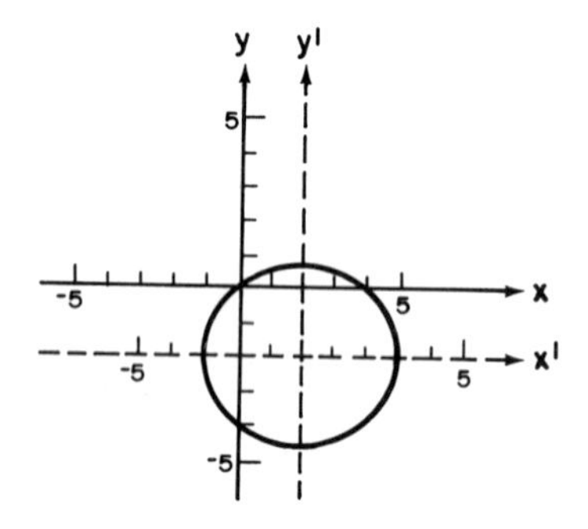

4.

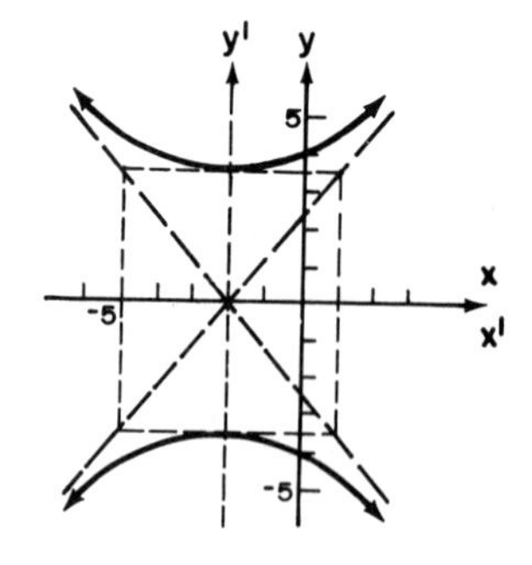

5. $x^2 - 6x + ? + y^2 + 4y + ? = -4$

$x^2 - 6x + 9 + y^2 + 4y + 4 = -4 + 9 + 4$

$(x - 3)^2 + (y + 2)^2 = 9$

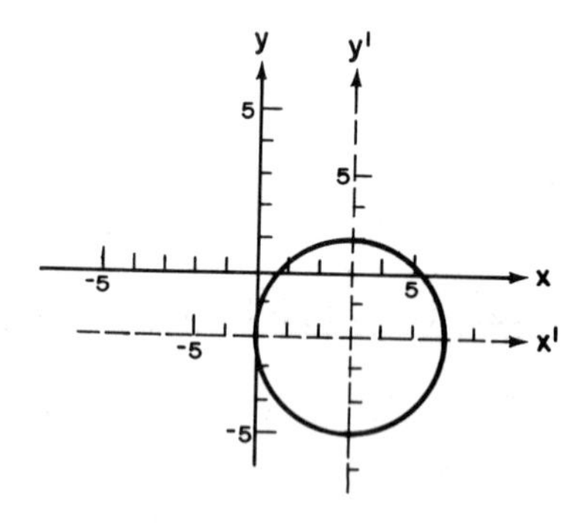

6. $y = -2x^2 + 4x - 2$

$y = -2(x^2 - 2x + ?) - 2$

$y = -2(x^2 - 2x + 1) - 2 - (-2) \cdot 1$

$y = -2(x - 1)^2$

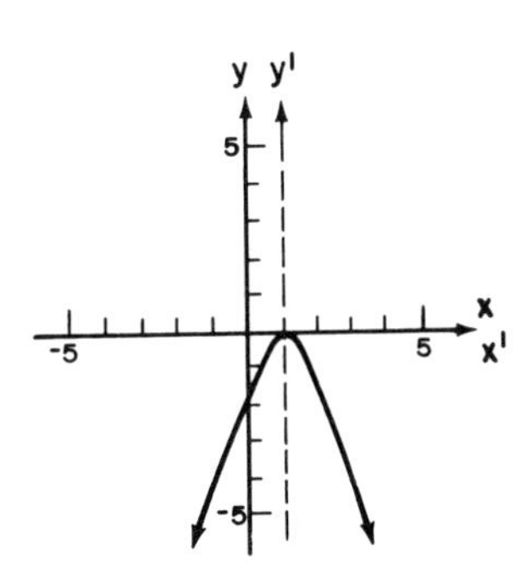

7. $y = x^2 - 2x + 1$
$x + y = 3.$

Substitute $x^2 - 2x + 1$ for y in $x + y = 3$.

$$x + (x^2 - 2x + 1) = 3$$
$$x^2 - x - 2 = 0$$
$$(x - 2)(x + 1) = 0$$
$$x = 2 \text{ or } -1.$$

From $x + y = 3$, when $x = 2$, we have $2 + y = 3$; $y = 1$. And when $x = -1$, we have $-1 + y = 3$; $y = 4$.

$\boxed{\{(2,1),(-1,4)\}}$.

8. $9x^2 + 16y^2 = 100$
$x^2 + y^2 = 8.$

$$9x^2 + 16y^2 = 100$$
$$-16x^2 - 16y^2 = -128.$$

$$-7x^2 = -28$$
$$x^2 = 4$$
$$x = \pm 2.$$

From $x^2 + y^2 = 8$, when $x = 2$, we have $4 + y^2 = 8$ and $y = \pm 2$. So we have (2,2) and (2,-2). Similarly, when $x = -2$, we obtain (-2,2) and (-2,-2).

$\boxed{\{(2,2),(2,-2),(-2,2),(-2,-2)\}}$.

9. $(3x^2 - 2xy + 3y^2) - 3(x^2 + y^2) = 34 - 3(17)$, $-2xy = -17$, $y = \dfrac{17}{2x}$.

Substituting in the second equation, $x^2 + \left(\dfrac{17}{2x}\right)^2 = 17$, $x^2 + \dfrac{289}{4x^2} = 17$,

$4x^4 - 68x^2 + 289 = 0$, $(2x^2 - 17)^2 = 0$, $2x^2 = 17$, $x = \pm\sqrt{\dfrac{17}{2}} = \pm\dfrac{\sqrt{34}}{2}$.

If $x = \dfrac{\sqrt{34}}{2}$, then $y = \dfrac{17}{2\left(\dfrac{\sqrt{34}}{2}\right)} = \dfrac{\sqrt{34}}{2}$; if $x = -\dfrac{\sqrt{34}}{2}$, then $y = -\dfrac{\sqrt{34}}{2}$.

9. cont'd.

The solution set is $\left\{\left(-\frac{\sqrt{34}}{2}, -\frac{\sqrt{34}}{2}\right), \left(\frac{\sqrt{34}}{2}, \frac{\sqrt{34}}{2}\right)\right\}$.

10. Factoring the left-hand member of the second equation, $(x + 4y)(x - 2y) = 0$, so $x = -4y$ or $x = 2y$.

Substituting $-4y$ for x in the first equation, $(-4y)^2 - (-4y)y + y^2 = 21$ or $21y^2 = 21$ so $y = \pm 1$. If $y = 1$, then $x = -4(1) = -4$; if $y = -1$, $x = 4$.

Substituting $2y$ for x, $(2y)^2 - (2y)y + y^2 = 21$ or $y = \pm\sqrt{7}$. If $y = \sqrt{7}$, $x = 2\sqrt{7}$; if $y = -\sqrt{7}$, $x = -2\sqrt{7}$.

The solution set is $\{(-4,1),(4,-1),(2\sqrt{7},\sqrt{7}),(-2\sqrt{7},-\sqrt{7})\}$.

11.

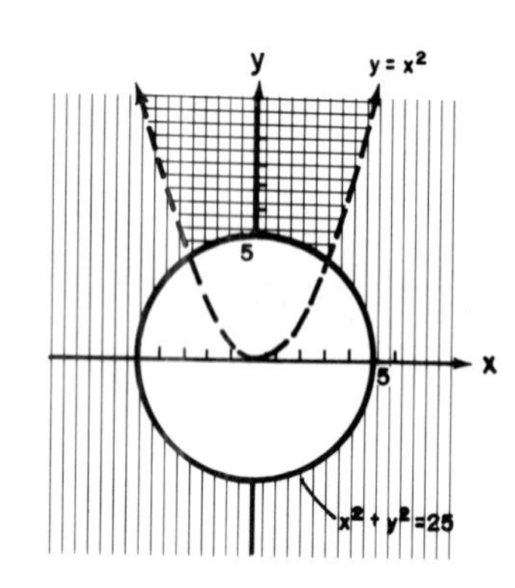

12. The width: x.
The length: $32 - x$.

$$x(32 - x) = 240$$
$$x^2 - 32x + 240 = 0$$
$$(x - 12)(x - 20) = 0$$
$$x = 12 \text{ or } 20.$$

When $x = 12$, $32 - x = 20$; and when $x = 20$, $32 - x = 12$. In any case, the dimensions are 20 feet by 12 feet.

13. Substitute (1,1), (-1,-5), and (2,10) into $y = ax^2 + bx + c$.

$$1 = a + b + c$$
$$-5 = a - b + c$$
$$10 = 4a + 2b + c.$$

The solution of this system yields $a = 2$, $b = 3$, $c = -4$.

Hence, the equation of the parabola is $y = 2x^2 + 3x - 4$.

CHAPTER 11

1. a. $\boxed{\text{Domain is } \{3,4,5,6\}. \quad \text{Range is } \{1\}. \quad \text{A function.}}$

   b. $\boxed{\text{Domain is } \{1\}. \quad \text{Range is } \{3,4,5,6\}. \quad \text{Not a function.}}$

2. a. $h(3) = 2(3)^2 + 3(3) - 2 = 2(9) + 9 - 2 = \boxed{25}$.

   b. $h(1) - h(0) = [2(1)^2 + 3(1) - 2] - [2(0)^2 + 3(0) - 2] = 3 - (-2) = \boxed{5}$.

3. a. $f(x + h) = (x + h)^2 + 2(x + h) = \boxed{x^2 + 2hx + h^2 + 2x + 2h}$.

   b. $$\frac{f(x+h) - f(x)}{h} = \frac{(x^2 + 2hx + h^2 + 2x + 2h) - (x^2 + 2x)}{h} = \frac{2hx + h^2 + 2h}{h}$$

   $$= \frac{h(2x + h + 2)}{h} = \boxed{2x + h + 2}\ .$$

4. $x^2 + y^2 = 49$, $y^2 = 49 - x^2$, $y = \pm\sqrt{49 - x^2}$.

   y is a real number if and only if $49 - x^2 \geq 0$. Solving this inequality, we obtain $|x| \leq 7$. Hence the domain is $\boxed{\{x \mid |x| \leq 7\}}$.

5. a. $$y = \frac{k}{t^2}$$

   $$2 = \frac{k}{2^2}$$

   $$k = 8.$$

   Then $\boxed{y = \frac{8}{t^2}}$.

   b. $y = \frac{8}{3^2} = \boxed{\frac{8}{9}}$.

6. a. $$R = \frac{kl}{d^2}$$

   $$100 = \frac{k(25)}{(0.006)^2}$$

   $$k = \frac{100(0.006)^2}{25} = 4(0.006)^2.$$

   Then, $\boxed{R = \frac{4(0.006)^2 l}{d^2}}$

   b. $$R = \frac{4(0.006)^2(100)}{(.012)^2} = 4(100)\left(\frac{.006}{.012}\right)^2$$

   $$= 400\left(\frac{1}{2}\right)^2 = 100.$$

   Hence, the resistance is $\boxed{100 \text{ ohms}}$.

7. a. Replace x by y and y by x in the given equation and obtain $\boxed{2y - 3x = 6}$.

b. The graph is

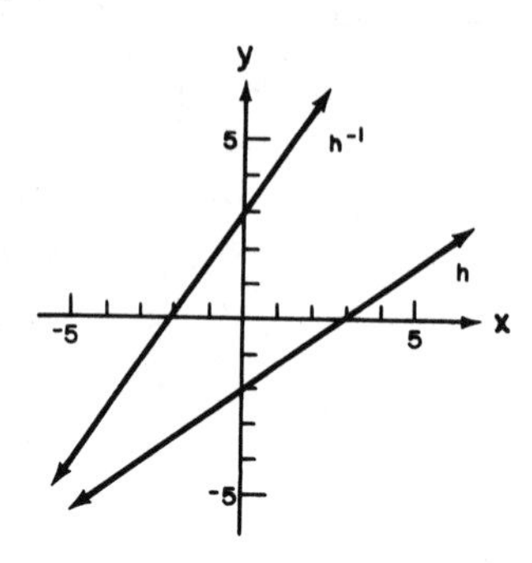

c. Since each x is paired with one and only one y, $\boxed{h^{-1} \text{ is a function}}$.

8. Some ordered pairs are

$(-3,\frac{1}{27}), (-2,\frac{1}{9}), (-1,\frac{1}{3}), (0,1),$
$(1,3), (2,9), (3,27)$.

The graph is

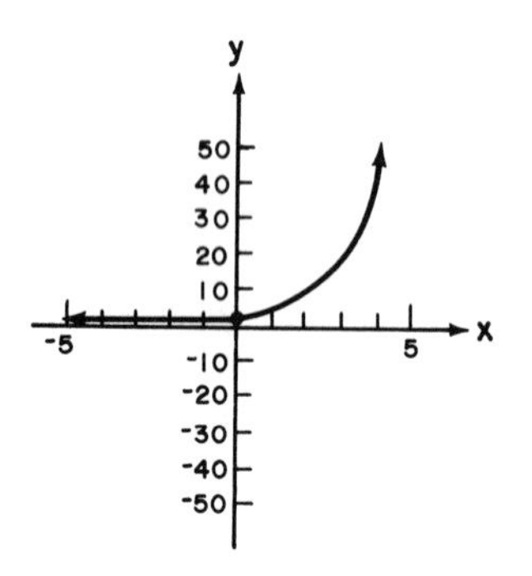

9. Some ordered pairs are

$(-3,27), (-2,9), (-1,3), (0,1),$
$(1,\frac{1}{3}), (2,\frac{1}{9}), (3,\frac{1}{27})$.

The graph is

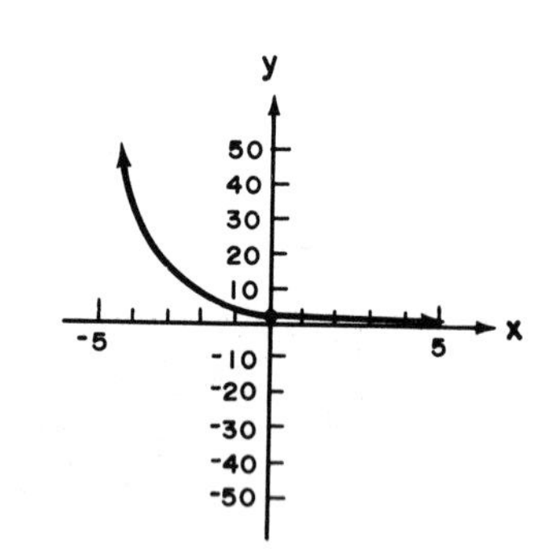

10. a. $10^{-4} = \frac{1}{10^4}$

$= \boxed{\frac{1}{10,000} \text{ or } 0.0001}$

b. $\left(\frac{2}{3}\right)^{-2} = \frac{1}{(2/3)^2}$

$= \frac{1}{4/9} = \boxed{\frac{9}{4}}$

11.

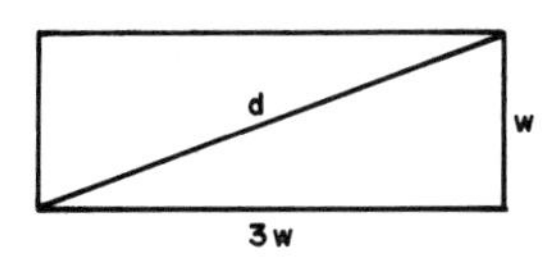

a. $A = 3w \cdot w = 3w^2$

From the Pythagorean theorem,

$$(3w)^2 + w^2 = d^2$$
$$10w^2 = d^2$$
$$w^2 = \frac{1}{10}d^2$$

Substituting $\frac{1}{10}d^2$ for w^2 in $A = 3w^2$, we have

$$A = 3 \cdot \frac{1}{10}d^2$$

$$\boxed{A = \frac{3}{10}d^2}$$

b. When $d = 10$, $A = \frac{3}{10}(10^2) = 30$.

The area is $\boxed{\text{30 square inches}}$.

12. a. $h = 4r$

Substitute $4r$ for h in $V = \pi r^2 h$ and obtain

$$V = \pi r^2(4r)$$

$$\boxed{V = 4\pi r^3}.$$

b. When $h = 5$,

$$V = 4\pi(5^3)$$
$$= 500\pi \approx 1570.8$$

The volume is $\boxed{\text{1570.8 cubic centimeters}}$ to the nearest tenth.

13.

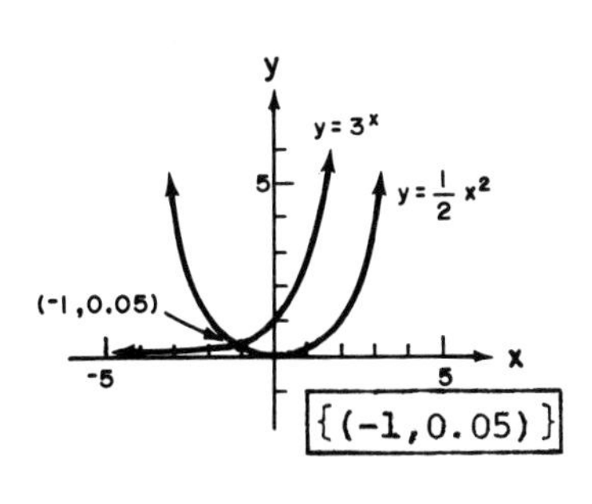

$\boxed{\{(-1, 0.05)\}}$

CHAPTER 12

1. a. $\log_2 16 = \boxed{4}$. b. $\log_{10} 0.01 = \boxed{-2}$.

2. a. $10^{-5} = \boxed{0.00001}$. b. $\left(\frac{1}{4}\right)^{-2} = \boxed{16}$.

3. a. $b^3 = 64$
$b^3 = 4^3$
$b = \boxed{4}$.

b. $\left(\frac{1}{3}\right)^{-3} = x$

$x = \frac{1}{\left(\frac{1}{3}\right)^3} = \boxed{27}$.

4. a. $\log_b \frac{x^3y^2}{z} = \log_b x^3y^2 - \log_b z = \log_b x^3 + \log_b y^2 - \log_b z$

$= \boxed{3 \log_b x + 2 \log_b y - \log_b z}$.

b. $\log_{10}\left(\frac{xy^2}{x - y}\right)^{1/4} = \frac{1}{4} \log_{10}\left(\frac{xy^2}{x - y}\right) = \frac{1}{4}[\log_{10}(xy^2) - \log_{10}(x - y)]$

$= \frac{1}{4}[\log_{10} x + 2 \log_{10} y - \log_{10}(x - y)]$

$= \boxed{\frac{1}{4} \log_{10} x + \frac{1}{2} \log_{10} y - \frac{1}{4} \log_{10}(x - y)}$.

5. a. $3 \log_{10} x + 4 \log_{10} y - \log_{10} z = \log_{10} x^3 + \log_{10} y^4 - \log_{10} z$

$= \boxed{\log_{10} \frac{x^3y^4}{z}}$

b. $\frac{1}{2}(\log_b x^3 + \log_b y^5 - \log_b z^3) = \frac{1}{2}\left(\log_b \frac{x^3y^5}{z^3}\right) = \log_b\left(\frac{x^3y^5}{z^3}\right)^{1/2} = \boxed{\log_b\sqrt{\frac{x^3y^5}{z^3}}}$.

6. $\log_{10}[\log_2(\log_3 9)] = \log_{10}[\log_2(2)] = \log_{10}[1] = \boxed{0}$.

7. $\log_{10}(x + 3)(x) = 1.$ $10^1 = (x + 3)(x).$ $x^2 + 3x - 10 = 0.$
$(x + 5)(x - 2) = 0.$ $x = -5, 2.$ -5 causes $\log_{10}(x + 3)$ or $\log_{10} x$ to be undefined and so must be rejected. $\boxed{\{2\}}$.

8. a. $\boxed{1.3304}$ b. $\boxed{7.3304 - 10}$.

9. Using Table II,

a. $10^{3.9258} = \text{antilog}_{10}\ 3.9258$. $\text{Antilog}_{10}\ 0.9258 = 8.430$.

Thus, $10^{3.9258} = \boxed{8430}$.

b. $10^{2.8156-4} = \text{antilog}_{10}\ (2.8156 - 4)$. $\text{Antilog}_{10}\ 0.8156 = 6.54$.

Thus, $10^{2.8156-4} = \boxed{0.0654}$.

10. Using Table III,

a. $e^{0.47} = \boxed{1.6000}$ b. $e^{-2.2} = \boxed{0.1108}$

11. Using Table IV, ln 20 = 2.9957.

12. a. $x \log_{10} 5 = \log_{10} 15$. $x = \boxed{\dfrac{\log_{10} 15}{\log_{10} 5}}$.

12. cont'd.

b. $(3 - x)\log_{10} 3 = \log_{10} 1000$

$$(3 - x)\log_{10} 3 = 3$$

$$3 - x = \frac{3}{\log_{10} 3}$$

$$-x = -3 + \frac{3}{\log_{10} 3}$$

$$x = \boxed{3 - \frac{3}{\log_{10} 3}} .$$

13.

$$E = 2ke^{1+t}$$

$$\ln E = \ln 2 + \ln k + (1 + t)\ln e$$

$$\ln E - \ln 2 - \ln k = (1 + t) \cdot 1$$

$$\boxed{-1 + \ln E - \ln 2 - \ln k = t} .$$

14.

$$\log_{10} \frac{1}{[H^+]} = 6.3$$

$$\log_{10} 1 - \log_{10}[H^+] = 6.3$$

$$0 - \log_{10}[H^+] = 6.3$$

$$\log_{10}[H^+] = -6.3$$

$$= [10 + (-6.3)] - 10$$

$$= 3.7 - 10$$

$$[H^+] = 10^{3.7-10}$$

To the nearest table entry,

$$[H^+] = \boxed{5.01 \times 10^{-7}} .$$

15.

$$N = N_0 10^{0.2t}$$

$$180 = 3 \cdot 10^{0.2t}$$

$$60 = 10^{0.2t}$$

$$0.2t \log_{10} 10 = \log_{10} 60$$

$$(0.2t)(1) = 1.7782$$

$$t = \frac{1.7782}{.2} = \boxed{8.891}\ .$$

16.

$$y = y_0 e^{-0.3t}$$

$$6 = 30e^{-0.3t}$$

$$\frac{6}{30} = e^{-0.3t}$$

$$e^{-0.3t} = 0.2$$

$$-0.3t \ln e = \ln 0.2$$

$$(-0.3t)(1) = (2.303)\log_{10} 0.2$$

$$-0.3t = 2.303(9.3010 - 10) = 2.303(-0.6990) = 1.6098$$

$$t = \frac{-1.6098}{-0.3} = \boxed{5.37}\ .$$

CHAPTER 13

1. $s_n = \dfrac{n}{n^2 + 1}$. $s_1 = \dfrac{1}{1^2 + 1} = \dfrac{1}{2}$. $s_2 = \dfrac{2}{2^2 + 1} = \dfrac{2}{5}$. $s_3 = \dfrac{3}{3^2 + 1} = \dfrac{3}{10}$.

   $s_4 = \dfrac{4}{4^2 + 1} = \dfrac{4}{17}$. So we have $\boxed{\frac{1}{2}, \frac{2}{5}, \frac{3}{10}, \frac{4}{17}}$.

2. $$\sum_{j=0}^{4} \frac{(-1)^j 3^j}{j + 1} = \frac{(-1)^0 3^0}{0 + 1} + \frac{(-1)^1 (3)^1}{1 + 1} + \frac{(-1)^2 (3)^2}{2 + 1} + \frac{(-1)^3 (3)^3}{3 + 1} + \frac{(-1)^4 (3)^4}{4 + 1}$$

   $$= \boxed{1 - \frac{3}{2} + 3 - \frac{27}{4} + \frac{81}{5}}\ .$$

3. $1 - 8 + 27 - 64 + 125 = 1^3 - 2^3 + 3^3 - 4^3 + 5^3$
$$= (-1)^2(1^3) + (-1)^3(2^3) + (-1)^4(3^3) + (-1)^5(4^3) + (-1)^6(5^3).$$

From this, we observe that $s_n = (-1)^{n+1}(n^3)$. Therefore, we have

$$\boxed{\sum_{j=1}^{5} (-1)^{j+1}(j)^3}.$$

4. a. $d = 3$. Therefore, the next three terms are $\boxed{x + 6, x + 9, x + 12}$.

   b. Use $s_n = a + (n - 1)d$ and obtain $s_n = x + (n - 1)(3) = \boxed{x + 3n - 3}$.

5. $-5, -2, 1, \ldots$ is an arithmetic progression with $a = -5$, $d = 3$. Therefore, using $s_n = a + (n - 1)d$, we have

$$s_n = -5 + (17 - 1)(3) = \boxed{43}.$$

6. Using $s_n = a + (n - 1)d$, we have $-46 = a + (20 - 1)d$ and $-30 = a + (12 - 1)d$. These can be written equivalently as $a + 19d = -46$ and $a + 11d = -30$. To solve this system of equations, we subtract the second equation from the first and obtain $8d = -16$, from which we have $d = -2$. Using this result in $a + 11d = -30$, we have $a + 11(-2) = -30$ and obtain $a = -8$. Then,

$$s_5 = -8 + (5 - 1)(-2) = \boxed{-16}.$$

7. $$\sum_{j=10}^{20} (2j - 3) = [2(10) - 3] + [2(11) - 3] + [2(12) - 3] + \cdots + [2(20) - 3]$$
$$= 17 + 19 + 21 + \cdots + 37.$$

This series is arithmetic with $a = 17$, $d = 2$, $n = 11$, and $s_{11} = 37$.

Using $S_n = \frac{n}{2}(a + s_n)$, we find

$$S_{11} = \frac{11}{2}(17 + 37) = \frac{11}{2}(54) = \boxed{297}.$$

8. a. To determine r, we divide -1 by $\frac{x}{a}$; that is, we have

$$-1 \div \frac{x}{a} = -1\left(\frac{a}{x}\right) = -\frac{a}{x}. \text{ So, } r = -\frac{a}{x}. \quad s_4 = s_3\left(-\frac{a}{x}\right) = \frac{a}{x}\left(-\frac{a}{x}\right) = -\frac{a^2}{x^2}.$$

$$s_5 = s_4\left(-\frac{a}{x}\right) = \left(-\frac{a^2}{x^2}\right)\left(-\frac{a}{x}\right) = \frac{a^3}{x^3}.$$

$$s_6 = s_5\left(-\frac{a}{x}\right) = \left(\frac{a^3}{x^3}\right)\left(-\frac{a}{x}\right) = -\frac{a^4}{x^4}.$$

8. cont'd.

Therefore, the next three terms are $\boxed{-\frac{a^2}{x^2}, \frac{a^3}{x^3}, -\frac{a^4}{x^4}}$.

b. $s_n = \left(\frac{x}{a}\right)\left(-\frac{a}{x}\right)^{n-1} = \frac{x}{a}(-1)^{n-1}\left(\frac{a^{n-1}}{x^{n-1}}\right) = \boxed{(-1)^{n-1}\frac{a^{n-2}}{x^{n-2}}}$.

9. For this geometric progression, $a = -81$ and $r = \frac{-27}{-81} = \frac{1}{3}$. Using $s_n = ar^{n-1}$ with $n = 9$, we have

$$s_9 = (-81)\left(\frac{1}{3}\right)^8 = -(-3)^4 \frac{1}{(3)^8} = -\frac{1}{3^4} = \boxed{-\frac{1}{81}}.$$

10. In $s_n = ar^{n-1}$, take $n = 6$, $s_6 = 60$, and $r = 3$ and obtain $60 = a(3)^5$. $60 = 243a$. $a = \frac{60}{243} = \frac{20}{81}$. Then $s_2 = \frac{20}{81}(3) = \boxed{\frac{20}{27}}$.

11.
$$\begin{aligned}\sum_{j=5}^{8}(2^j - 5) &= (2^5 - 5) + (2^6 - 5) + (2^7 - 5) + (2^8 - 5)\\ &= 2^5 + 2^6 + 2^7 + 2^8 - (5 + 5 + 5 + 5)\\ &= 2^5 + 2^6 + 2^7 + 2^8 - (20).\end{aligned}$$

$2^5 + 2^6 + 2^7 + 2^8$ is a geometric progression with $a = 2^5$, $r = 2$, and $n = 4$.

Using $S_n = \frac{a-ar^n}{1-r} = \frac{2^5 - 2^5(2)^4}{1-2} = \frac{2^5(1-2^4)}{-1} = \frac{32(1-16)}{-1} = -32(-15) = 480$.

Therefore, $\sum_{j=5}^{8}(2^j - 5) = 480 - 20 = \boxed{460}$.

12. The diagram 3, __, __, __, 243 suggests that there are 4 multiplications by r between 3 and 243. Hence,

$$r^4 = \frac{243}{3},\ r^4 = 81,\ r = 3.$$

Therefore, the required means are $\boxed{9, 27, 81}$.

13. Since $r = -\frac{3}{4}$, that is $|r| < 1$, the series has a sum given by $S_\infty = \frac{a}{1-r}$.

Taking $a = 2$ and $r = -\frac{3}{4}$, we obtain $S_\infty = \frac{2}{1-\left(-\frac{3}{4}\right)} = \frac{2}{1+\frac{3}{4}} = \frac{2}{\frac{7}{4}} = \boxed{\frac{8}{7}}$.

14. $\dfrac{(12!)(8!)}{16!} = \dfrac{\cancel{12!}\,8!}{16 \cdot 15 \cdot 14 \cdot 13 \cdot \cancel{12!}} = \dfrac{\cancel{8} \cdot \cancel{7} \cdot 6 \cdot \cancel{5} \cdot \cancel{4} \cdot \cancel{3} \cdot \cancel{2} \cdot 1}{\cancel{16} \cdot \cancel{15} \cdot \cancel{14} \cdot 13} = \boxed{\dfrac{12}{13}}$.

15. The first four terms of $[x + (-2y)]^8$ are

$$x^8 + \frac{8x^7(-2y)^1}{1!} + \frac{7 \cdot 8x^6(-2y)^2}{2!} + \frac{6 \cdot 7 \cdot 8x^5(-2y)^3}{3!} .$$

These can be simplified to $\boxed{x^8 - 16x^7y + 112x^6y^2 - 448x^5y^3}$.

16. The 7th term of $(a^3 - b)^9$ is $\dfrac{4 \cdot 5 \cdot 6 \cdot 7 \cdot 8 \cdot 9(a^3)^3(-b)^6}{6!}$.

This can be simplified to $\boxed{84a^9b^6}$.

17. $1.027\overline{027} = 1 + .027 + .000027 + .000000027 + \cdots$. Beginning with the second term, .027, we have an infinite geometric series with $a = .027$ and $r = .001$. So using

$$S_\infty = \frac{a}{1 - r}$$

on this infinite geometric series, we have

$$1.027\overline{027} = 1 + \frac{.027}{1 - .001} = 1 + \frac{.027}{.999} = 1 + \frac{27}{999} = 1 + \frac{1}{37} = \boxed{\frac{38}{37}} .$$

18. $P_{5,5} = 5! = 5 \cdot 4 \cdot 3 \cdot 2 \cdot 1 = \boxed{120}$.

19. Eight positions remain to be filled and nine boys are available to fill them.

$$P_{9,8} = 9 \cdot 8 \cdot 7 \cdot 6 \cdot 5 \cdot 4 \cdot 3 \cdot 2 = \boxed{362{,}880} .$$

20. Since the letters and digits may be used more than once, the letters may each be chosen in 8 ways and the digits may each be chosen in 10 ways. So, we have

$$8 \cdot 8 \cdot 8 \cdot 10 \cdot 10 \cdot 10 = \boxed{512{,}000} .$$

21. $P = \dfrac{9!}{2!3!2!} = \dfrac{9 \cdot \cancel{8} \cdot 7 \cdot \cancel{6} \cdot 5 \cdot 4 \cdot 3 \cdot \cancel{2} \cdot 1}{\cancel{2} \cdot 1 \cdot \cancel{3 \cdot 2} \cdot 1 \cdot \cancel{2} \cdot 1} = \boxed{15{,}120}$.

22. 3 men can be chosen from 6 men in $\binom{6}{3}$ ways. $\binom{6}{3} = \dfrac{\overset{1}{\cancel{6}} \cdot 5 \cdot 4}{\cancel{3 \cdot 2} \cdot 1} = 20.$

3 women can be chosen from 8 women in $\binom{8}{3}$ ways. $\binom{8}{3} = \dfrac{8 \cdot 7 \cdot \overset{1}{\cancel{6}}}{\underset{1}{\cancel{3 \cdot 2}} \cdot 1} = 56.$

Then a committee can be chosen in 20 x 56 = $\boxed{1120}$ ways.

23. This is equivalent to being asked to determine how many nonempty subsets there are in a four-member set. This is given by $2^4 - 1 = \boxed{15}$. An alternative solution is obtained by computing $\binom{4}{1} + \binom{4}{2} + \binom{4}{3} + \binom{4}{4} =$ $4 + 6 + 4 + 1 = \boxed{15}$.

24. 2 red balls can be chosen from 5 red balls in $\binom{5}{2} = 10$ ways. 3 black balls can be chosen from 4 black balls in $\binom{4}{3} = 4$ ways. Each of the choices of a pair of red balls can be paired with 4 choices of a combination of black balls. Therefore, there are 10 x 4 = $\boxed{40}$ ways in total.

APPENDIX

1. P(1) is equal to the remainder when P(x) is divided by x - 1.

1⌋	2	-3	1	1
		2	-1	0
	2	-1	0	1

Hence, $\boxed{P(1) = 1}$.

P(-1) is found similarly.

-1⌋	2	-3	1	1
		-2	5	-6
	2	-5	6	-5

Hence, $\boxed{P(-1) = -5}$.

2. $x^3 - x^2 + 3x$ is divided successively by -3, -2, -1, 0, 1, 2, 3.

-3⌋	1	-1	3	0
		-3	12	-45
	1	-4	15	-45

-2⌋	1	-1	3	0
		-2	6	-18
	1	-3	9	-18

-1⌋	1	-1	3	0
		-1	2	-5
	1	-2	5	-5

0⌋	1	-1	3	0
		0	0	0
	1	-1	3	0

1⌋	1	-1	3	0
		1	0	3
	1	0	3	3

2⌋	1	-1	3	0
		2	2	10
	1	1	5	10

2. cont'd.

3	1	-1	3	0
		3	6	27
	1	2	9	27

Hence, the following ordered pairs correspond to points on the graph:

(-3,-45), (-2,-18), (-1,-5), (0,0), (1,3), (2,10), (3,27) .

The graph is:

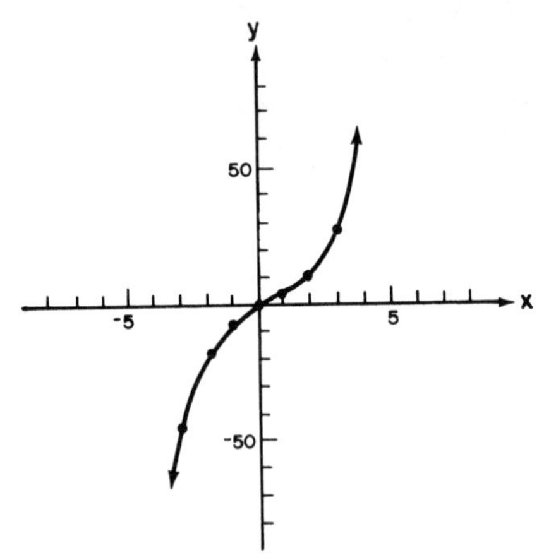

3. a.

-1	1	1	-4	-4
		-1	0	4
	1	0	-4	0

Since the remainder is 0, **x + 1 is a factor** .

b.

1	1	1	-4	-4
		1	2	-2
	1	2	-2	-6

Since the remainder is not 0, **x - 1 is not a factor.** .

4. In the solution to 3(a) above, we found that x + 1 is a factor of $x^3 + x^2 - 4x + 4$. Hence, from the synthetic division

$x^3 + x^2 - 4x + 4 = (x + 1)(x^2 - 4) = (x + 1)(x + 2)(x - 2) = 0.$

Then, by inspection the required solutions are x = -1, x = -2, x = 2 and the solution set is $\{-1,-2,2\}$.

5. Vertical asymptote: $x = -2$. Since the degree of the numerator is equal to the degree of the denominator, $y = 1$ is a horizontal asymptote.

   When $x = 0$, we have $y = \frac{3}{2}$; y-intercept is $\frac{3}{2}$.

   When $y = 0$, $x = -3$.

   Computing the coordinates of a few other points on the graph leads to the figure at the right.

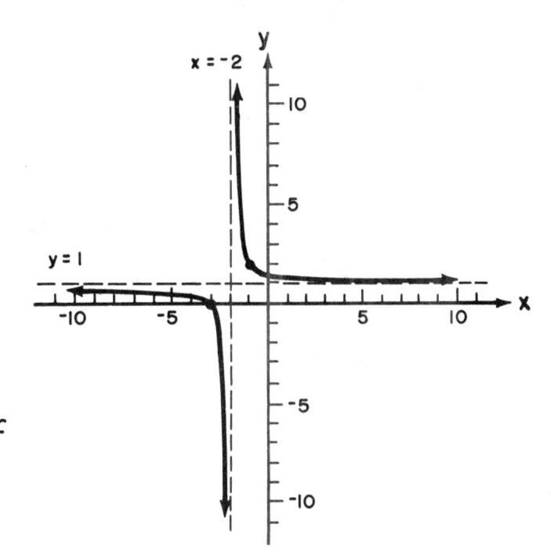

6. For $2x - 2 \geq 0$, we have $y = 2x - 2$.
   For $2x - 2 < 0$, we have $y = -(2x - 2)$.
   $2x - 2 \geq 0$ is equivalent to $x \geq 1$.
   $2x - 2 < 0$ is equivalent to $x < 1$.
   Therefore, we graph $y = 2x - 2$ for $x \geq 1$ and $y = -2x + 2$ for $x < 1$.

7. For $-3 \leq x < -2$, we have $-2 \leq x + 1 < -1$ and $[x + 1] = -2$, and we graph $y = -2$.

   For $-2 \leq x < -1$, we have $-1 \leq x + 1 < 0$ and $[x + 1] = -1$, and we graph $y = -1$.

   For $-1 \leq x < 0$, we have $0 \leq x + 1 < 1$ and $[x + 1] = 0$, and we graph $y = 0$.

   For $0 \leq x < 1$, we have $1 \leq x + 1 < 2$ and $[x + 1] = 1$, and we graph $y = 1$.

   For $1 \leq x < 2$, we have $2 \leq x + 1 < 3$ and $[x + 1] = 2$, and we graph $y = 2$.

   For $2 \leq x < 3$, we have $3 \leq x + 1 < 4$ and $[x + 1] = 3$, and we graph $y = 3$.

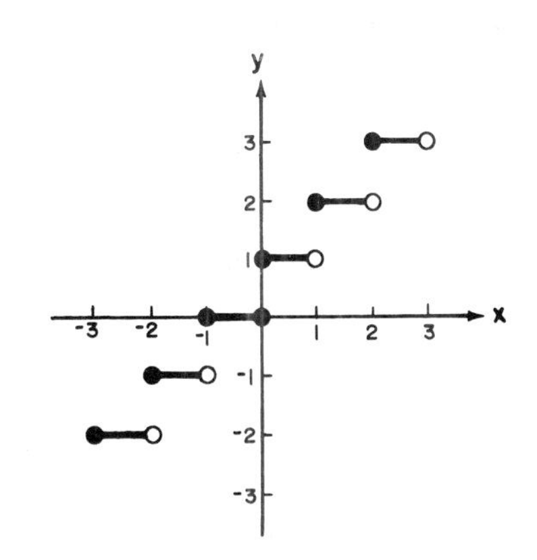

8.

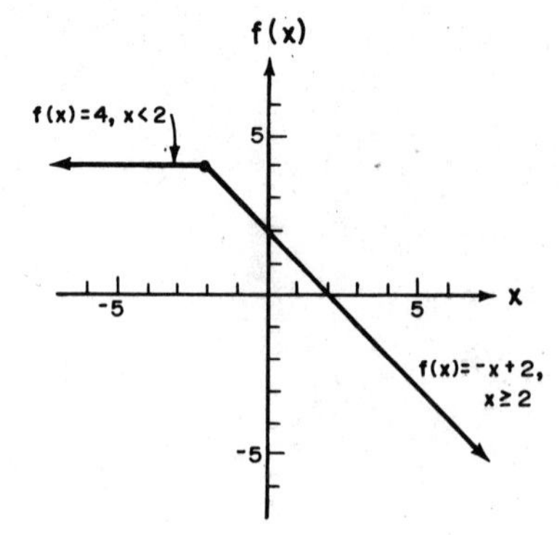